From
Edison
to LEDs

The Science and Story
of Light Sources

From
Edison
to LEDs

The Science and Story
of Light Sources

Faiz Rahman

Ohio University, USA

World Scientific

NEW JERSEY · LONDON · SINGAPORE · BEIJING · SHANGHAI · HONG KONG · TAIPEI · CHENNAI · TOKYO

Published by

World Scientific Publishing Co. Pte. Ltd.

5 Toh Tuck Link, Singapore 596224

USA office: 27 Warren Street, Suite 401-402, Hackensack, NJ 07601

UK office: 57 Shelton Street, Covent Garden, London WC2H 9HE

Library of Congress Cataloging-in-Publication Data
Names: Rahman, Faiz, author.
Title: From Edison to LEDs : the science and story of light sources /
    Faiz Rahman, Ohio University, USA.
Description: New Jersey : World Scientific, 2023. | Includes index.
Identifiers: LCCN 2022045160 | ISBN 9789811267581 (hardcover) |
    ISBN 9789811268274 (paperback) | ISBN 9789811267598 (ebook for instituions) |
    ISBN 9789811267604 (ebook for individuals)
Subjects: LCSH: Electric lamps--History. | Light sources--History.
Classification: LCC TK4310 .R | DDC 621.32/6--dc23/eng/20221128
LC record available at https://lccn.loc.gov/2022045160

British Library Cataloguing-in-Publication Data
A catalogue record for this book is available from the British Library.

For any available supplementary material, please visit
https://www.worldscientific.com/worldscibooks/10.1142/13178#t=suppl

Desk Editors: Logeshwaran Arumugam/Amanda Yun

Typeset by Stallion Press
Email: enquiries@stallionpress.com

Preface

During our waking moments, light — the enabler of vision — plays a central role in our lives. Sight is, arguably, the most important of our five senses, and its deprivation — whether due to an eye disease or due to the absence of light — can completely incapacitate us. This is probably the most ancient of human observations and so is, thus, the importance of light in our lives. This book deals with how we produce light that makes our surroundings visible to us. While this has been done for much of human existence on this planet, here we examine artificial light generation mainly through electrical means. To enhance broader understanding, we do cover some other techniques for light generation, and even some instances of the production of radiations that are invisible to us.

This book follows developments in a roughly chronological order and, thus, moves from inventions of the 19th century to developments in the 21st. The treatment of various topics is somewhat encyclopedic, describing both accounts of developments and the science behind particular light sources. The coverage — as the title implies — progresses from the earliest days of electrical lighting to the current frontiers. I have had the good fortune to be involved in some of these developments during this century and, thus, some of the accounts draw from my personal experience as a researcher in semiconductor-based light sources.

I began writing this book several years ago, but other ongoing professional commitments made me take a long time to actually finish it, with all eight planned chapters in place. The text is supported by a large number of illustrations — many of which are historic and some of which are rare. Substantial amount of material present in this book came from various historic archives, article depositories and museums. My endeavor has been to piece it all together into a holistic account of the development of modern lighting technologies. I have tried to do this in an engaging manner, so as to keep the interest alive as the reader progresses through the various chapters. My hope is that this book will help deepen the knowledge base of both students and professional practitioners of optical science and technology. Knowing about the many different kinds of light sources and where each fits in the larger scheme, as well as the technology that underpins them, should make for better-informed scientists and engineers.

The book begins with an introductory chapter on the physics of light as an electromagnetic phenomenon. The second chapter deals with incandescent light sources where electrically-heated solid forms are used to generate heat — and some visible light. Incandescents are on their way out but are still available for use in legacy equipment. The next chapter describes electric discharge-based light sources. There are an exceptionally large number of lamps in this category. Some of these are used for space illumination while others are employed in scientific applications. Chapter 4 makes a logical progression to electrically-generated plasma sources. These are of increasing importance in contemporary science and technology, so a separate chapter has been devoted to their description. Chapter 5 describes light-emitting diodes (LEDs), i.e. semiconductor junction-based non-coherent light emitters. The next chapter moves on to coherent light sources, i.e. lasers. These form their own distinct class, and I have tried to include some of the more prominent types in this chapter. This is followed by a chapter on solid-state lighting systems — based on both semiconductor lasers and LEDs. The last chapter of the book covers miscellaneous light sources and technologies that do not neatly fall into areas that are covered in earlier chapters. Some of these may not be familiar to many readers, so this chapter nicely rounds out our coverage of light sources.

I have dedicated this book to all members of my family — past and present — from my grandparents and parents to my wife and son. Each of them has contributed to my making. I would also like to take this opportunity to thank notable people who have helped with the production of this book. James Hooker of lamptech.co.uk was generous with providing access to many historical illustrations that appear here. His website is well worth a visit by all who are interested in the historical development of light sources. Staff personnel at the British Museum in London and at the Smithsonian Museum in Washington D.C. were also of great help in gathering a good amount of historic information on relevant UK and US companies and their products. Chris Davis, executive editor at World Scientific Publishing (WSP) Corporation, enthusiastically got the publication project started and has been a great help throughout. Amanda Yun, senior editor at WSP Singapore took care of all editing requirements and was extremely helpful in the publication process. My thanks also go to Logesh at Academic Consulting and Editorial Services (ACES) who worked on the manuscript to transform it into its final book form.

Faiz Rahman
Athens, Ohio
United States of America

About the Author

Faiz Rahman is an associate professor of Electrical Engineering at Ohio University. His research is based on investigating new semiconductor materials and devices in order to develop functionally superior components for electronic/photonic applications. He holds a bachelor's degree in Physics, a master's in Physics and Electronics, another master's degree in Semiconductor Science & Technology and a Ph.D. in Electrical Engineering, the last two from Imperial College, London. After getting his academic degrees, he did postdoctoral work on spintronics at the University of Nottingham in England, before moving to the California Institute of Technology. For three years, at Caltech, Dr. Rahman was involved with developing the sensor microelectronics for NASA and ESA's Herschel and Planck Surveyor space telescopes. At the conclusion of that project, he moved to Cypress Semiconductor's Fab-4 in Bloomington, MN, where he worked on miniaturizing integration technology for logic and memory devices. In 2002, he joined the University of Glasgow in Scotland as a faculty member in Electrical Engineering. Later, he also founded Electrospell Ltd. — a technology company for developing, manufacturing and selling advanced LEDs. In 2013, Dr. Rahman joined Ohio University as a member of the Electrical Engineering faculty. His current research interests are focused on developing visible and ultraviolet radiation emitters from wide band gap semiconductors. Dr. Rahman is a fellow of the Institute of Physics and a senior member of Optica and the IEEE.

Contents

Contents

1

The Nature and Properties of Light

1.1 Introduction

Artificial lighting is central to our existence on this planet. These days, our lives would be unimaginable without some form of artificial lighting. Electrically-powered lighting fixtures light up our indoor and outdoor environments at all times of day and night. A widely publicized picture of the earth's surface during night-time, composed from a number of separate images taken from near-earth orbit, shows the concentration of lighting all over the globe. This picture, reproduced here as Figure 1.1, shows the distribution of human activity and major population centers over the Earth's surface, highlighted through the emanation of night-time artificial lighting. This picture is sufficient to show the over-sized role man-made illumination plays all over the world.

The proliferation of outdoor illumination creates a rich tapestry of light that is both visually appealing and is useful for various scientific and technical studies. Figure 1.2 here shows a view of the Italian peninsula taken by astronauts on the International Space Station, showing the spread of lighting throughout the boot of Italy.

After discussing the nature and general origin of light in this chapter, we shall look at incandescent and discharge lamps in the next two chapters. Light sources based on artificially-generated plasmas will be discussed in Chapter 4. Coherent light sources, i.e. lasers form the subject

Fig. 1.1.   Distribution of artificial lighting on the earth's surface, as seen from earth-orbiting satellites.

Fig. 1.2.   Night-time view of Italy, captured by cameras on board the International Space Station.

matter of Chapter 5. This will be followed by a chapter on solid-state junction-based light emitters, i.e. light-emitting diodes (LEDs). Each of those chapters will begin by looking at the physics of a particular technique for generating light. Afterward, the discussions will include the development and features of various kinds of light-generating devices. Chapter 7 deals with solid-state lighting while miscellaneous other sources — not discussed elsewhere in the book — are covered in the last chapter.

In what follows, we are going to take a look at some of the physics that underlies our understanding of what is light, and some of its properties. We start by looking at the development of the science of electromagnetism in the 19th century. Initial work on the science of electricity and magnetism was carried out by experimenters, who studied the nature of electric and magnetic phenomena, and established connections between them — all through experimental observations. Then came a slew of analytical physicists who re-casted experimentally obtained wisdom into mathematical language. This created a firm theoretical foundation that led to many further advances.

1.2 The Development of Electromagnetism

Electric and magnetic phenomena have been known since antiquity. The attractive or repulsive state acquired by certain objects, when rubbed, was a well-known observation since ancient times. Benjamin Franklin studied the phenomenon of thunderstorms and arbitrarily labeled certain objects as positively or negatively charged. Natural magnets, called loadstones were known in many parts of the world. These rocks, composed of magnetic iron oxide, were mined in places such as Asia minor (now part of Turkey). The attractive tendencies between pieces of loadstone were, thus, also quite well-known wherever these objects were available, and the Chinese had even invented the compass using magnetized pieces of iron to tell direction. Electric and magnetic objects were a standard repertoire of tricksters and magicians during the middle-ages. By the 18th century it was recognized that electric charges can flow through conducting objects, and steady electric currents could be established by the use of electro-chemical batteries. Serious scholarly studies of electric and magnetic phenomena began during the later half of the 18th century, but these disciplines remained as separate sciences until the year 1819 when the Danish scholar Hans Christian Ørsted noticed that a current flowing through a metal wire could deflect a magnetic compass needle. Ørsted was much ahead of his time and had conjectured that there may be some connection between electricity and magnetism. The French scientist, André-Marie Ampère soon came to know of Ørsted's discovery, and within a week put forward a rough explanation in 1820. A year later, he presented a full theory, explaining how an electric current can generate magnetic effects. Figure 1.3 shows pictures of Ørsted (left) and Ampère (right).

In 1822, Ørsted's discovery was put to practical use with the invention of the first galvanometer to measure electric currents. Galvanometers turned out to be an extremely important invention, as they allowed precise quantitative measurement of electric currents. By the first quarter of the 19th century, a basic understanding of electric and magnetic phenomena was in place and the stage was, thus, set for subsequent work on the combined disciplines of electricity and magnetism.

Fig. 1.3.   (Left) Hans Christian Ørsted, (Right) André-Marie Ampère.

During the quarter century, from 1830 to 1855, Michael Faraday, a self-taught experimentalist from a poor background, carried out thousands of experiments to better elucidate the connection between electricity and magnetism. Faraday was born to a poor family in south London, and was apprenticed to a London bookbinder when he was only 14 years old. There, he read voraciously and educated himself on a variety of scientific topics. Later, he obtained an appointment as a technical assistant at the Royal Institution in London, where he worked for scientists, such as Sir Humphery Davy. While at the Royal Institution, Faraday was able to meet many notable scientists of that era, and to investigate many topics in chemistry and physics. As an over-zealous scientist, he went to great lengths to perform detailed experiments on many aspects of electric and magnetic phenomena, including their connection to each other. Among his many discoveries was the observation that a moving magnet in the vicinity of an electrical conductor can produce an electrical effect in it. This was the opposite of Oersted's and Ampere's observations. Faraday's work, therefore, hinted at a beautiful symmetry inherent in electric and magnetic phenomena. Faraday made many other valuable contributions to science and engineering, including the invention of the first electric motor and the first dynamo (generator). His observations, however, remained

Fig. 1.4.   (Left) Michael Faraday, (Right) James Clerk Maxwell.

descriptive in nature as he neither had the inclination nor the requisite mathematical training to cast his observations in precise quantitative terms. This task eventually fell to James Clerk Maxwell — an astute mathematical physicist, Scottish by birth and English by domicile — who took Faraday's systematic, though descriptive, observations and tried to assemble them into a coherent mathematical framework. Figure 1.4 shows pictures of Faraday (left) and Maxwell (right).

In stark contrast to Michael Faraday, Clerk Maxwell was from a rich intellectual family, and had attended Edinburgh and Cambridge Universities. He, thus, had the right training in physics and mathematics to enable him to cast Faraday's observations into standard mathematical notation of his period. Maxwell began by introducing the concepts of electric and magnetic fields. Prior to his involvement, electric and magnetic effects were thought to be mediated instantaneously through 'lines of force' permeating the space around charged objects and magnetized bodies. Faraday was particularly fond of this concept, and assigned more reality to them than they actually deserved. This is not surprising because it is easy to visualize lines of force, especially around a magnet, by sprinkling some iron fillings around it. Iron particles get magnetized and assembled in a characteristic pattern, as seen in Figure 1.5 here.

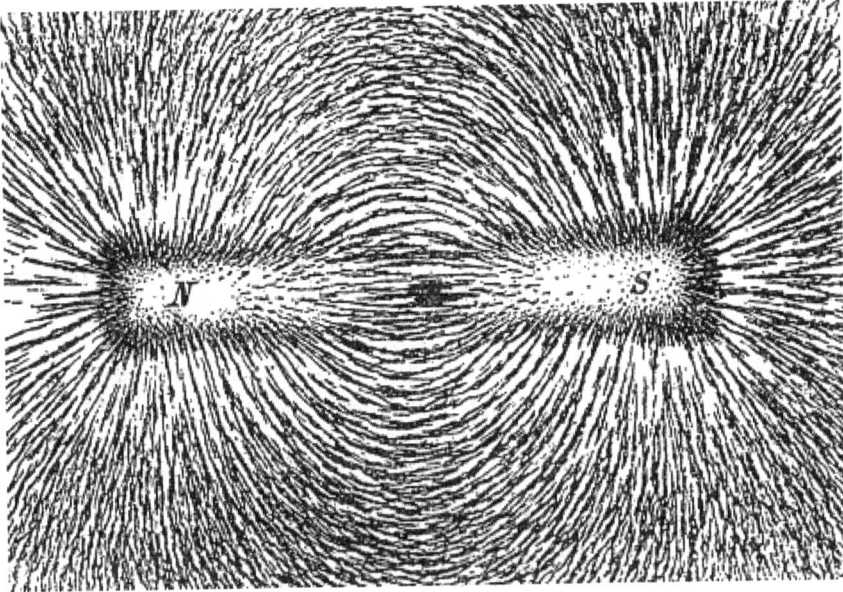

Fig. 1.5. Pattern of magnetic 'lines of force', as traced by iron filings around a bar magnet.

Faraday thought that the lines of force actually represented the loci of magnetic forces. Maxwell's superior physical intuition and mathematical sophistication enabled him to dispel this notion and instead led him to describe the space around charged or magnetized bodies as containing electric or magnetic fields. He assigned vector quantities to describe the strength and direction of these fields. The familiar notations for various types of electric and magnetic fields that we see today, **E, D, H** and **B** were all introduced by Maxwell. We now understand that electric and magnetic fields are central concepts in any description of electric and magnetic phenomena. Vector fields, like these, assign both a strength and a characteristic direction at each point in space. Lines of force are then revealed as simply the integral curves to these vector fields. Having introduced the concept of electric and magnetic fields, Maxwell then used the notations of calculus to describe the way electric and magnetic fields depend on each other. It should be noted that Maxwell's work went beyond just systemizing the observations of Faraday and other experimentalists. He recognized the symmetries that exist between electric and magnetic fields, and, thus, introduced the concept of a displacement current, i.e. that a changing electric field can generate a magnetic field, just as much as a changing magnetic field can generate an electric field. It is interesting to note that Maxwell did not outline his theory of electromagnetism in the form of the four equations that today go by his name. In addition to using scalars and vectors, he made extensive use of mathematical entities called quaternions in his equations. Maxwell's theory, later published in his book, *A Treatise on Electricity and Magnetism*, in the year 1873, contained 20 equations in 20 variables to describe all the electric and magnetic phenomena, as well as their interdependencies. This book, now a classic, is still available in print and a picture of its title page appears here in Figure 1.6. Maxwell may have further refined his equations had he not died prematurely from abdominal cancer at the age of 48.

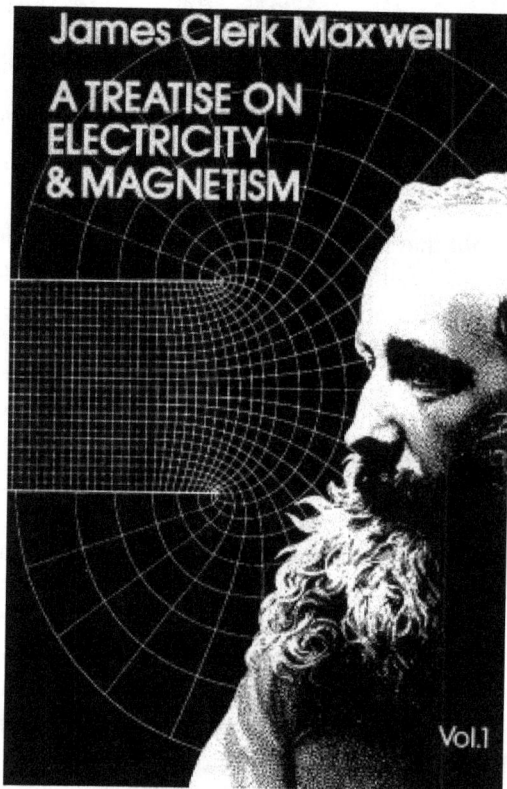

Fig. 1.6.   Cover of James Clerk Maxwell's Treatise on Electricity and Magnetism.

The 'original' Maxwell's equations bear little resemblance to the set of equations we are now familiar with. The modern versions of Maxwell's equations were developed by the English physicist Oliver Heaviside, some 20 years after the publication of Maxwell's book. Somewhat like Michael Faraday, but with a clear mathematical inclination, Oliver Heaviside was a self-taught engineer and mathematical physicist. Born in London in 1850, and mostly home-schooled, he spent a few years of his life working for the Great Northern Telegraph Company. Heaviside came across Maxwell's two-volume Treaties on Electricity and Magnetism soon after its publication in 1873, and was immediately struck by their great scope and applicability to electric telegraphy and other emerging electric technologies. However, he found their mathematical structure to be unnecessarily complicated and thought that Maxwell's use of quaternions had obscured much of the physical reality that his theory was describing. Heaviside, thus, set himself the task of cleaning up Maxwell's original equations in order to make them more explicit. He introduced modern vector differential operators to recast twelve of Maxwell's twenty equations into the four equations that we know today as Maxwell's equations. Heaviside made many other contributions to the theory of electromagnetism, such as introducing the use of complex numbers in analyzing alternating current circuits. In this connection, he also introduced the now familiar terms, such as impedance and conductance. Some of his work in mathematical physics was controversial at his time, as professional mathematicians sometimes objected to the lack of rigor in some of his statements. Once counteracting some of his critics he is said to have put forward the argument: "I do not refuse my dinner simply because I do not understand the process of digestion".

Returning back to Maxwell's contributions, not only did he develop a coherent theory of electromagnetic phenomenon, his subsequent analysis of the mathematical relationships between electric and magnetic fields, using the tools of vector calculus, led to another startling conclusion. On combining some of these equations he found that a wave equation resulted, which predicted the existence of oscillating electric and magnetic fields traveling at a well-defined speed, given by a combination of electric and magnetic constants. On evaluation, this speed turned out to be identical to that of light. This was indeed an extremely important deduction, for it firmly hinted that light itself was an electromagnetic phenomenon. At the time this conclusion was published by Maxwell in 1862, it did not attract much attention, and in fact Maxwell himself paid scant

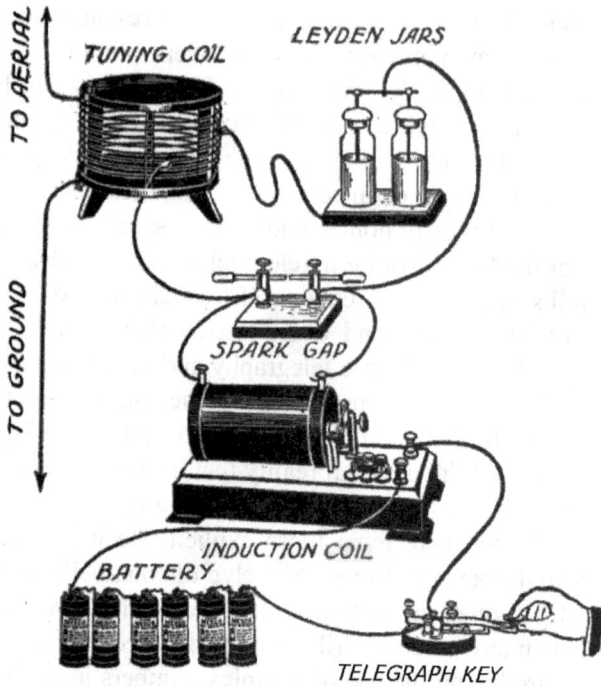

Fig. 1.7.   Spark gap apparatus, similar to the arrangement used by Heinrich Hertz, for generating and transmitting radio waves.

attention to it. In 1887, the German physicist Heinrich Rudolf Hertz experimentally discovered the existence of traveling waves, which clearly had an electromagnetic origin. Hertz generated these waves using a spark gap apparatus, and demonstrated that these waves were transverse in character, and, thus, can be polarized just like light waves. An illustration of equipment similar to what Hertz used appears in Figure 1.7 here. When the telegraph key was briefly pressed down and released, a high voltage was generated by the induction coil, which caused a bright spark to appear in the attached spark gap. This generated a wide spectrum of ultrahigh frequency (UHF) radio waves. The waves traveled outward from the transmitting apparatus, and were received by a similar receiving apparatus a few feet away, where they caused small sparks to appear in the spark gap. The tuning coil (inductor) and the Leyden jars (capacitors) were used as a tuned circuit on the receiver side of the apparatus.

The properties of these radio waves also resembled ordinary light very closely. For instance, the waves could be reflected and polarized, just

Fig. 1.8. (Left) Oliver Heaviside, (Right) Heinrich Hertz.

as well as light waves. Thus, it became clear that Maxwell's theory had correctly predicted the existence of, so-called, electromagnetic waves. More importantly, Hertz's discovery firmly established the fact that light was nothing but a form of electromagnetic waves, whose velocity could be both measured experimentally and calculated theoretically from purely electrical and magnetic measurements. Figure 1.8 shows pictures of Oliver Heaviside (left) and Heinrich Hertz (right). It should be mentioned here that nine years prior to Hertz's work, the English engineer and musician, David Edward Hughes, had carried out similar experiments with spark gap apparatus to demonstrate the existence of radio waves. He called them aerial waves, but his work was somewhat less extensive and was not widely recognized by his peers.

1.3 Maxwell's Equations and Electromagnetic Waves

Maxwell's equations, as were first written by Oliver Heaviside, consist of a set of four equations which can be written as follows in modern notation:

$$\nabla.\mathbf{D} = \rho \tag{1.1}$$

$$\nabla.\mathbf{B} = 0 \tag{1.2}$$

$$\nabla \times \mathbf{E} = \frac{-\partial \mathbf{B}}{\partial t} \qquad (1.3)$$

$$\nabla \times \mathbf{H} = \mathbf{J} + \frac{\partial \mathbf{D}}{\partial t} \qquad (1.4)$$

The first equation is Gauss's law for electric charges. It relates the divergence of electric displacement, **D**, at a point to the electric charge density at that point. One of the principal consequences of this relation is that electric charges only reside at the outer surface of charged conducting objects.

The second equation here is the analogous relation for magnetic fields. It says that the divergence of the magnetic induction (also called magnetic flux density), **B**, at a point is always zero. This is because free magnetic charges (magnetic monopoles) do not exist, and, thus, the divergence of magnetic induction always vanishes identically.

The third equation above is a statement of Faraday's law and simply states that the curl of the electric field, **E**, at a given point is equal to the rate at which a magnetic field there changes in time. This phenomenon forms the basis of all electric generators, as well as some charged particle accelerators, such as the betatron.

The last equation here is the mathematical statement of Ampère's law, as modified by Maxwell. It relates the divergence of the magnetic field strength, **H**, at a point to the current density at that point, as well as to the time rate of change of electric displacement at that point. The latter term was added by Maxwell, as has already been pointed out. This equation clearly shows that magnetic fields can be created by the presence of electric currents or by changing electric fields.

Note that the electric field strength and electric displacement vectors, **E** and **D**, respectively, are related as:

$$\mathbf{D} = \varepsilon\, \mathbf{E} = \varepsilon_r\, \varepsilon_0\, \mathbf{E} \qquad (1.5)$$

Here, $\varepsilon = \varepsilon_r\, \varepsilon_0$, where ε is electric permittivity, ε_r is relative electric permittivity (also called dielectric constant, κ), and ε_0 is the absolute electric permittivity of free space. Its value is 8.85×10^{-12} Farad/m.

Similarly, the magnetic field strength and magnetic induction vectors, **H** and **B**, respectively, are related as

$$\mathbf{B} = \mu\, \mathbf{H} = \mu_r\, \mu_0\, \mathbf{H} \qquad (1.6)$$

Here, $\mu = \mu_r \mu_0$, where μ is magnetic permeability, μ_r is relative magnetic permeability, and μ_0 is the absolute magnetic permeability of free space. Its value is $4\pi \times 10^{-7}$ Henry/m.

These four equations are remarkable in their scope, as essentially almost all our classical knowledge of electric and magnetic phenomena is contained in these four relations. The operating principles of all electromagnetic devices including motors, generators, radars, particle accelerators, microwave ovens and radio communication systems are based on the physics contained in these equations. The compilation of this set of four equations was a crowning glory of classical physics, and laid the foundation of much of the engineering that followed in later years.

Maxwell's equations, as presented above, are written in differential form, i.e. in terms of vector differential operators and apply point-wise in a given region of space. It is possible to re-write them in an integral form so that these relations apply to finite volumes of space instead. The integral forms of Maxwell's equations are written as

$$\oint \boldsymbol{D} \cdot \boldsymbol{ds} = \rho \tag{1.7}$$

$$\oint \boldsymbol{B} \cdot \boldsymbol{ds} = 0 \tag{1.8}$$

$$\oint \boldsymbol{E} \cdot \boldsymbol{dl} = -d\phi_B/dt \tag{1.9}$$

$$\oint \boldsymbol{H} \cdot \boldsymbol{dl} = J + \varepsilon_0 d\phi_E/dt \tag{1.10}$$

Equations (1.3) and (1.4) (or (1.9) and (1.10)) clearly show that the electric and magnetic fields are related to each other in dynamical situations, i.e. change in one type of field creates the other type of field. In order to go further, we can start by writing down equations (1.1) to (1.4) in vacuum, defined as a region free of electric charges and currents ($\rho = 0$ and $J = 0$). Additionally, in a vacuum we have, $\varepsilon_r = 1$ and $\mu_r = 1$, i.e. relative electric permittivity and relative magnetic permeability are both equal to unity. Then, we get:

$$\nabla . \boldsymbol{E} = 0 \tag{1.11}$$

$$\nabla . \boldsymbol{H} = 0 \tag{1.12}$$

$$\nabla \times \mathbf{E} = -\mu_{o} \frac{\partial H}{\partial t} \qquad (1.13)$$

$$\nabla \times \mathbf{H} = \varepsilon_{o} \frac{\partial E}{\partial t} \qquad (1.14)$$

Now, applying the vector differential operator 'curl' ($\nabla \times$) to equation (1.14), we get:

$$\nabla \times (\nabla \times \mathbf{H}) = \varepsilon_{o} \nabla \times \left(\frac{\partial E}{\partial t} \right)$$

using the vector calculus identity which holds for any vector **G**:

$$\nabla \times (\nabla \times \mathbf{G}) = \nabla(\nabla \cdot \mathbf{G}) - \nabla^{2}\mathbf{G} \qquad (1.15)$$

we obtain:

$$\nabla(\nabla \cdot \mathbf{H}) - \nabla^{2}\mathbf{H} = \varepsilon_{o} \nabla \times \left(\frac{\partial E}{\partial t} \right) = \varepsilon_{o} \frac{\partial(\nabla \times E)}{\partial t}$$

now, $\nabla.\mathbf{H} = 0$ and $\nabla \times \mathbf{E} = -\mu_{o} \frac{\partial H}{\partial t}$ from equations (1.12) and (1.13) above, so, we get:

$$\nabla^{2}\mathbf{H} - \frac{1}{c^{2}} \frac{\partial^{2} H}{\partial t^{2}} = 0 \,;\, \text{with } \frac{1}{c^{2}} \text{ written in place of } \varepsilon_{0}\mu_{0} \qquad (1.16)$$

In a similar manner, from equation (1.13), we can write:

$$\nabla \times (\nabla \times \mathbf{E}) = -\mu_{o} \nabla \times \left(\frac{\partial H}{\partial t} \right)$$

which, on account of the identity (1.15) becomes:

$$\nabla(\nabla \cdot \mathbf{E}) - \nabla^{2}\mathbf{E} = -\mu_{o} \nabla \times \left(\frac{\partial H}{\partial t} \right) = -\mu_{o} \frac{\partial(\nabla \times H)}{\partial t}$$

now, $\nabla.\mathbf{E} = 0$ and $\nabla \times \mathbf{H} = \varepsilon_o \frac{\partial E}{\partial t}$ from equations (1.11) and (1.14) above, so that:

$$\nabla^2 \mathbf{E} - \frac{1}{c^2}\frac{\partial^2 \mathbf{E}}{\partial t^2} = 0; \text{ with } \frac{1}{c^2} \text{ written in place of } \varepsilon_o\mu_o \qquad (1.17)$$

Equations (1.16) and (1.17) are mathematically of the same form, one involving the electric field $\mathbf{E}$ and the other involving the magnetic field $\mathbf{H}$. Furthermore, these equations have the same mathematical structure as the general wave equation which describes the un-attenuated propagation of any physical disturbance $\varphi(x,t)$ in space and time, with velocity V:

$$\nabla^2 \varphi - \frac{1}{V^2}\frac{\partial^2 \varphi}{\partial t^2} = 0 \qquad (1.18)$$

This leads to the conclusion that electric and magnetic fields can propagate through vacuum (free space) at speed given by:
$c = 1/\sqrt{\varepsilon_o\mu_o}$, which is very close to the experimentally measured speed of light, 2.9979×10^8 m/s.

Now, the wave equation (1.18) can have its solution written as

$$\varphi = \varphi_0 e^{-i(\omega t - k.r)} \qquad (1.19)$$

and, thus, the solutions of equations (1.17) and (1.16) can be analogously written as

$$\mathbf{E} = \mathbf{E}_0 e^{-i(\omega t - k.r)} \qquad (1.20)$$

$$\mathbf{H} = \mathbf{H}_0 e^{-i(\omega t - k.r)} \qquad (1.21)$$

Here $\mathbf{k}$ is the so-called wave vector given by

$$\mathbf{k} = k\mathbf{n} = (2\pi/\lambda)\mathbf{n} = (2\pi\nu/c)\mathbf{n} = (\omega/c)\mathbf{n}$$

where $\mathbf{n}$ is a unit vector in the direction of propagation of the wave, and, thus, the vector $\mathbf{k}$ also points in the direction of wave propagation. λ and ν are the wavelength and frequency of the wave, respectively. $\omega = 2\pi\nu$ is called the circular frequency of the wave. We, thus, establish that the $\mathbf{E}$ and $\mathbf{H}$ vectors, that form an electromagnetic wave, oscillate in a sinusoidal fashion.

On substituting equations (1.20) and (1.21) into equations (1.16) and (1.17) it becomes clear that applying the space differential operator is equivalent to multiplying the field vectors by $i\mathbf{k}$, whereas applying the time derivative operator is equivalent to multiplying the field vectors by $-i\omega$. Thus, equations (1.11) to (1.14) can be re-written as:

$$\mathbf{k}.\mathbf{E} = 0 \qquad (1.22)$$

$$\mathbf{k}.\mathbf{H} = 0 \qquad (1.23)$$

$$\mathbf{k} \times \mathbf{E} = \omega\mu_0\mathbf{H} \qquad (1.24)$$

$$-\mathbf{k} \times \mathbf{H} = \omega\varepsilon_0\mathbf{E} \qquad (1.25)$$

These relations are true for a plane electromagnetic wave traveling in free space, and lead to the following conclusions:

According to equation (1.22), the $\mathbf{E}$ vector is perpendicular to the direction of travel of the wave and according to equation (1.23), the $\mathbf{H}$ vector is also perpendicular to the direction of travel of the wave. Thus, in an electromagnetic wave both the electric field vector $\mathbf{E}$ and the magnetic field vector $\mathbf{H}$ remain at right angle to the direction of propagation of the wave. Therefore, electromagnetic waves are transverse in character and can, therefore, be polarized, i.e. their $\mathbf{E}$ and $\mathbf{H}$ vectors could be made to lie in planes that are perpendicular to each other. Furthermore, taking into account equations (1.24) and (1.25) above, $\mathbf{E}$ and $\mathbf{H}$ are also perpendicular to each other. Thus $\mathbf{E}$, $\mathbf{H}$, and the direction of wave propagation, together constitute an orthogonal system for an electromagnetic wave propagating in free space, as shown in Figure 1.9.

We can now go back to equation (1.24):

$$\mathbf{K} \times \mathbf{E} = \omega\mu_0\mathbf{H}$$

and rewrite it as

$$\mathbf{H} = k/\omega\mu_0 \, (\mathbf{n} \times \mathbf{E})$$

This follows because $\mathbf{k} = \mathbf{n}k$.
Thus, we have:

$$\mathbf{H} = \mathbf{n} \times \mathbf{E}/c\mu_0 = c\varepsilon_0(\mathbf{n} \times \mathbf{E}) \qquad (1.26)$$

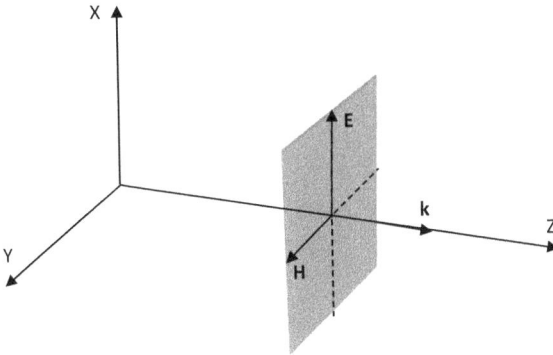

Fig. 1.9. Schematic illustration, showing the electric and magnetic field vectors at right angles to the direction of propagation of an electromagnetic wave.

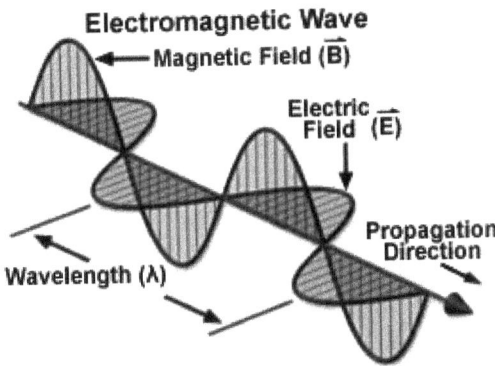

Fig. 1.10. Sinusoidal and in-phase variation of electric and magnetic field intensities in a propagating electromagnetic wave.

using $k = \omega/c$ and $1/c^2 = \varepsilon_0\mu_0$, we obtain:

$$|\mathbf{E/H}| = E_0/H_0 = c\mu_0 = 1/\, c\varepsilon_0 = \sqrt{(\mu_0/\varepsilon_0)} = Z_0 \qquad (1.27)$$

The ratio of $\mathbf{E}$ to $\mathbf{H}$ being real and positive implies that these vectors remain in phase as an electromagnetic wave propagates through free space, i.e. $\mathbf{E}$ and $\mathbf{H}$ reach their maxima and minima at the same time. This is illustrated in Figure 1.10 here.

From equation (1.27) it also follows that the magnitude of the electric vector **E** is Z_0 times the magnitude of the magnetic vector **H**. The quantity Z_0 has the dimension of impedance. It is called the characteristic impedance of free space and it has the value:

$$Z_0 = 120\pi = 376.73 \ \Omega$$

1.4 Energy Carried by Electromagnetic Waves

Electromagnetic waves transport energy. This is given by the so-called Poynting vector:

$$\mathbf{S} = \mathbf{E} \times \mathbf{H} \qquad (1.28)$$

As **S** clearly is orthogonal to **E** and **H**, so it lies in the direction of propagation of the electromagnetic wave, as one would expect.

With **H** given by equation (1.26) above, we can write:

$$\mathbf{S} = \mathbf{E} \times \mathbf{H} = \mathbf{E} \times \mathbf{n} \times \mathbf{E}/c\mu_0 = [(\mathbf{E}.\mathbf{E})\mathbf{n} - (\mathbf{E}.\mathbf{n})\mathbf{E}]/c\mu_0 = \mathbf{n}(E^2/c\mu_0) \qquad (1.29)$$

Here the identity for the vector triple product, $\mathbf{A} \times (\mathbf{B} \times \mathbf{C}) = (\mathbf{A}.\mathbf{C})\mathbf{B} - (\mathbf{A}.\mathbf{B})\mathbf{C}$, has been used, and the fact that $\mathbf{E}.\mathbf{n} = 0$ as $\mathbf{E} \perp \mathbf{n}$.

As the quantity in the bracket on the extreme right in equation (1.29) is a scalar so it is clear that the energy carried by an electromagnetic wave flows in the direction of wave propagation.

A further result can be obtained by taking the ratio of electric and magnetic energy densities, i.e. u_E and u_H. It can be shown that in free space (vacuum), $u_E = \frac{1}{2}\varepsilon_0 E^2$ and $u_H = \frac{1}{2}\mu_0 H^2$. We, therefore, get:

$$u_E/u_H = \frac{1}{2}\varepsilon_0 E^2 \Big/ \frac{1}{2}\mu_0 H^2 = \varepsilon_0 /\mu_0 (E/H)^2 = 1 \qquad (1.30)$$

this has used equation (1.27) above.

Equation (1.30) shows that in an electromagnetic wave propagating in free space, the electric and magnetic fields carry equal energy densities.

1.5 The Generation of Electromagnetic Radiation — Classical Description

As should be clear from the discussion in the previous sections, the classical theory of electromagnetism provides a firm mathematical and physical foundation for explaining the nature of electromagnetic waves. Depending on their frequency, these can range from AC voltages on our household mains lines to gamma rays that are emitted by radioactive materials. In between exist radio waves, microwaves, infrared radiation, visible light, ultraviolet radiation and X-rays — in increasing order of frequency. It is remarkable that a single theory encompasses such a very wide spectrum of wave phenomena. The entire electromagnetic spectrum is so wide that very different techniques are used for generating and detecting electromagnetic radiation in different regions of the spectrum. Even the way the different types of electromagnetic waves are specified is different. Radio and microwaves, for instance, are identified with respect to their frequency, infrared radiation is often specified in terms of wave numbers, visible light and ultraviolet radiation are identified with their wavelengths while X-rays and gamma rays are specified with regard to the energy in electron volt units, carried by their photons.

The discussion of previous sections can be extended to explain also the generation of electromagnetic radiation. Stationary electric charges create a static electric field around them, charges in uniform rectilinear motion create both an electric field and a magnetic field around them, whereas charges in non-uniform motion, such as accelerated charges, can radiate self-sustaining electromagnetic fields that detach from charges and can freely propagate in space. This last phenomenon is the origin of all electromagnetic radiation, including the light that we see all around us. Lower frequency electromagnetic waves — up to the microwave range — is usually generated by accelerating charge distributions in electrical conductors, such as antennas and resonant cavities. Higher frequency radiation, on the other hand, is generated by charge acceleration in molecules, atoms and atomic nuclei. The detailed electromagnetic field theory of radiation generation from accelerated charges was developed by the French physicist Alfred–Marie Liénard in 1898, and independently by the German scientist Emil Wiechert in 1900. Their pictures appear in Figure 1.11. Liénard and Wiechert's theory describes the so-called Liénard–Wiechert potentials that can be shown to give rise to freely propagating electromagnetic waves.

That accelerating charges give rise to electromagnetic radiation is not hard to verify. Charged particles, accelerating in both linear and circular accelerators, lose energy by generating electromagnetic waves.

In fact, this phenomenon is used to advantage for generating highly intense beams of radiation in the short wavelength electromagnetic regions, such as ultraviolet radiation and X-rays. Synchrotron radiation

Fig. 1.11.   (Left) Alfred-Marie Liénard, (Right) Emil Wiechert.

Fig. 1.12.   Advanced Photon Source (APS) synchrotron at the Argonne National Lab.
Courtesy: Argonne National Lab.

sources, for instance, accelerate charged particles and extract electromagnetic energy from accelerating streams at periodic locations, called beam
lines, around the circumference of the accelerator. Figure 1.12 shows an
inside view of the synchrotron at the Advanced Photon Source (APS)

located at the Argonne National Lab in Lemont, Illinois. This facility provides 65 X-ray beam lines around its circumference, where experiments on materials science, biology, physics etc. could be carried out with intense X-ray photon flux. The current system operates with maximum electron beam current of 100 mA, but this will be raised considerably in the near future as the source is modified to generate much higher X-ray flux. Similar synchrotron radiation sources are also operated in several other countries.

Synchrotron-based radiation sources accelerate electrons to high energies and then pass high-energy electrons through a set of periodic 'undulator' magnet poles, as shown in the schematic illustration in Figure 1.13. The strong cyclic field causes the electrons to bunch and wiggle, and the resulting rapid accelerations and decelerations result in the emission of X-ray photons. This mechanism generates extremely high X-ray intensities. The APS beam lines can be configured for X-ray flux as high as 6×10^{19} photons per second per square millimeter. In the future, this will be

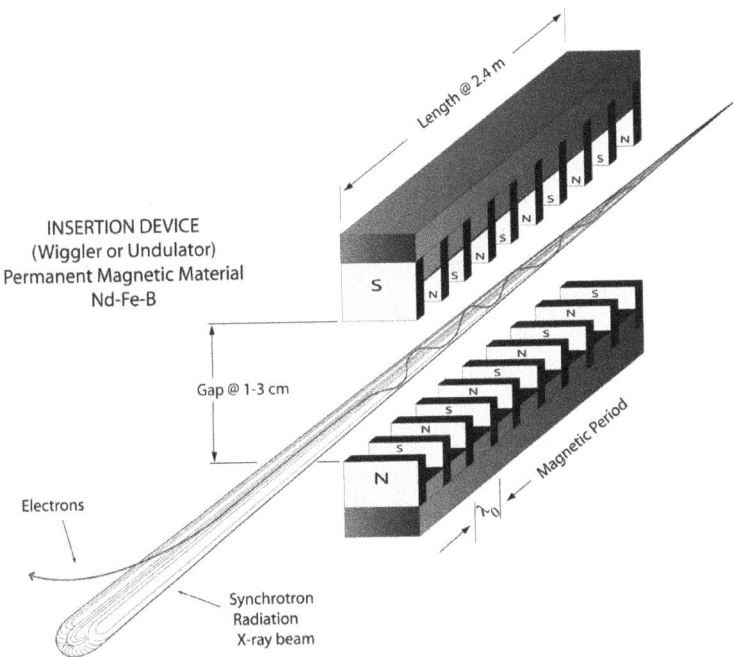

INSERTION DEVICE
(Wiggler or Undulator)
Permanent Magnetic Material
Nd-Fe-B

Length @ 2.4 m

Gap @ 1-3 cm

Magnetic Period

Electrons

Synchrotron
Radiation
X-ray beam

Fig. 1.13.   Schematic illustration of the use of a magnetic wiggler (undulator) used to generate X-rays from an accelerated electron beam.

increased by several hundred times, once modifications to the synchrotron have been completed.

A further development of this concept is the free electron laser (FEL) where synchrotron-emitted X-rays are made to interact with accelerated electron bunches in a resonant cavity. This process generates extremely intense, short duration, X-ray pulses through stimulated emission. Unlike conventional lasers, which are discussed in Chapter 6, in an FEL there is no material gain medium, and the laser radiation is generated by the inter-action of X-rays with relativistic electrons (having speeds close to that of light). The absence of a material gain medium means that FELs are freely tunable over the region of their operation. Besides their high flux output, wavelength tuneability is one of the most prominent, and useful, charac-teristics of these sources. The wavelength of the radiation emitted can be tuned by adjusting either the energy of the electron beam or the period of the undulator magnet poles.

The output wavelength, λ, from an FEL is closely given by

$$\lambda = L/2\gamma^2 \tag{1.31}$$

Here, L is the undulator period (typically, 1–2 cm) and γ is the Lorentz factor $1/\sqrt{(1 - v^2/c^2)}$, with v being the velocity of the electrons injected into the undulator. As usual, c stands for the speed of light.

FEL operation was first demonstrated at Stanford University in 1977. The device built there, operated in the infrared region at 3.4 microns. Operation in the X-ray region was achieved later with more energetic electron beams. Due to their desirable properties, FELs are now being developed by several organizations around the world as valuable research resources. Figure 1.14 shows an X-ray FEL linear accelerator beam line at the 3.4 km long European X-ray FEL facility in Hamburg, Germany.

Radiation generated by synchrotrons and X-ray FELs is transported to experimental stations through long photon delivery tubes. There, the radiation is made to interact with experimental samples under controlled conditions to investigate various aspects of their structure, and other prop-erties of interest. Figure 1.15 shows synchrotron radiation (visible through its fluorescence) exiting a photon delivery tube at an experimental station.

So far, we have seen that accelerated charged particle beams can be used as sources of electromagnetic radiation. It is also possible to produce radia-tion from the deceleration, i.e. negative acceleration of beams of charged

Fig. 1.14.   The linear accelerator beam line at the European free electron laser facility.
Courtesy: European XFEL.

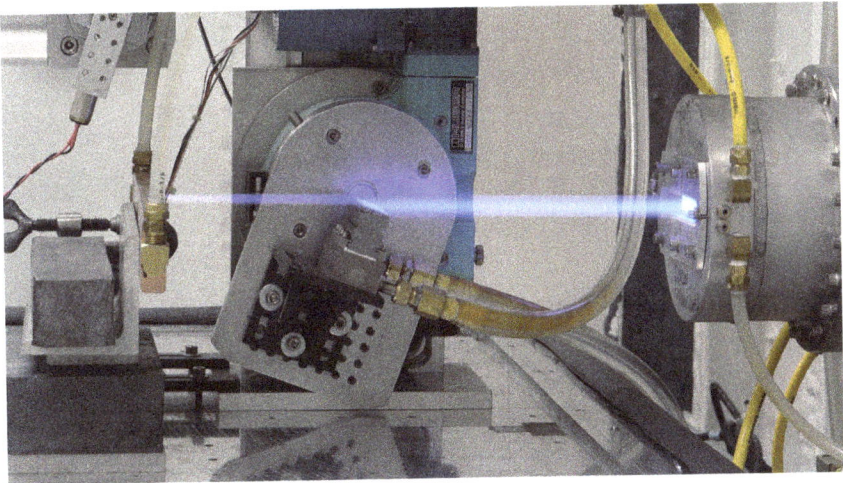

Fig. 1.15.   Synchrotron radiation exiting from a photon delivery tube.
Courtesy: European XFEL.

Fig. 1.16. Wilhelm Conrad Röntgen.

Fig. 1.17. A commercial 50–75 W output X-ray tube.
Courtesy: Oxford Instruments.

particles. One example of a source that operates on this principle is an X-ray tube. In a typical device, electrons are accelerated by a set of high-voltage electrodes inside an evacuated glass tube, and the accelerated beam is made to strike a heavy metal target, made of tungsten or molybdenum. The sudden deceleration experienced by electrons, as they enter the metal block, causes the emission of X-rays. This mechanism was discovered by Wilhelm Conrad Röntgen (see Figure 1.16) in Germany, and he was awarded a Nobel Prize in 1901 for the discovery of X-rays. This was the first ever Nobel Prize for physics. Figure 1.17 shows a commercial 50–75 W X-ray tube.

Finally, an interesting way in which fast particles can produce electromagnetic radiation is if their speed exceeds that of the speed of light in a dielectric medium. Of course, nothing can travel faster than light in vacuum, but in material media, where light travels at a slower speed, it is possible for sub-atomic particles to have speeds in excess of the local speed of light. If this condition is satisfied then particles lose energy by producing electromagnetic radiation, and gradually slow down to speeds below that of light. This is the optical counterpart to the sonic boom generated when a solid object travels in air at a speed exceeding that of sound. This radiation is, thus, analogous to acoustic shock waves. Radiation emitted in such manner is called Cherenkov radiation, after its discoverer, the Soviet physicist Pavel Alekseyevich Cherenkov. He received a joint Nobel Prize in 1958, with Ilya Frank and Igor Tamm, for the discovery of Cherenkov radiation. Nuclear reactors moderated by liquid water often display a characteristic blue glow which is caused by the emission of Cherenkov radiation from fast neutrons, emitted by the reactor's core, traversing through the surrounding mass of water (see Figure 1.18).

While classical electromagnetic theory provides a clear and straightforward description of radiation generation from accelerated or decelerated

Fig. 1.18.   Cherenkov radiation, seen in the moderator water pool of a nuclear reactor.

charges (whether in free space or inside conductors), it is not as easily applicable to higher-frequency radiation, which is not usually treated in terms of electric and magnetic field distributions. Infrared, visible light and other higher-frequency radiation is more easily treated as collection of energy- and momentum-carrying non-material particles, called photons. In this 'optical' regime, the generation and absorption of radiation is best treated through quantum mechanical treatments that involve either discrete energy levels or energy bands that are occupied by electrons in atoms, molecules and ions. In this formulism, it is the change of electronic energy between allowed energy levels that is behind the absorption and emission of light. The physics of photon generation and absorption through quantum mechanical mechanisms will be discussed where appropriate in later chapters.

2 Incandescent Lamps

2.1 Introduction

Solids heated to incandescence have long provided sources of heat and light. The most familiar electrical example being the ordinary tungsten lamp. This kind of light source is usually simple and cheap, but quite inefficient in converting electrical energy to visible light. The luminous intensity and spectrum of the emitted radiation is a function of the surface temperature of the heated element and its material properties. This chapter is devoted to a detailed description of filament-based incandescent lamps. We start by looking at the physics of light generation from heated solid objects and the various measures used to quantify light generated by different light sources. This is followed by a review of the development of incandescent lamp technology during the 19th and early 20th centuries. This was a period marked by many inventions and equally numerous inventors. We also discuss the Nernst glower which was a different incandescent lamp technology that briefly made its appearance during the early years of the 20th century. By the end of the first quarter of that century, the tungsten filament lamp had taken the form that we are now familiar with. The prominent features of standard tungsten lamps are described next, followed by those of tungsten-halogen lamps which were developed during the 1950s. Initially, developed for specialist lighting applications, halogen lamps have gradually proliferated to all kinds of lighting applications, and are now also widely used for domestic lighting. Until quite recently, incandescent lamps remained the most widely manufactured electrical products, despite their pitiful efficiency in converting electrical energy to light. This is a testament to the engineering effort that, over the years, has gone into extracting as much efficiency as reasonably possible from this seemingly trivial technology.

2.2 Generation of Heat and Light from Electrically-heated Conductors

Electrically-generated incandescence arises from the heating of a solid or liquid material when a current is passed through it. Resistive heating raises the temperature of all conductors, except superconductors. Due to obvious practical reasons, liquids melted and heated to incandescence are, generally, not used in practical light sources. This is in spite of the fact that liquid incandescence can generate much higher brightness and that too at a higher energy conversion efficiency.

2.2.1 *Resistive heating due to the passage of electric current*

It is a common observation that the passage of an electrical current serves to heat the conductor that it flows through. Another common observation is that as a material is heated it first begins to give off heat (infrared radiation), and on further heating this changes to a visible red glow. Heating even further causes both the intensity of the emitted radiation to increase and the emitted color to gradually change from red to orange and then to yellow; finally turning to white heat. We'll now look at the mechanism behind electrical heating and the physics that underlies the trend that gives rise to changing colors, as the temperature of a heated object changes. The flow of an electrical current through a metallic conductor is simply the movement of freely mobile electrons through the conductor. The electrons move from the high electrical potential end of the conductor to the low potential end, and in many respects the flow resembles the flow of water through an obstructed pipe, due to difference in water pressure at the two ends of the pipe. The electrons travel through the conductor at a velocity that is determined by the amount of current, the geometry of the conductor and the density (number per unit volume) of free (conduction) electrons present in the material, according to the relation:

$$v_d = J/ne = i/Ane$$

Here, v_d is the drift velocity of electrons, J is the current density, i.e. the current per unit cross-sectional area of the conductor (i/A), n is the number density of electrons in the conductor and e is the electron charge (1.6021×10^{-19} Coulomb).

Calculation of the drift velocity, for a typical material, such as a copper wire, will show that drift velocities are usually of the order of a few millimeters per second. Thus, electrons don't move particularly fast through ordinary conductors. The chief reason for their sluggish movement is their continual collisions with the ion cores that make up the structure of the conductor. Negatively charged electrons experience strong coulomb scattering from positively charged ion cores. Each ion core may carry one, two, three or even four units of positive charge, depending on the number of valence electrons that have been shed by the metal atoms. The free electrons permeate the spaces between the ion cores and serve to bind them together in a stable metallic structure. This type of bonding, where the free electrons act as a kind of glue, is called metallic bonding. The free electrons stabilize an otherwise extremely unstable system of mutually repelling positive ion cores. Besides keeping the metal intact, the free electrons also shield the ion cores so that their effective charge is reduced from the values they will have without the intervening electrons. Free electrons are also responsible for most of the thermal conductivity exhibited by metals, and it is their presence that makes most metals such good conductors of heat.

It is an interesting fact that if we had a perfect metallic crystal, with no disorder or thermal vibrations, then electrons will be able to traverse it without undergoing any collisions at all. This is not an obvious conclusion but can be rigorously proved from quantum mechanical considerations. However, we never observe collision-less electron transport in practical conductors (the only exceptions being superconducting materials and situations where electrons travel over very short distances — so-called ballistic conductors). This is because all conductors have both structural disorder and non-stationary ion cores due to thermal vibrations. Both these imperfections tend to scatter electrons; greatly impeding their progress and giving rise to electrical resistance. The higher the structural disorder due to factors such as grain boundaries, ion cores out of lattice positions, presence of impurity atoms or ions etc., the higher the electrical resistance of the conductor. Metallic conductors encountered in everyday life have plenty of these structural faults and, thus, present considerable resistance to the flow of electrical current.

Each time an electron scatters, it starts on a fresh journey; accelerated by the electric field generated by the potential difference (voltage) across the length of the conductor. After a short time, it suffers another collision and then gets accelerated again. The energy gained by the electron during its acceleration is promptly lost as it suffers a scattering event. This energy shakes the ion cores which gain kinetic energy and start to vibrate more vigorously. The vibrating cores, thus, get their energy from the electric field through the intermediary of conduction electrons. The vibration of ion cores is perceived as heat and is experienced as an elevated temperature. The vibrating cores shed their energy by emitting long wavelength infrared photons. This is what happens when a conductor gets hot to touch due to the passage of an electrical current. We say that electrical energy has been dissipated as heat. This energy has to be supplied by the source that maintains a voltage, i.e. potential difference between the ends of the conductor. The dissipated power (energy consumed per unit time) can be calculated as

$$E = VI = I^2R = V^2/R$$

Here, V is the potential difference across the conductor, I is the current through the conductor and R is the resistance of the conductor.

The jostling of metallic ion cores results in the generation of electromagnetic radiation with a continuous spectrum. Such continuum radiation is characteristic of all incandescent light sources.

2.2.2 *Thermodynamics of incandescence*

As the temperature of an incandescent body increases, the radiation from it becomes more intense. This is a consequence of the excitation of ionic cores in accordance with the classical Maxwell–Boltzmann distribution. The probability of finding a core at an energy E is given by the Maxwell–Boltzmann distribution function as

$$f(E) = (1/kT) \, e^{-E/kT}$$

If N is the total number of cores, then the number of cores in the differential energy interval E to $E + dE$ can be written as

$$N(E)dE = (N/kT)e^{-E/kT}dE$$

The number of excited ion cores in the energy range E_a to E_b can then be calculated as:

$$N_{a-b} = \int_{Ea}^{Eb} N(E)dE = (N/kT)\int_{Ea}^{Eb} e^{-E/kT} \, dE$$

This equation can be used to investigate the temperature required to raise the energy of ion cores in metallic solids to values that will result in the emission of visible radiation.

Ordinary metals used for making incandescent elements, such as nickel-chromium alloys (nichrome), tungsten etc. have an ion core density of around 5×10^{28} m^{-3}. In order to emit visible light (wavelength band: 700–400 nm) ion cores in solids must be excited enough to have energies of at least 1.7 eV, that corresponds to a photon energy of 700 nm. Thus, we are interested in finding the number of cores with energies in excess of this lower limit. If a sufficiently large number of cores possess energies above this threshold, then enough photons will be emitted to make the heated object visible to us. Taking the number density of ion cores to be

$N = 5 \times 10^{28}$ m^{-3}, the lower energy limit as $E_a = 1.7$ eV and the upper energy limit as $E_b = \infty$, we can evaluate the integral above for room temperature $(T = 300\text{K})$ as

$$N_{\text{vis/300K}} = (N/kT)\int_{1.7eV}^{\infty} e^{-E/kT}\, dE \approx 5\times10^{28}\times3\times10^{-29} = 1.5\text{m}^{-3}.$$

Practically, this means that at room temperature (300 K) there are almost no ion cores excited enough to emit radiation visible to human eyes.

Repeating the same calculations for $T = 1000$ K we get:

$$N_{\text{vis/1000K}} = 1.2 \times 10^{20} \text{ m}^{-3}.$$

Thus, at 1000 K a very substantial number of ion cores are excited enough to visible radiation in the red to orange region.

At a still higher temperature of 2500 K, which is typical for an operating incandescent filament lamp, the number of excited ion cores approaches 10^{24} m^{-3}. With so many ion cores excited and emitting visible radiation, the filament appears extremely bright and, thus, acts as a source of bright light.

2.3 Characteristics of Radiation from Heated Surfaces

In order to understand the nature of radiation emitted from hot surfaces it is necessary to have an understanding of a hypothetical ideal emitter that can emit radiation at the highest theoretical efficiency. While such an object doesn't exist in nature, it can be closely approximated by surfaces that are very good absorbers of radiation. An idealized perfect absorber is called a black body. It is defined as any surface that absorbs all the radiation that falls on it, irrespective of the intensity or spectral makeup of the radiation. No physical object behaves like this, as all real materials reflect at least some of the radiation that falls on them. However, certain artificial materials, such as carbon black (a form of carbon soot), make very good absorbers of electromagnetic radiation and, thus, are very good approximations of an ideal black body. A more recent example is that of vertically oriented random single-walled carbon nanotubes. Surfaces composed of this material absorb more than 98% of incident radiation and, thus, make extremely good examples of physically accessible black bodies. Yet another example of a black body is a small hole drilled in a solid block of metal or graphite, containing an internal cavity. Any radiation that enters such a hole is multiply reflected inside and is almost completely absorbed. The entrance to the cavity then acts as an excellent approximation to a black body. The chief reason for the usefulness of the concept of a black body lies in the fact that black bodies are not only perfect absorbers but, conversely, also perfect emitters of radiation. This means that, when heated, a black body will emit more radiant power per unit area than any other surface at the same temperature. Thus, if a solid material block with a cavity open to the outside through a small hole, as described above, is heated, then radiation will first be observed to come out of the hole rather than the outer surface of the block. Even when the outer surface begins to emit radiation, that emitted from the hole remains significantly higher in intensity. We explain such observations by saying that every surface has a characteristic emissivity, with a black body having the highest emissivity. Moreover, the emissivity of a black body is independent of wavelength. The emissivity, ε, of a black body is defined as 1. For real surfaces, emissivities are wavelength-dependent and are always less than unity.

The radiation that comes out of a black body cavity is termed black body radiation or cavity radiation. This radiation has a unique spectral makeup at any given temperature which is completely independent of the shape of the cavity or the material that it is made of. This is a striking fact, and forms the basis of much of radiation physics. Typical black body spectral curves for several temperatures in the 100 K to 10,000 K range are shown in Figure 2.1.

This figure shows emitted optical power per unit area per unit wavelength interval in the 0.1 μm (100 nm) to 100 μm range. With rise in temperature, the emission increases at all wavelengths, the peak of the emission shifts toward shorter wavelengths and so does the short wavelength cut-off. Thus, increase in temperature causes the spectrum of thermal radiation from heated bodies to show a blue shift, i.e. the light changes from a reddish to a bluish hue. All light sources show these trends, to various degrees of agreement with this figure, depending on how closely the light-emitting source approximates an ideal black body.

In general, radiation emission from heated tungsten filaments closely agrees with the spectral intensity distributions shown in this figure.

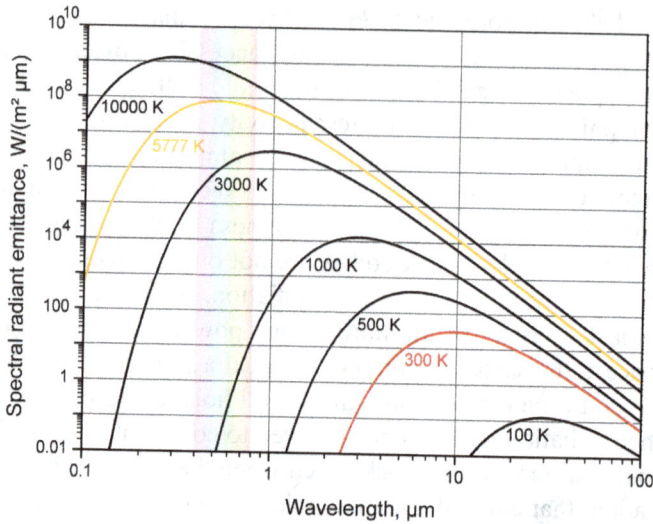

Fig. 2.1.   Ideal black body spectral emission curves for several temperatures in the 100 K to 10,000 K range.

2.4 Incandescent Lamps

During the late 19th century, and most of the 20th century, incandescent filament lamps were the most prominent sources of artificial visible radiation around. Their technology was perfected by the 1940s, and thereafter a more-or-less standard form of this lamp kept being produced in very large numbers, all over the world. The later development of the tungsten-halogen lamp, during the 1950s, served to keep incandescent lamps in the forefront of indoor lighting applications for many more decades. This lighting technology also survived after the development of electric discharge lamps, but gradually met its demise with the appearance of solid-state lighting devices based on the light-emitting diode (LED). In the rest of this chapter, we look at the history and technology of most kinds of incandescent lamps, before going on to examine other lighting technologies, in later chapters.

2.4.1 *Early history of the development of incandescent lamps*

Tungsten filament lamps owe their genesis to early efforts in the 19th century for developing electrically heated sources of visible radiation.

Fig. 2.2.   Humphrey Davy.

The thermal effect produced when an electrical current flowed through a resistive material was well known by the early years of the 19th century and it was appreciated that by heating a resistor to white hot heat it would be possible to create an electrical source of light. As early as 1802, Humphrey Davy (see Figure 2.2), in London, had demonstrated that when an electrical current was passed through thin strips of platinum, they could be made to glow white hot. Platinum, however, has always been very expensive, and its relatively low melting point of 1768°C meant that thin platinum conductors did not survive for long when resistively heated to high temperatures. Nevertheless, efforts to develop platinum strip-based incandescent light sources continued for several decades. Notable among these was the invention of an enclosed lamp by the Englishman, Warren De La Rue in 1840. He placed a platinum filament inside a partially evacuated glass bulb, making a safer and slightly longer lasting electric light.

In the following year, another English inventor, Frederick de Moleyns, patented a somewhat similar lamp. The very early work on incandescent lamps was carried out with platinum wire filaments because it is very easy to draw platinum into a fine wire, owing to its outstanding ductility. Platinum also is very resistant to oxidation at any temperature.

Other workers also tried filaments made of iridium — another metal, very similar to platinum. The prohibitive cost of platinum and iridium meant that this route was never going to result in a successful commercial invention. Efforts then turned toward finding cheaper materials for making electric lamps. In the 1840s John Wellington Starr of Cincinnati, Ohio, developed a 'continuous carbon burner' which consisted of strips of conductive carbon placed inside an evacuated glass bulb. Starr filed a patent for his lamp in 1844. This invention was preceded by that of carbon arc lamps which consisted of two carbon rods with an intervening gap. Because the carbon conductor in Starr's lamp, in contrast, was a continuous strip of graphitic carbon so it was named continuous burner. During his investigations, Starr quickly found out that a carbon conductor in the open, or inside an air-filled glass bulb, oxidized quickly at the high-operating temperatures. In order to protect the element, he next enclosed his carbon strip in a glass envelope and tried to remove as much of the air from the bulb as possible; mostly using crude chemical methods that were only marginally effective. Starr's lamp was the first serious step toward producing continuous conductor incandescent lamps but suffered from short conductor life, bulb blackening, and high-current requirements. As a note of historical interest, Starr's work was funded by the philanthropist George Peabody. Shortly after developing his lamp, Starr traveled to England with his attorney associate Edward Augustin King, where the latter filed for a patent for the lamp. Starr demonstrated his lamp in London but died a short time later in Birmingham, due to tuberculosis. The patent was granted to King on November 4, 1845, as British patent 1,0919 — a year after Starr's death.

The next major step in the evolutionary development of electric lamps was taken by Joseph Swan of England. A contemporary of John Starr, Swan was well aware of Starr's invention and set out to develop an even better lamp. After considerable experimentation he realized that the form of the conductive element in Starr's lamp was the cause of many problems. Until that time, both Starr and Swan had used thin carbon rods as the incandescent elements in their lamps. These rods had low resistance and, thus, the lamps had to use long rods to reduce the currents required to manageable values. This resulted in physically large lamps that were good for demonstration purposes, but were clearly not suitable for widespread use. Swan reasoned that a high-resistance element would be better as it would require less current to heat up to the same temperature. Accordingly, he introduced filamentary resistive elements in his lamp designs. At first, these were made of carbonized paper, but were latter replaced by carbon-coated collodion fibers. By the 1850s and 1860s, good mechanical pumps had become available and Swan was also quick to put them to good use in evacuating his light bulbs. The resulting Swan lamps were much more durable than its predecessors and gave off bright light with economical usage of electricity. These lamps were the first commercial luminaries to have been used on large scales for public space lighting.

At the same time as Swan's developments, Thomas Edison in the United States had also developed a very similar lamp. His lamps first used carbonized thread, but he later started using carbonized bamboo filaments. In the quest to build a better lamp, he even sent his technicians to Japan to find the very finest bamboo for use in filament lamps. The bamboo filaments were attached to lead in wires in his lamps using a carbon paste. Edison was somewhat ahead of Swan in that he had devised not only carbon filament lamps with sufficient brightness and longevity to be a commercial success but had also developed much of the infrastructure for implementing a proper electrical lighting system. This included things like generators and a power transmission system. Edison's bulbs were, arguably, also better evacuated than Swan's bulbs because he had discovered that materials inside glass bulbs, including the inner surface of the bulb itself, gave off gases after the bulb had been evacuated. This 'out gassing' spoiled the vacuum inside the bulb and, thus, shortened the lamp's life. Edison developed a better way of evacuating light bulbs using the then state-of-the-art Sprengel pumps. He also started the practice of heating the bulb and the filament mount before and during evacuation. This resulted in a much better ultimate vacuum and, thus, greatly prolonged the life of his incandescent lamps. Figure 2.3 shows one of Joseph Swan's lamps (left) with an Edison lamp (right).

For a few years, a controversy raged as to whether Swan or Edison was the true inventor of incandescent filament lamps. By the 1880s, it was widely recognized that both Swan and Edison had independently developed practical light bulbs, and the two went on to establish a joint company (Ediswan) in 1883, to manufacture and install electric lamps. The company went on to sell incandescent lamps for many years. A catalog and price list of Ediswan light bulbs, dating from 1893 appears in Figure 2.4. The Ediswan Company was later incorporated into Thorn Lighting Limited. Figure 2.5 shows the portraits of Joseph Swan (left) and Thomas Edison (right).

Fig. 2.3.   Swan lamp (left), Edison lamp (right).

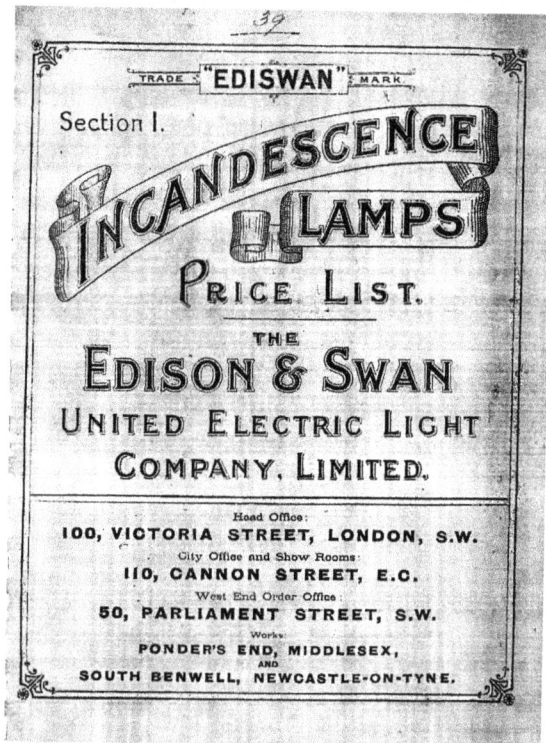

Fig. 2.4.   Catalog, with price list, of Ediswan light bulbs.

Fig. 2.5.   Joseph Swan (left), Thomas Edison (right).

Edison was not only a skilled inventor but also a shrewd businessman. He once said that "I have never perfected an invention that I did not think about in terms of the service it might give to others". By the year 1890, he had consolidated his various business interests into the Edison General Electric Company. However, soon afterwards, Edison faced stiff competition from a rival company called the Thomson-Houston Company. This company, headed by Charles A. Coffin — a former shoe manufacturer from Lynn, Massachusetts — had gradually formed from the merger of a number of smaller manufacturing companies. For a period, both companies existed independently and competed with each other, but it was eventually decided by Thomas Edison and Charles Coffin to merge the businesses together so as to take advantage of their combined strengths in manpower, patents and manufacturing facilities. In 1892, the two companies combined together to form General Electric (GE), which became a dominant player in the further development of incandescent lamp technology.

By the end of the 19th century, incandescent lamps were a common sight in the major cities in both North America and Western Europe. Electrical lighting installations were to be found in both public places, such as roads and parks, and in the homes of rich individuals. At that time, the traditional carbon filament lamp was a mature product, with production estimated at several million units a year. The lamps, however, had woefully low efficacy of only around 3 lumens per watt. Not only were these lamps inefficient, they also did not provide light that was bright enough. The last significant development in carbon filament technology was the work by Willis Whitney of GE in developing high temperature-baked carbon filaments. Whitney developed this new variety of carbon filament using an electric resistance furnace at GE's Schenectady laboratory. This filament material (pyrolytic graphite) had a superior morphology to the existing carbon filaments, which gave it a metal-like appearance. It also had a positive temperature coefficient of resistance like metals, instead of the negative coefficient that other forms of carbon have. For some time, it was used in commercial lamps that were sold under the trade name, 'General Electric Metallized' (GEM) lamp. Figure 2.6 shows a GE advertisement for GEM lamps.

The low efficiency and short lifetime of carbon filament lamps forced inventors, especially in Europe with its high energy costs, to search for

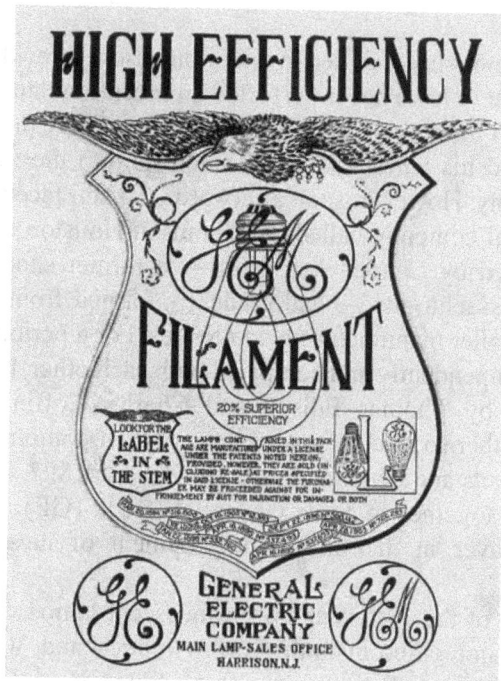

Fig. 2.6.   Advertisement for General Electric Metallized (GEM) lamp.
Courtesy: General Electric Company.

new alternatives to carbon filaments. The main problem with carbon-based lamps was the propensity for carbon to evaporate and coat the inside of the glass envelope. This was due to the high vapor pressure of carbon, and caused both the filament to weaken and the bulb to blacken over a period of time. It was obvious that replacing carbonized elements with metal wires was the answer to this problem. Metal wires could sustain much higher temperatures without appreciable evaporation and produce brighter, more energy efficient lamps. The Austrian physicist, Carl Ritter Auer van Welsbach, was the first to use a metal filament in an electric lamp in around 1898. The very brittle osmium filaments that he used, yielded brighter light than carbon filament lamps but were difficult and expensive to manufacture. This metal had the advantage of having a high melting point (2700°C), combined with good ductility that allowed it to be drawn into fine wires. Osmium filament lamps were manufactured in Austria and Germany and were sold in Austria, Germany and Britain under the trade names of Osmin, Auer-Os and Osmi, respectively. Due to the very high cost of Osmium, these lamps were usually rented out by the manufacturer instead of being sold. Once a lamp was burnt out, its osmium filament was recycled for making new lamps. Osmium filament lamps went obsolete by 1905, as osmium was replaced by tantalum — a cheaper alternative with superior strength and an even higher melting point (2996°C). Tantalum lamps were principally developed by Dr Werner von Bolton and Dr Otto Fuerlein of Siemens and Halske Company, in Berlin–Moabit, in Germany. Later, Siemens established a subsidiary called Osram (the name comes from the combination of Osmium and Wolfram — an alternative name for tungsten) for the development of commercial incandescent lamps. Tantalum lamp technology was later also licensed to GE in the United States for $250,000 and for several years tantalum filament was exported from Germany for the manufacture of tantalum filament lamps in the United States. Figure 2.7 shows a German tantalum filament lamp, dating from 1907. Glass bulbs were evacuated through an opening at the top, which was then fused to form the characteristic fused tip at the top. All vintage bulbs have this feature. Modern bulbs are evacuated and gas filled through the bottom, using a dedicated exhaust tube and, thus, these are described as tip-less bulbs.

Tantalum lamps produced by GE in the United States were sold under the trade-name MAZDA (after the Persian god of light), around 1909. Figure 2.8 shows a GE tantalum filament lamp. Note the similarity in design to the German lamp. Because of the lower resistivity of metals,

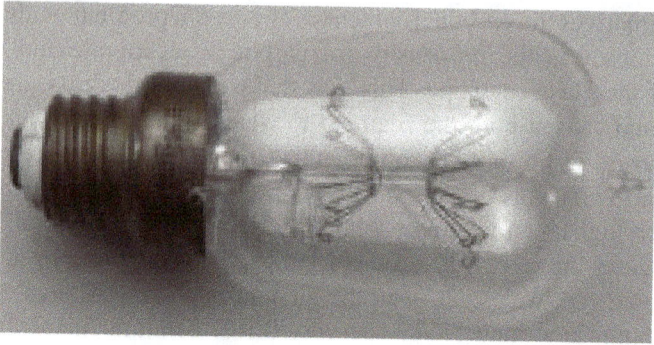

Fig. 2.7. Osram tantalum filament lamp.
Courtesy: Osram A.G.

Fig. 2.8. GE tantalum filament lamp.
Courtesy: General Electric Company.

metallic filament lamps had much longer filaments than carbon filament lamps of the same resistance. These long straight filaments required extensive support and, as seen in the figures here, tantalum lamps used to have numerous upper and lower hooks to guide and support the filament. These lamps were the most advanced of their times and allowed electric lighting to really flourish for the first time. Due to their reliability and high light output, tantalum lamps were extensively used on the ill-fated vessel Titanic, where they were one of the many high-end features on-board. In 1907, the Siemens Company also started marketing lamps with filaments made from an alloy of tungsten and tantalum. These were sold as WOTAN lamps — the name coming from a combination of wolfram (tungsten) and tantalum.

By the year 1911, however, tantalum too was replaced by tungsten — a metal even superior to tantalum for use in lamp filaments. Its usefulness in this regard arises from its very high-melting point of 3422°C (the highest of all metals, and second only to carbon among all elements), and a very low vapor pressure (1 Pascal at 3204°C). Combine this with its high density, high tensile strength, and extremely low coefficient of thermal expansion and tungsten becomes impossible to beat for use as incandescent lamp filament wire. The remarkable properties of tungsten originate from the very strong covalent bonding mediated by 5d valence electrons in its crystals.

Commercially, the very first tungsten filament lamps were marketed by the Hungarian company, Tungsram, in 1904. In that year, Hungarian Sándor Just and Croatian Franjo Hanaman were jointly awarded a Hungarian patent for a tungsten filament lamp much superior to the carbon filament lamp of the day. Figure 2.9 shows an advertising poster for Tungsram light bulbs. The Tungsram Company has been a successful light bulb and vacuum tube company throughout these years. It was acquired by GE in 1990 and is now a subsidiary of GE Lighting.

While the early developments in tungsten filament lamps took place almost entirely in Europe, the Americans were quick to catch on by developing industrial processes for the manufacture of ductile tungsten wires. This was an essential development because it enabled industrial scale manufacture of superior lamps at economical prices. Before the advent of drawn tungsten wires, only sintered tungsten wires were available, which formed brittle filaments. William Coolidge of the General Electric Company, seen here in Figure 2.10, was awarded several patents for this important development. For several years, he worked on the metallurgy of tungsten, and, in the end, succeeded in developing an industrial process for making uniform diameter, die-drawn, ductile tungsten wire.

Filaments made from ductile tungsten wire quickly became the mainstay of all incandescent light bulbs, and remain so to this day. By the time of this development, electric lamps had reached an efficacy figure of 10 lumens per watt. Coolidge's colleague, Irving Langmuir, soon after discovered that by coiling the tungsten filament and filling an inert gas like nitrogen inside the bulb he could make an even better lamp. Coiling the filament allowed a much longer and higher resistance element to be

Fig. 2.9.   Advertisement for Tungsram tungsten filament lamp.

placed inside the bulb, whereas the inert gas prevented excessive tungsten evaporation. This prolonged the lamp life significantly while allowing it to operate at still higher temperature and increasing the efficacy to 12 lumens per watt. These developments are described in more detail where we discuss the modern incandescent lamp further on in this chapter.

Fig. 2.10.   William Coolidge.

Table 2.1   Properties of various filament materials used in incandescent lamps.

Filament Material	Year Introduced	Efficacy (Lumens/watt)	Mean Lifetime (hours)
Carbonized paper	1880	1.7	600
Carbonized cellulose	1884	3.4	400
Osmium	1889	5.5	1500
Tantalum	1902	5.0	700
Metallized carbon	1904	4.0	600
Sintered tungsten	1904	7.9	800
Drawn tungsten	1910	10	1000
Coiled tungsten	1916	12.5	1000
Coiled-coil tungsten	1936	15.3	1000

Table 2.1 lists the various filament materials used in the evolution of the incandescent light bulb, together with the efficacies and lamp lifetimes attained.

2.4.2 *Component features of modern incandescent lamps*

Modern tungsten filament lamps are the product of many years of painstaking developments. A traditional general lighting service (GLS)

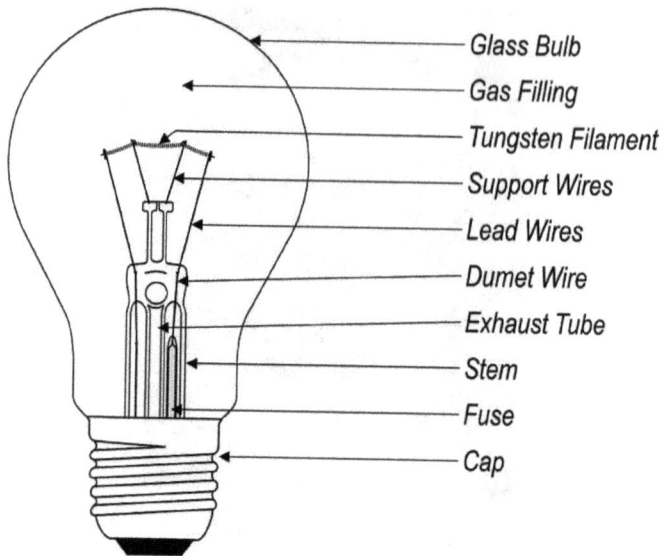

Fig. 2.11.   Schematic diagram of a modern tungsten filament lamp.

incandescent bulb is shown in Figure 2.11. Let's examine its construction in some detail, so as to appreciate the technological developments that have resulted in such a widespread, mass-produced commodity product.

2.4.2.1 *Glass bulb*

The glass bulb is the outer protective enclosure that also houses the lamp filament on its mount assembly. It is most commonly made from low-cost soda-lime glass, although in high wattage lamps it may be made of borosilicate glass. Soda-lime glass can withstand temperatures up to 300°C whereas borosilicate glass can be safely used at temperatures up to 450°C. Bulbs are formed on a separate glass forming line before these are united with filament mount assemblies. A ribbon of glass is extruded on to a moving conveyor belt that contains holes at periodic intervals. Air is blown through molten glass at the site of these holes so that the glass blows into mould cavities that have the desired bulb shape. After this blow moulding process, the bulbs are annealed inside a furnace and then cooled to ambient temperature. Initially the bulbs have a long neck which is later trimmed during the sealing process, after the filament mount assembly has been fitted inside the bulb. For over 90 years, the pear-shaped glass bulb has remained the standard envelop shape for most common incandescent lamps. Popular wattage ratings of 40, 60 and 100 W all come in this shape with an outer diameter of 60 mm. For several decades now, mushroom-shaped bulbs have also been available. Other commonly available bulb shapes include the candle and round shapes. All of these lamps are available in both clear and frosted glass finishes. The latter cut down on the glare from a visible filament. Frosted bulbs were conceived of quite early, but at first only bulbs frosted on the outside were available. Such bulbs had a rough outer surface due to the acid etching that had to be carried out to give the bulb a frosty appearance. The rough surface attracted dust and marred the appearance in decorative applications. Efforts were started to develop inside frosted bulbs but at first were met with only limited success. This was because traditional acid etching caused the inner surface to weaken and, thus, the bulbs became fragile. This problem was solved in 1925 at GE by the chemist Marvin Pipkin. He developed a double etching technique that resulted in internally frosted bulbs that retained their structural strength. Soon afterwards, GE began marketing internally frosted bulbs that had an attractive, blemish-free, smooth outer surface. Nearly a quarter of a century later, in 1947, Pipkin also developed the now-familiar soft-white bulb which has an electrostatically-applied inside coat of fine silica powder that spreads the brilliant filament's light diffusely through Mie scattering. This produces gentler and evenly distributed light.

2.4.2.2 *Lamp filament*

The filaments of incandescent lamps are now almost universally made from tungsten. This metal only became the material of choice for lamp filaments after several other metals had been tried over many years. The chief problem in using tungsten as a filament material was the fact that tungsten has low ductility and, thus, it is very difficult to obtain tungsten wires using the ordinary wire drawing process. However, as it appeared to have many desirable properties for use as incandescent lamp filaments, so interest in it started as the 19th century ended and a new century started. Alexander Just and Franz Hanaman in Austria were one of the first people who gave tungsten a try in 1903. They made a hybrid carbon-tungsten filament by depositing tungsten on a carbon filament. This proved somewhat successful in operation but they were not able to develop an industrial scale process for manufacturing such filaments. In order to make proper tungsten filaments, Hans Kuzel, also in Austria, experimented with a technique that was similar to that used for making osmium filaments. This method involved making tungsten filaments from pressed tungsten powder. These sintered filaments were very fragile, and could not tolerate even mild shocks. Their manufacture was a long and involved process that consisted of mixing fine black tungsten metal powder with starch or dextrin as a binder and then extruding this mixture through small diameter diamond dies. The filament thus produced was shaped like a hairpin, cut to length and then heated in a hydrogen atmosphere to burn off the organic binder, leaving a pure, sintered tungsten filament. A number of these hairpin segments were connected in series to obtain the appropriate resistance for the line voltage and lamp wattage needed. A vintage lamp with sintered tungsten filaments is shown in Figure 2.12, where the characteristic 'U' shaped tungsten hairpin filaments can be seen. Due to the brittleness

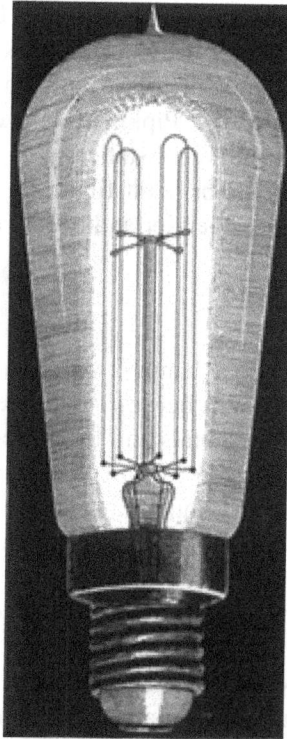

Fig. 2.12.   Incandescent lamp with sintered tungsten filament.

of sintered tungsten filament, tantalum filament lamps remained at the forefront. Until work, mainly carried out in the United States, resulted in the development of drawn tungsten wire filaments. However, at first, these were prone to offset and sagging effects. As the filaments were heated during lamp operation and then cooled when the lamp was switched off, the tungsten tended to recrystallize with the grains growing as wide as the diameter of the wire itself. These grains were oriented with boundaries perpendicular to the wire axis, thus giving the wire the segmented appearance of a bamboo. During subsequent operation, these sections had a tendency to slip (offset) relative to each other. This sliding at transverse grain boundaries led to the fracture of filaments at room temperature. Sag, on the other hand, refers to the droop of a long filament under its own weight. Sag caused the filament to elongate and the efficiency of the lamp to degrade. Grain slippage was initially overcome with the addition of thoria to tungsten, but very quickly it was found that thoriated tungsten was even more prone to sagging than un-thoriated tungsten. For a number of years, these problems beset the nascent tungsten incandescent bulb industry until Aladar Pacz, around 1915, discovered that the addition of potassium to tungsten solved both offset and sag problems. It is now known that the large potassium atoms occupy interstitial sites in the tungsten crystals in such a way as to lock the grains together, making filaments resistant to both offset and sag. Modern lamps utilize filaments made from a tungsten-rhenium alloy. The addition of rhenium — a very rare metal — improves the properties of tungsten by making it more ductile for the wire drawing process and also improves its high temperature stability. Rhenium can be alloyed into tungsten up to the solubility limit of 27%, but the material used for making lamp filaments contains a much smaller amount of this metal. In ordinary argon-nitrogen-filled incandescent lamps, the filament is operated at a temperature in the range of 2400–2600°C. Such lamps convert 5–8% of the electrical energy input into visible radiation — the rest appearing as heat. Thus, tungsten filament lamps are mainly devices that produce light as a by-product of heat generation. Lamps with krypton and xenon fillings, as well as tungsten-halogen lamps, can operate at much higher filament temperatures because of reduced rate of tungsten evaporation in such lamps. Higher temperature operation produces more visible light compared to infrared radiation, and, thus, leads to higher lamp efficacy. However, even at temperatures of 3000°C, only about 10% of the emitted radiation falls in the visible region, making tungsten lamp technology inherently inefficient for producing visible light. It is interesting to note that the highest theoretical efficacy for incandescent tungsten lamps is predicted to be 53 lumens/W, as this is the amount of visible light emitted by molten tungsten at its melting point, when supplied with 1 W of power to maintain its temperature. This figure provides a useful reference to compare incandescent lamps with other technologies, such as discharge lamps and LEDs.

2.4.2.3 *Gas fill*

While good, high quality drawn tungsten wires solved many of the problems that the electric bulb industry initially faced, bulb blackening still persisted as tungsten evaporated from hot filaments and was deposited all around the glass envelope. This phenomenon is exactly the same as that encountered in metal evaporators where material is evaporated on purpose in a high vacuum, in order to coat a substrate. As the early lamps were evacuated to quite low pressures, so in the absence of any collisions from gaseous molecules, the evaporated tungsten atoms traveled in straight lines and were deposited on all exposed bulb surfaces. This caused both the filament to thin down quickly, shortening the life of the lamp, and to gradually reduce the light output from the bulb as more and more light was absorbed by the thickening tungsten deposit. Several people found a solution to this problem quite early in the evolution of tungsten incandescent bulb technology — fill the bulb with an inert gas after removing all the air from it. The atoms of the fill gas serve to bounce back evaporating tungsten atoms on to the filament, thus significantly retarding the rate of tungsten evaporation. The gas used must be chemically inert and very pure so as not to react chemically with the filament material at high temperature. Pure nitrogen was first used for this purpose but it was found that it conducted away too much heat from the filament. This caused the filament to require too much current in order to glow at the same brightness as a similar filament in vacuum. The heat was removed from the filament by both conduction and convection through the gas. This is seen in Figure 2.13 here which shows tungsten filament bulbs with (right), and without (left) fill gas.

The evacuated bulb on the left glows brightly, while the gas-filled bulb on the right, operating with the same current, glows only feebly. The fact that heat is being lost rapidly from the filament on the right is easily established because the glass envelope of the bulb on the right is much hotter than that of the bulb on the left despite the fact that the bulb appears

Fig. 2.13.   Tungsten filament lamp with gas fill (left) and without gas fill (right).

to be glowing only faintly. The heat in this case is being efficiently carried away from the filament to the bulb surface, making the latter much hotter to touch. From Figure 2.13, one can also observe the effect of convective heat transfer, as only the top part of the filament on the right glows, being heated both by the flowing current and from the heat carried to it from the lower part of the filament through convection currents in the fill gas. If the lamp is inverted, then it is the other end of the filament that starts to glow.

It was speculated that argon would be a better choice for the fill gas, as it has a lower thermal conductivity compared to nitrogen. Irving Langmuir and his colleague Lewi Tonks at GE laboratories in Schenectady carried out a series of experiments to verify the superiority of argon as the fill gas for incandescent lamps. However, the commercial use of argon as lamp fill gas only started once inert gas distillation plants became widely available during the Second World War. Argon is also a better choice over pure nitrogen because its larger and heavier atoms are more effective at preventing the evaporation of tungsten. The filling used in modern tungsten light bulbs is a mixture of argon and nitrogen (usually in the ratio of 85 parts of argon to 15 parts of nitrogen). Nitrogen is added because with pure argon there is a tendency for arcing to take place if the filament breaks for any reason. This happens because of the low ionization potential of argon which allows an arc discharge to form easily even at relatively low voltages. In the United States, the lower operating voltage of 120 volts enables the use of argon with lower dilution ratio of 95 parts of argon with 5 parts of nitrogen. This composition is more effective at preventing heat loss from filament, as well as in reducing tungsten evapo- ration, and, thus, tungsten incandescent lamps in America operate at slightly higher luminous efficiency. The gas mixture is filled in bulbs at lower than atmospheric pressure, usually 70 kPa (0.7 atm). During lamp operation, as the internal temperature rises, the fill gas pressure increases to a little above atmospheric pressure. This can be readily appreciated by applying the gas law equation in the form, PV/T = constant. The volume of the bulb remains the same in cold and hot states but as the temperature changes, the pressure of the fill gas changes, in accordance with the relation, $P_{cold}/T_{cold} = P_{hot}/T_{hot}$. With T_{cold} = 296 K (23°C), T_{hot} = 448 K (175°C) and P_{cold} = 70 kPa, we get, $P_{hot} = P_{cold} (T_{hot}/T_{cold})$ = 106 kPa. This pressure, a little above atmospheric pressure (101.325 kPa), is very effec- tive at retarding the evaporation of tungsten from hot filaments, while at the same time ensuring that if the bulb breaks accidentally during its operation, glass fragments are not blown out with any significant force to cause injury.

While argon is a good fill gas for tungsten incandescent lamps, kryp- ton is an even better choice as it has bigger atoms and a still lower ther- mal conductivity. In fact, this trend continues such that xenon turns out to be the best fill gas with even bigger atoms and lower thermal conduc- tivity than krypton. The reason krypton and xenon are not used in mass- produced incandescent lamps is simply the prohibitive cost of these very rare gases. The properties of various lamp fill gases are given here in Table 2.2.

Krypton and xenon are now widely used in small high-performance lamps, such as those used for film production and projection, automotive headlights, and lamps for mining operations. These fill gases allow such lamps to operate at high temperature (and, thus, high brightness and lumi- nous efficacy), producing light that is significantly whiter due to its higher color temperature. The use of premium fill gases is justified in such situations because of the higher admissible cost of such bulbs, and also because such lamps are usually physically small, requiring only small amounts of fill gas. Historically, krypton-filled lamps were first developed in 1930 by the Hungarian Physicist Imre Bródy and his colleagues at Tungsram in Hungary. Figure 2.14 here shows modern miniature

Table 2.2   Properties of various fill gases used in incandescent lamps.

Gas	Symbol	Molecular Weight	Thermal Conductivity	% in Atmosphere
Hydrogen	H_2	2	0.1805	N/A
Helium	He	4	0.1513	0.0005
Neon	Ne	20	0.0491	0.0015
Nitrogen	N_2	28	0.02583	78.03
Argon	Ar	40	0.01772	0.94
Krypton	Kr	84	0.00943	0.0001
Xenon	Xe	131	0.00565	0.000009

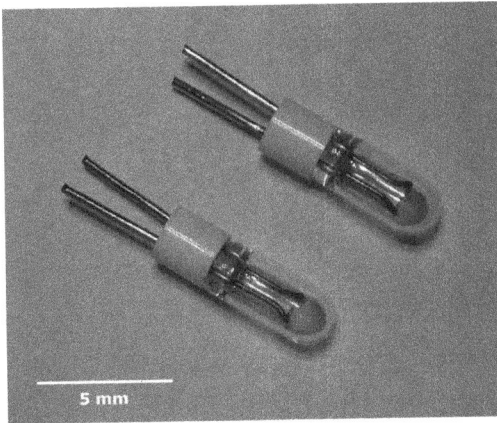

5 mm

Fig. 2.14.   Miniature krypton-filled tungsten filament bulbs.

krypton-filled light bulbs for use in high-performance flash lights. Such bulbs can operate at filament temperature of up to 3000°C, producing an intense white light.

The only other gases which have been commercially employed in incandescent lamps are hydrogen and helium. Both of them are character-ized by very high-thermal conductivities, making them inefficient for general lighting applications. However, this property is utilized in signal-ing lamps which are switched on and off quickly. These fill gases serve to cool the filament more rapidly between pulses, enabling higher signaling speeds.

The ambient environment inside a lamp must be exceptionally well-controlled during the bulb evacuation and fill process, in order to ensure a long lamp life. This was poorly understood during the early years of lamp production when even highly evacuated lamps had short lifetimes of at most a few tens of hours. It was only when Edison discovered that filaments and other materials inside lamps outgas when first operated, spoiling the vacuum inside the bulbs, that a proper manufacturing process with lamp component heating during manufacture was adopted; greatly prolonging lamp lifetimes. The presence of even trace amounts of water vapor can greatly shorten the life of incandescent lamps through a process called the water cycle. This process operates as follows. The very high filament temperature causes water molecules to break into oxygen and hydrogen:

$$2H_2O \rightarrow 2H_2 + O_2$$

The oxygen then attacks the filament, forming tungsten trioxide:

$$2W + 3O_2 \rightarrow 2WO_3$$

Due to its volatility, the tungsten oxide leaves the filament and deposits on the interior of the glass bulb surface. This is usually seen as a brown-colored deposit in imperfectly sealed bulbs. Over a period of time, tungsten oxide, in turn, is attacked by hydrogen formed in the decomposition reaction above:

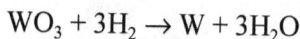

$$WO_3 + 3H_2 \rightarrow W + 3H_2O$$

The oxide is, thus, reduced to a metallic tungsten deposit, and the oxygen is set free to attack the filament again. This destructive cycle causes rapid bulb blackening from the formation of tungsten deposits, reducing the light output. The thinned down filament also breaks after a short time, causing premature lamp failure. From this description it is clear that residual water vapor can have a very deleterious effect on the longevity of tungsten incandescent bulbs, and thus, every precaution has to be taken to remove even slight traces of water vapor from incandescent lamps. A quantity of water vapor as small as eight parts per million can cause a lamp to fail within 15 min, while even smaller amounts of water vapor will greatly reduce the operating life of a tungsten filament lamp.

It is also extremely important to ensure that the glass to metal seal in lamps is as perfect as possible, because air leakage can also cause lamps to fail early. Oxygen in the leaked air can react with the hot filament to form either milky yellow-white deposits of tungsten trioxide (WO_3) or dark blue deposits of tungsten pentoxide (W_2O_5). For the removal of residual atmospheric gases and other contaminants, manufacturers often place a getter, such as phosphorus pentanitride (P_3N_5), inside the bulb. This material — a safer alternative to red phosphorus or phosphorus oxides that were used in the past — serves to absorb remaining traces of oxygen and water vapor, and thus helps prolong lamp life.

2.4.2.4 *Coiled filament*

A straight filament is prone to high heat loss during lamp operation. This was not a problem as long as bulbs had a vacuum in them, but as soon as these started being filled with gases to suppress filament evaporation, the conductive and convective heat loss became a substantial problem. Irving Langmuir, working at the Schenectady, New York, laboratories of the General Electric Corporation, worked on this problem for several years. He discovered that whereas convective currents were set up once a gas-filled bulb was operating, there was always a stationary region of hot gas around the filament. This region, now called the Langmuir sheath, is typically two millimeters thick. Langmuir also found that by winding the filament into an elongated coil such that the spacing between adjacent windings is less than the thickness of the Langmuir sheath, the stationary gas layer around one coil merges with the next. Making the filament into a coil, thus, greatly reduces the effective filament area exposed to convecting gas flows. This was a momentous discovery that allowed filaments to run much hotter than before and, thus, resulted in a significant increase in the luminous efficacy of incandescent lamps. GE commercialized the first coiled filament gas-filled lamps in 1913, and these were immediately a great success. Figure 2.15 is a photograph of Irving Langmuir.

Tungsten filaments are formed into coils by winding them on a thin molybdenum mandrel. The coiled filament is then annealed in a hydrogen atmosphere to remove the coiling strain and the mandrel is subsequently removed by dissolving it in an acid bath. In early practical tungsten filament lamps, the tungsten filament coil was supported between lead wires in a horizontal configuration called the Wreath or C-9 configuration, as shown in Figure 2.16. This ensured that the entire coil saw the same convective cooling effect, and, thus, remained at a uniform temperature. A vertical configuration will result in the top of the coil getting hotter

Fig. 2.15.   Irving Langmuir.

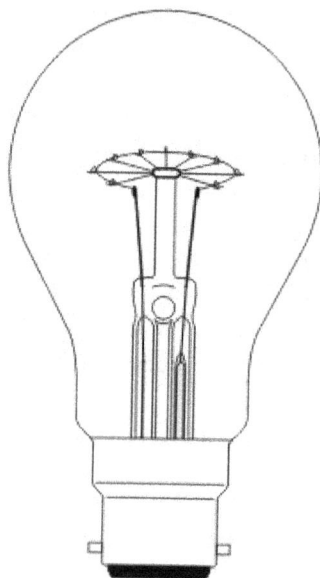

Fig. 2.16.   Wreath or C-9 configuration of tungsten filament as used in early tungsten filament lamps.

because of rising convection currents, compared to the bottom of the coil. The excess temperature would lead to premature filament failure at its top. This is the reason, for best longevity, tungsten filament lamps are operated in a horizontal position so that the entire filament remains at a uniform temperature. However, some lamps do use vertically oriented filaments, because this configuration can be used with physically small tubular bulbs, for use in restricted spaces. Another advantage of the coiled filament is that it simplifies lamp manufacture, because owing to its reduced length a number of support wires can be eliminated, and the overall bulb volume is reduced.

A few years after the development of the coiled filament, Iris Runge and Ellen Lax working at Osram in Germany, demonstrated that coiling the filament into a coil had an interesting effect on the light radiated from the filament. Tight coiling, they showed, begins to form a loosely enclosed space — essentially an approximation of an elongated tungsten cavity — where light photons are multiply reflected between individual windings. This creates a black body effect and results in the intensification of light inside the winding compared to that outside the coil. The outer surface of the coiled filament radiates less light than the inner surface that is seen through the openings in the turns. This is essentially because the emissivity of the inner surface is enhanced by the black body effect. As a result, the amount of light that is radiated by a coiled filament is somewhat increased compared to that which would be radiated by a straight filament of the same total wire length. This is yet another benefit of a coiled filament.

If a coiled filament offers so many benefits then what about a double-coiled filament, i.e. where the coil is wound a second time to form an even tighter filament? This question was asked a short time after the commercial introduction of single coil tungsten filament lamps. It was clear that a double-coiled (also called coiled-coil) filament would be an improvement over a single coil filament by reducing heat losses to the fill gas even further, reducing the number of filament supports required, and shrinking the size of the bulb itself. While the idea was self-evident, its practical implementation had to await the development of fully automatic production-line machines to form such coiled filaments, and to incorporate them into lamps in a consistent manner. Burnie Lee Benbow of General Electric's Cleveland, Ohio, facility was awarded a patent for a double coiled tungsten filament in 1917 although mass-produced lamps with coiled coil filaments appeared much later. Figure 2.17 is a view of his patent document.

B. L. BENBOW.
FILAMENT.
APPLICATION FILED OCT. 4, 1913.

1,247,068.

Patented Nov. 20, 1917.

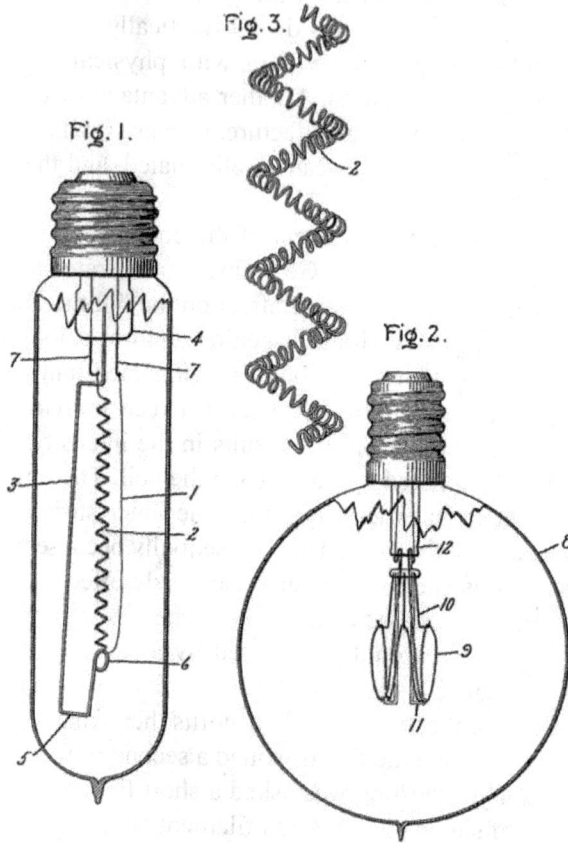

Fig. 3.

Fig. 1.

Fig. 2.

WITNESSES:

Benjamin B. Hall
Helen Orford

INVENTOR :
BURNIE LEE BENBOW,
BY Albert S. Davis
HIS ATTORNEY.

Fig. 2.17.   Patent diagram of double-coiled tungsten filament.
Courtesy: General Electric Company.

Across the Atlantic, in Europe, research and development on double-coiled filament lamps started in the 1920s, and the first such lamp, developed in 1924 at the Osram-GEC labs in London, is shown in Figure 2.18 here. The double-coiled lamp appeared commercially in 1933.

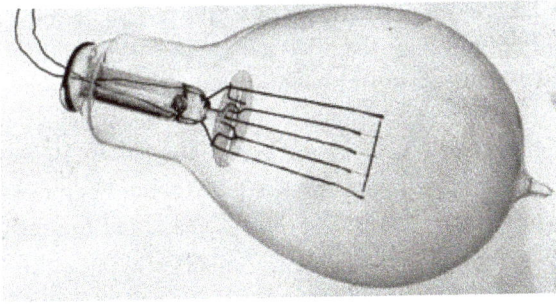

Fig. 2.18.   Osram-GEC double-coiled tungsten filament lamp.

Courtesy: General Electric Company.

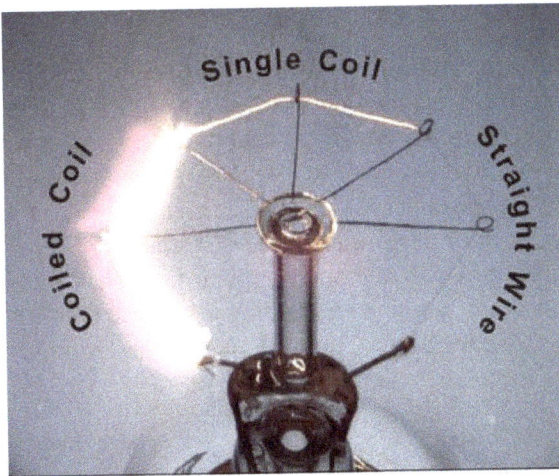

Fig. 2.19.   Comparison of luminous output from un-coiled, single-coiled and double-coiled tungsten filament.

Double coiling is carried out by winding a pre-coiled filament a second time on a retractable steel mandrel, annealing the assembly and then releasing the steel mandrel.

The effect of single and double coiling on heat transfer loss and, thus, on luminous output from incandescent lamps can be seen in Figure 2.19. Here a double coiled filament has been stretched first to a single coil configuration and then further stretched to a straight un-coiled state. This procedure divides a continuous filament into three equal length segments with different coiling configurations, which is then enclosed in a bulb with the standard 85:15 argon:nitrogen gas fill. When a current is passed through this lamp, the double-coiled segment glows brightly whereas the single-coiled segment glows only faintly, and no light comes out of the straight filament segment. If this segmented filament is placed in an evacuated bulb and driven by the same current then all three segments light up at the same luminosity. This is an excellent demonstration of the fact that heat loss from filament to the fill gas is the lowest for a double-coiled filament.

A typical double-coiled filament lamp has a coiled filament less than an inch long between its end supports. If fully unwound, however, this will yield a tungsten wire around 20 inches in length. This illustrates how tightly the wire is wound in its double-coiled configuration. A scanning electron micrograph (SEM) of a double-coiled tungsten filament is shown in Figure 2.20.

There has been further development in this direction, in that Toshiba in Japan has commercialized triple-coiled lamps that feature three levels of tight coiling. The main benefit of such a lamp is in the compact form factor for the lamp which enables it to be economically filled with premium fill gases, such as krypton and xenon. The triple coiling and use of heavier atomic weight gases allows these lamps to be operated at much higher filament temperatures than usual lamps, thus producing more light with higher efficacy. Another benefit is that the smaller filament volume helps create a near-point source of light which improves optical

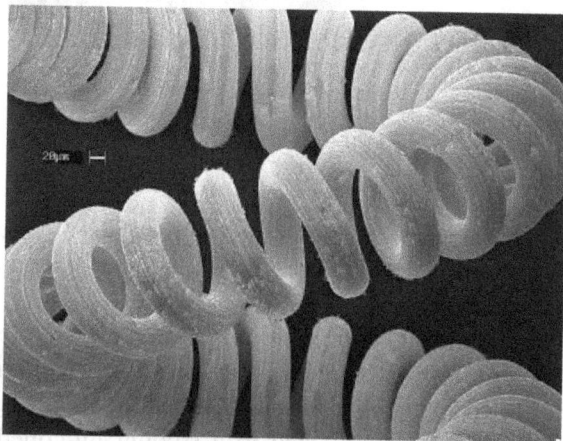

Fig. 2.20.   Scanning electron micrograph of a double-coiled tungsten filament.

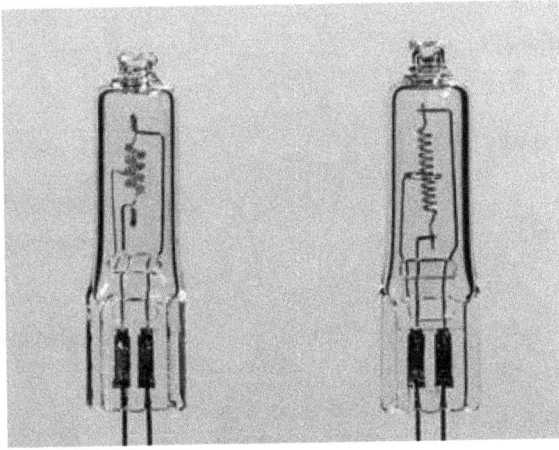

Fig. 2.21.   Triple-coiled (left) and double-coiled (right) tungsten filament bulbs.
Courtesy: Toshiba Corporation.

characteristics when such lamps are used in applications, such as film projection systems. Figure 2.21 shows a triple-coiled filament lamp (left) compared with a double-coiled filament lamp (right).

2.4.2.5 *Filament mount assembly*

Another important component of all incandescent lamps is the filament mount. This consists of several sub-components. Its function is to hold the filament in a fixed position and to provide a housing for electrical connections to the filament, internal fuses, and gas exhaust tube. The main structural elements of the assembly consist of a glass 'pinch' tube which contains a secondary tube called the exhaust tube. These parts are made from either soda lime glass or lead glass. As seen in Figure 2.22, lead-in wires, that hold the two ends of the filament, pass through the glass mount assembly. The portion that goes through the glass is made of nickel-iron wire (called 'dumet' because it uses two metals). During assembly, the dumet part of the wire is dipped into a borax solution to make the wire more adherent to glass. Dumet wire is used because this bi-metallic wire has a coefficient of thermal expansion close to that of glass, which simplifies its union with glass. The lead-in wires may also be coated with a paint containing zirconium and aluminum metals, which act as a getter by absorbing any remaining traces of oxygen and water vapor inside the

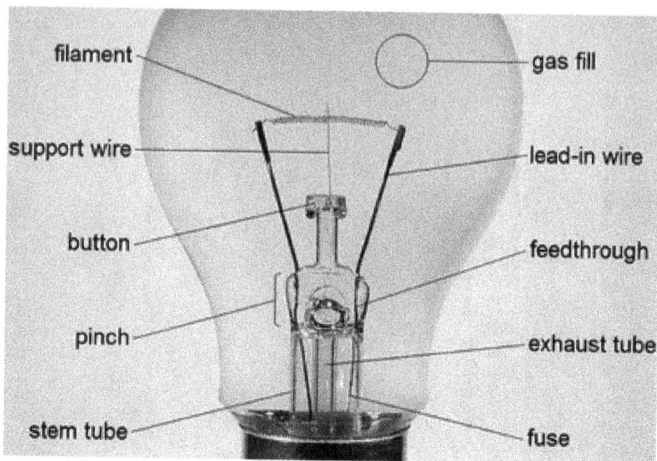

Fig. 2.22.   Various component parts of a modern tungsten filament lamp.

bulb. One or both dumet wires then connect to a fuse, as described below. Apart from the ends, the filament is also supported in its mid-section with a supporting mechanism. This is also illustrated in Figure 2.22, as a molybdenum or nickel-iron wire terminating in a loop that supports the coiled tungsten filament and protect it from both mechanical shocks and from sagging under its own weight. Some lamps have only one support wire, while others may have two or even three supports. It should be noted that each support loop locally reduces the filament temperature at the point of contact, decreasing the lamp efficiency by about 1%. Thus, the lower the number of support loops, the higher the efficiency of the lamp. Double coiling the filament reduces the effective length of the filament, so that in most lamps only two supporting loops are needed — this is yet another advantage of coiling the lamp filament.

The lower end of the exhaust tube is connected to an evacuation and gas manifold during lamp assembly. Air inside the bulb is removed through this tube, and then it is flushed with dry nitrogen gas. This procedure is repeated several times before the glass bulb is filled with its fill gas. As mentioned earlier, a getter, usually phosphorus pentanitride P_3N_5 is placed in modern incandescent lamps. It absorbs any oxygen and water vapor remaining inside the bulb, thus removing the atmospheric constituents that can shorten filament life.

Some lamp mount assemblies are designed to accommodate two separate filaments instead of just one filament. Such dual filament lamps have been around from the time of carbon filament lamps. Either filament can be made to light up, or both filaments can be lit up simultaneously, thus providing a choice of three different light intensities from the same lamp. For this reason, such lamps are popularly known as 3-way bulbs. Dual filament lamps have been available in the US from Westinghouse since 1933.

2.4.2.6 *Lamp fuse*

Most modern incandescent lamps incorporate a fuse for safety reasons. When a lamp is switched on from its cold state, it draws a much higher current than its operational current rating. This is because, when cold, a filament has a much lower resistance than when it is hot. The resistance of the filament in the hot state can be determined by dividing the square of the operating voltage by the wattage rating of the lamp. The lamp draws a heavy surge current for a few milliseconds when switched on, and then gradually stabilizes to its nominal operating current, as the filament reaches its operating temperature. The electrical supply circuit must be capable of providing this surge current, which can be ten to twenty times the normal operating current of the lamp. The fuse inside a lamp can provide protection, in case an abnormally high surge current is drawn for some reason. More importantly, however, the built-in fuse is meant to protect from high current draws when a filament blows and an arc succeeds in forming between the broken ends of the filament. If formed, such an arc can draw destructive amounts of current, and the internal fuse then blows to prevent further damage to the electrical system that powered the lamp. The fuse itself can be either a simple fuse wire made of monel, cupro-nickel, or a more elaborate 'Ballotini' fuse consisting of a thin wire enclosed in a sealed glass capillary and surrounded by tiny glass beads. This type of fuse is more expensive but has the advantage that in the event it blows the heat causes the glass beads to melt and form a non-conducting medium between the bare ends of the fuse wire. This prevents any potential arcs from forming and causing further damage. Figure 2.23 shows a Ballotini fuse.

2.4.2.7 *Lamp bases*

Lamp bases seal the ends of glass bulbs and provide the means to connect the lamp to a source of electrical power. Edison's lamps had a screw-in base with male screw threads. It was originally made of brass and

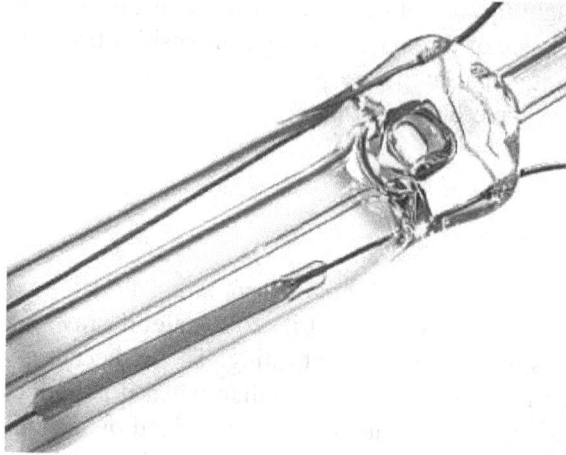

Fig. 2.23.   Ballatoni fuse.

insulated with plaster of Paris and, later, with porcelain. Nowadays, the base is made of formed aluminum, with either a low melting point glass-like black enamel (vitrite) or a thermosetting resin material, used to seal and insulate the base. The lamp fuse is internally connected to the center 'eye' contact, which makes connection with the live wire. The threaded portion of the base internally connects to the other end of the filament, with or without an additional fuse. This section makes contact with the ground wire of the electrical supply. This type of screw-in base is still called the Edison base. Another widely used style of lamp base is the bayonet base, that has two short pins for fitting the base into spring-loaded lamp sockets. The bayonet base is mainly used in the UK and the commonwealth countries. Figure 2.24 shows drawings of both Edison (right) and bayonet (left) style lamp bases.

3-way lamps with Edison base have an additional contact ring surrounding the center contact at the base. This contact connects to the live side of the lower wattage filament, while the center contact connects to the live side of the higher wattage filament. The threaded portion of the base connects to the neutral side of both filaments.

2.4.3 *Lifetime and failure of tungsten incandescent lamps*

Modern incandescent lamps have a very much longer lifetime than their early versions. This is a result of more than a century's careful engineering

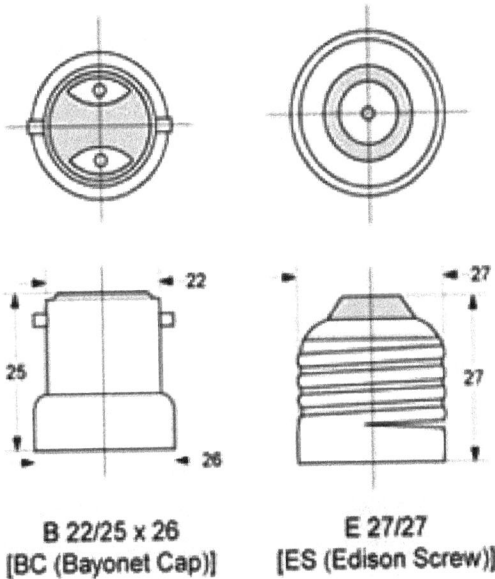

B 22/25 x 26
[BC (Bayonet Cap)]

E 27/27
[ES (Edison Screw)]

Fig. 2.24. Edison (right) and bayonet (left) style lamp bases.

advances in improving all aspects of the tungsten filament lamp. Ordinary double-coiled, argon-nitrogen-filled, domestic light bulbs are manufactured for a nominal lifetime of 1000 hours. Of course, some bulbs will burn out after only a few hundred hours while some may last for 2000 hours or more, so 1000 hours is only a manufacturing specification. The chief failure mechanism remains the loss of tungsten from the filament, which gradually thins down with operation to such an extent that it eventually breaks, rendering the lamp useless. Figure 2.25 shows a typical mortality curve for early generation of gas-filled double-coiled tungsten filament lamps. It shows the percentage of lamps surviving after several hours of continuous operation. It is notable that because of the low resistance of lamps in their cold state, and the consequent surge of current on switching them on, lamps with weakened filaments tend to blow when switched on rather than during their stable operating state.

For any sufficiently large collection of lamps, failure is a random occurrence. A tungsten incandescent lamp could fail in as few as a few hours or could last for thousands of hours of continuous operation. The many variables that go into the manufacture of each lamp make it a statistical process, which cannot be predicted with any degree of accuracy. While the vast majority of commercial light bulbs last for a few hundred

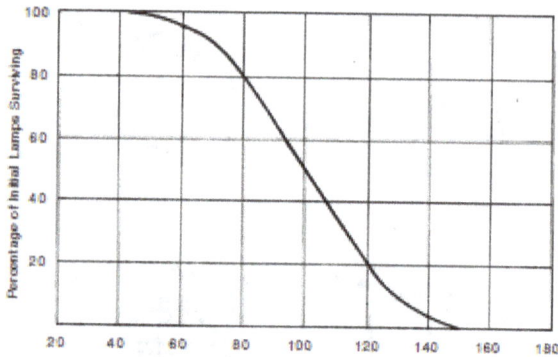

Fig. 2.25.   Longevity of double-coiled tungsten filament lamps as a function of operating time in hours.

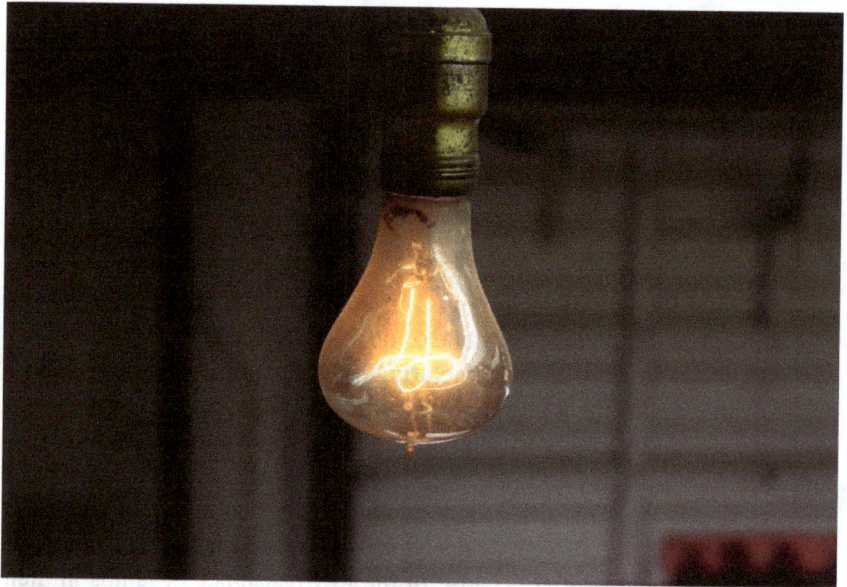

Fig. 2.26.   The Centennial Light.

hours, some prove to be noticeably short-lived, and a precious few show remarkable longevity. One particular lamp has drawn worldwide attention in this respect, for being in continuous operation (with very few interruptions) from the year 1901. Manufactured by the Shelby Electric Company in Shelby, Ohio, it is a handblown bulb with a looped carbon filament. By now, it has greatly surpassed a million hours of operation. Known as the Centennial Light (see Figure 2.26), this tungsten filament light bulb is installed in Livermore-Pleasanton Fire Department's Fire Station #6, in Livermore, California. Due to its unbelievable longevity, it claims a global celebrity status, with a huge fan following and a real-time webcam feed, showing its continuing operation. Hordes of tourists visit the fire station to observe the lamp in operation.

2.5 Tungsten-halogen Lamps

Since the advent of the incandescent lamp, reducing bulb blackening and prolonging the life of the filament had been the major goals of people and companies involved with manufacturing electric lamps. Early attempts toward these goals involved replacing carbon filaments with metallic filaments and, later, filling bulbs with evaporation-suppressing gases. These, and other improvements, resulted in light bulbs that had mean lifetimes of over 1000 hours. But, the quest to further increase the longevity and efficiency of these lamps continued. During the 1930s and 1940s there were no significant changes in the design of the standard tungsten lamp. While its low efficiency was well-known, it was generally conceded that the poor efficiency was a consequence of the lamp's thermodynamics and there were no prospects for any significant increases in either its efficiency or longevity. However, a development took place in later years that resulted in a marked improvement in incandescent lamps, beyond the expectations of its proponents and critics during the first half of the 20th century.

2.5.1 *History of the development of tungsten-halogen lamps*

During the early 1950s, General Electric became interested in developing lamps for heating purposes. As incandescent lamps produce more heat than light so it was surmised that lamps, specifically meant for heating purposes, such as for industrial drying applications, could be developed with minor modifications to the existing tungsten incandescent lamp technology. It did not turn out to be as straightforward as was initially hoped for, because major problems with material properties had to be surmounted. High temperature operation of tungsten filament lamps necessitated the use of small bulbs made of fused quartz. In contrast to ordinary glass, fused quartz — a non-crystalline variety of quartz — can tolerate much higher temperature without melting or shattering. GE researchers at GE's NELA park research lab in Cleveland, Ohio, found that the lamps they created had an increased propensity to blacken because of the

Fig. 2.27.   Elmer Fridrich.

proximity of the filament to the bulb, and its high temperature operation. This quickly reduced the light and heat flux from the lamps. For many months, this remained an outstanding problem with no apparent solution. A breakthrough occurred when Elmer Fridrich — one of the researchers in the group — decided to investigate the effect of iodine vapor inside the experimental high intensity tungsten heating lamps. He was led to this idea from his knowledge of metal-halogen chemistry, used for refining some exotic metals. Figure 2.27 shows Elmer Fridrich at GE NELA park laboratory, in the year 1959.

After consultations within his group, Fridrich obtained permission from group leader Alton Foote to try the use of iodine in experimental lamps. Permission was granted, and with the help of his colleagues, William Hodge, Mary Jaffe and Emmet Wiley, he tested the first lamp with a dilute iodine fill. The experiment was an immediate success, and, thus, the first quartz-iodine lamp was born. By 1953, the scope of the ongoing work was broadened as Elmer Fridrich and Emmett Wiley also started working on long-life high-intensity tungsten lamps for general illumination. The first working prototypes of these lamps were made by the end of 1953. Their invention was the first major evolution in incandescent tungsten lamp technology since the developments of the early 20th century. An image of their patent application appears in Figure 2.28. The linear variety of tungsten-halogen lamp, first presented

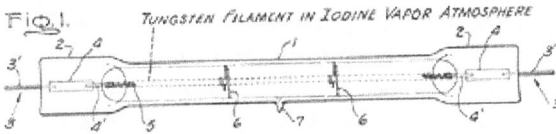

April 21, 1959   E. G. FRIDRICH ET AL   2,883,571

ELECTRIC INCANDESCENT LAMP

Filed March 3, 1958

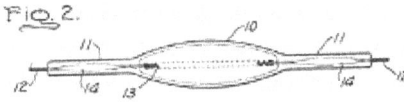

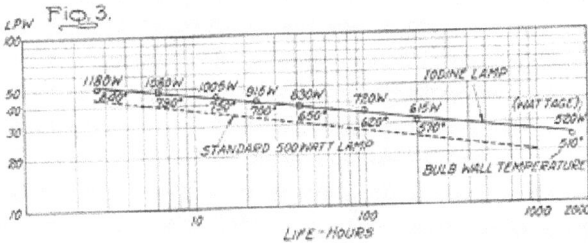

Fig. 2.28.   Tungsten-halogen lamp patent.

Courtesy: General Electric Company.

in this patent application, has remained virtually unchanged since its original design.

The earliest of these lamps employed fuzed quartz bulbs and iodine vapor and were, therefore, called 'quartz-iodine' lamps, but since other high temperature bulb materials and other halogens may also be used, the more generic term 'tungsten-halogen' lamp is now used. It took a few years, after the initial invention, to develop proper industrial procedures for manufacturing iodine-based tungsten-halogen lamps in large numbers. The very earliest lamps had inconsistent performance — a problem that was solved by Edward Zubler, a physical chemist, at the GE Cleveland, lab. Zubler's investigations revealed the essential role performed by oxygen in the working of the tungsten-iodine lamps. This then led to ways of introducing controlled amounts of oxygen in the lamps which resulted in marketable lamps. Although Elmer Fridrich and Emmet Wiley are credited with the invention of the modern tungsten-halogen lamp, they were certainly not the first to consider using halogens inside incandescent lamps. Edwin Scribner of the United States Lighting Company was issued a patent in 1882 for a carbon filament lamp containing a minute amount of chlorine gas. Scribner's lamp was the first demonstration of the effectiveness of a halogen in clearing the inner surface of incandescent bulbs from evaporated filament material. In his patent application, Scribner rightly conjectured that the chlorine reacted with carbon deposits at high temperature and, thus, served to remove such deposits from the bulb's inner surface. Figure 2.29 is an illustration of Scribner's lamp from his

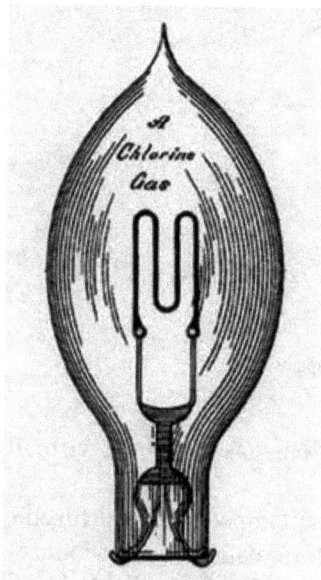

Fig. 2.29.   Scribner's chlorine-filled lamp.

patent application. The characteristic M-shaped filament in this lamp was developed by Hiram Maxim, the founder of the United States Lighting Company, where Scribner worked as an engineer.

Chlorine-containing lamps were never produced on an industrial scale, and, thus, remain of only historical interest. However, during 1893 and 1894 a lamp containing traces of bromine did appear on the market. This lamp, called the 'Novak' lamp, was the invention of John Warring of the Warring Electric Company. Warring was granted a patent in May 1893, for his invention. The Novak lamp had marked improvement in bulb blackening, compared with similar lamps without bromine, though bromine acted principally as a getter in this type of lamp. Warring's development was less a response to the problem of bulb blackening than an effort to circumvent Edison's patents on evacuated carbon filament lamps. The name 'Novak' was meant to signify the fact that his bulbs contained 'no vacuum' but instead had some bromine vapor. The lamp contained U-shaped, square cross-section, carbonized bamboo filaments, connected at both ends to platinum lead wires with a carbon paste. Figure 2.30 shows a Novak lamp.

Notwithstanding these previous attempts at including halogens inside filament lamps, it is undoubtedly the team at GE's Cleveland laboratory that developed the first proper tungsten-halogen lamps operating with the regenerative halogen cycle. The lamps were put on the market by GE in 1960, and over the years have become extremely popular. Fridrich

Fig. 2.30. Novak lamp.

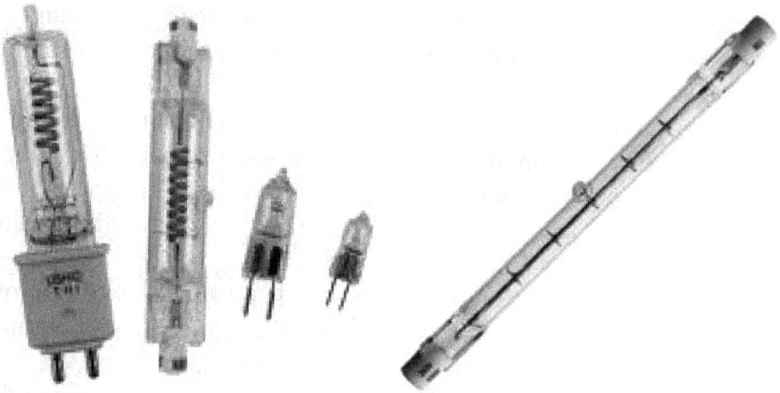

Fig. 2.31.   Various styles of modern tungsten-halogen lamps.

remained committed to the further development of tungsten-halogen lamps — making several later improvements. Figure 2.31 shows a variety of modern tungsten-halogen lamps.

2.5.2 *Operation of tungsten-halogen lamps — the tungsten-halogen regenerative cycle*

Tungsten-halogen lamps utilize a chemical vapor deposition (CVD) process to deliver higher brightness and longevity over ordinary tungsten lamps. This technology, developed during the 1950s, as described above, overcomes several of the shortcomings of conventional tungsten filament lamp technology. Almost all of the limitations of original tungsten lamp technology have their origin in the evaporation of tungsten from lamp filaments. The higher the operating temperature of the lamp, the higher is the loss of tungsten from filaments and, thus, the shorter the bulb life. The propensity for tungsten to evaporate from filaments can be reduced by the use of appropriate fill gases, as has been already described. This is a physical process that works by momentum reversal of evaporating tungsten atoms through collisions with fill gas atoms. Tungsten halogen technology, in contrast, depends on a chemical process to return evaporated tungsten back to the filament. This self-regenerative process, called the tungsten-halogen cycle, is an efficient mechanism for significantly reducing tungsten loss, and, thus, increasing the life of tungsten filament lamps.

The tungsten-halogen regenerative cycle depends on the fact that the halides and oxy-halides of tungsten (and other similar metals such as tantalum, niobium and molybdenum) are volatile compounds that decompose at high temperatures to form metallic tungsten and the corresponding halogen. A halogen lamp contains, in addition to its fill of argon and nitrogen gases, a small amount of a halogen such as iodine or bromine in a physically small bulb. The smaller volume of the bulb helps keep the wall temperature high during the operation of the bulb and also helps in resisting the high pressure of the fill gas. Another advantage of the smaller bulb is that it can be economically filled with premium high atomic weight gases, such as krypton and xenon, which confer advantages that have been described before. The gases are filled at much higher pressure than in non-halogen lamps, using liquid nitrogen cooling during the bulb fill process. The bulb is, therefore, thicker and stronger to contain the fill gases at high pressure. It is made of fused quartz or special grade glass, to be able to withstand the much higher operating temperature, as the filament is intentionally positioned much closer to the bulb surface than in ordinary lamps. As a tungsten-halogen lamp is switched on it initially works as an ordinary tungsten filament lamp with the temperature and pressure quickly rising to equilibrium operating values. In this case, however, the proximity of the filament to the bulb makes the bulb surface very hot and, consequently, the lamp achieves much higher temperature and pressure than non-halogen lamps. A few hundred milliseconds after switch on, the inner surface of the bulb reaches temperatures in excess of 250°C, at which point the tungsten-halogen cycle begins to operate. Tungsten evaporating from the filament moves out and begins to deposit on the inner surface of the bulb. The halogen, symbolized as X_2 below, (often with oxygen, if present) reacts with the tungsten deposit at high temperature to form volatile halides and oxy-halides.

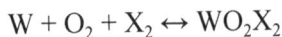

$$W + O_2 + X_2 \leftrightarrow WO_2X_2$$

These are immediately desorbed from the hot bulb surface and move back toward the filament. Close to the filament, the very high temperature causes these compounds to break down. The halogen is again released into the bulb to continue the cycle whereas the tungsten is deposited on the filament, so the amount lost is replenished. A combination of synergistic factors, the regenerative cycle, high pressure operation and the use of heavier fill gases such as krypton and xenon, means that the filament in a tungsten-halogen lamp can be operated at much higher temperature than in an ordinary incandescent lamp. Typical filament temperatures are in the range of 2600–3200°C. The high temperatures cause visible light to be generated more efficiently as a larger portion of the output lies in the visible region. The peak of the spectrum also lies at shorter wavelengths, giving the light a whiter appearance, with better color rendering properties.

For the regenerative cycle to operate successfully, the lamp wall has to reach a high enough temperature for tungsten halides to form. Thus, operating a tungsten-halogen lamp at too low power levels could stop the cycle from operating and lead to shortened lamp life. Theoretically, a tungsten-halogen lamp should last forever, as there is no net loss of tungsten from the filament, however, lamps do fail after a certain period of time. This happens because the returning tungsten is deposited at random points on the filament, rather than on locations from where most tungsten was lost. This causes the filament to thin down at certain 'hot spot' locations, and the filament to eventually break. All aged filaments from tungsten-halogen lamps show a dendrite-like random build-up of tungsten with thin and thick sections throughout the length of the filament. At the thin portions, the resistance is higher and, thus, the filament runs locally hotter, losing tungsten more rapidly and getting even thinner with time. This causes the filament to fail at such thin hot spots in both halogen and non-halogen lamps, except than in lamps with a halogen fill the failure occurs after a considerably longer time.

The first halogen-lamps were made with iodine, but gradually it was realized that it is not the best choice because of its low reactivity compared to other halogens, and also because iodine vapor has a deep purple color which can tint the light output from iodine-filled lamps. Philips later introduced lamps based on bromine, and nowadays almost all tungsten-halogen lamps contain bromine instead of iodine. Another reason for using bromine in preference to iodine stems from the fact that when there is a large difference in the molecular weights of the fill gas (argon) and the halogen the heavier component tends to accumulate near the bottom of the lamp due to diffusion-based segregation. This causes problems in the effective operation of the tungsten-halogen cycle. This problem can be greatly reduced if the molecular weight of the halogen is close to that of the fill gas. Bromine's lower molecular weight, thus, makes it a better choice. Modern tungsten-halogen lamps now often use krypton as the fill gas instead of argon, and then bromine becomes an even better choice as the two have comparable atomic and molecular weights. The actual material used for the bromine fill has ranged from elemental bromine to hydrogen bromide and bromo-phosphor-nitrile — a polymer of bromine, phosphorus and nitrogen. The polymer decomposes to its constituent elements when the lamp is first operated. Bromine then becomes the halogen fill, nitrogen acts as an arc suppressant and phosphorus acts as a getter. Chlorine has never been used in commercial tungsten-halogen lamps because of its toxicity and corrosive effect on lamp components. This leaves fluorine as a potential halogen for future lamps. Interest in fluorine-based tungsten-halogen lamps has existed for a long time but its high reactivity toward glass and quartz, coupled with the requirement for precise fill composition, has dissuaded attempts to use it in practical lamps. Some progress has been recently achieved by coating the inner surface of the bulb with a solution-deposited glass containing oxides of aluminum, titanium and phosphorus. A major postulated advantage of fluorine as the regenerative gas in halogen lamps is that tungsten fluoride dissociates at around 3030°C — a significantly higher temperature than for other tungsten halides. This means that tungsten will be selectively deposited at thin hot spots on the filament, potentially leading to a very long lamp life. Experimental lamps have shown lifetimes much in excess of commercial bromine-filled lamps, but further technical problems remain to be overcome if tungsten-fluorine lamps are to become a commercial product.

2.5.3 *Component features of modern tungsten-halogen lamps*

Just like the ordinary tungsten filament lamps that came before them, modern tungsten-halogen lamps are also the result of many years of development and continuous improvements. Careful material engineering and innovative processing techniques have resulted in a relatively high efficiency incandescent source, with several attractive features.

The bulbs of tungsten-halogen lamps have been traditionally made from an industrial grade of fused quartz. However, some lower cost, mass-produced tungsten-halogen lamp ampoules are now made from alumino-silicate glass, rather than quartz. This type of glass has been used for low wattage halogen lamps since the early 1970s. It is cheaper than fused quartz, while having a significantly higher melting point as compared to ordinary soda-lime glass, so that it can be used for lamp ampoule applications at temperatures as high as 750°C. Aluminosilicate glass contains about 20% of aluminum oxide together with smaller amounts of calcium oxide, magnesium oxide and boric oxide. There is very little soda in this type of glass. It is remarkably resistant to high temperatures and thermal shock. The introduction of aluminosilicate glass resulted in automated production of tungsten-halogen lamps on high-speed manufacturing lines, significantly reducing the cost of these lamps. Fused quartz, however, is still used for high power tungsten halogen lamps, with power ratings above 150 W, as it can be used at lamp wall temperatures up to 900°C. Both fused quartz and aluminosilicate glass require special handling precautions. This is because at high temperatures, such as those encountered during the operation of tungsten-halogen lamps, these amorphous materials tend to change into a more stable crystalline form. This devitrification is greatly accelerated in the presence of trace amounts of alkalis, such as those present in body oils, transferred as a result of touching the glass envelopes with bare fingers. Devitrified envelope material is porous to the passage of lamp fill gases from inside the lamp, and air and water vapor from outside the lamp, which drastically shortens the life of these lamps. Thus, exposed tungsten-halogen lamps and capsules must never be handled with bare hands. It is best to hold them using cloth or paper. If accidentally touched with human skin, the bulb should be wiped with isopropyl alcohol (propanol) and dried before further handling.

The very high temperatures attained by tungsten-halogen lamp ampoules pose significant fire-safety risks, if such lamps are improperly used. As discussed later, halogen lamps are also widely used for direct heating in several applications, on account of their heat radiating capability. Tungsten-halogen bulbs get much hotter than simpler non-halogen tungsten lamps and, thus, pose higher safety risks. These lamps must be used in proper fixtures, and must not be placed near flammable materials, such as paper or cloth lamp shades. Halogen lamps have been implicated in many fire accidents — both accidental and deliberate. In October 2018, a 6-year-old boy, Riley Jackson, died in Derbyshire, England, after his bed-side lamp — with a tungsten-halogen bulb — fell down and set fire to the lamp shade. This widely-reported incident was just one of many accidents — some fatal — that have been attributed to fires set alight by improper use of tungsten-halogen lamps. Although all incandescent light bulbs have been banned in the EU and many other places in the world, this ban has been based on environmental (power efficiency, hazardous construction materials etc.) grounds rather than due to safety concerns. Their fire risks must be properly taken into account, as long as these lamps are commercially available.

Tungsten-halogen lamps use a coiled or double coiled filament supported by tungsten supports. At each end, the filament is held by a tungsten clip formed out of a short section of tungsten wire. This wire is welded to a thin strip of molybdenum which passes through the glass or quartz envelope, making a hermetic pinch seal. Molybdenum is the metal of choice for this application because of its low coefficient of thermal expansion. Outside the bulb, the molybdenum strip is welded to platinum-clad molybdenum pins. The platinum coating prevents oxidation of molybdenum at high temperatures. Some lamps use a lead borate glass seal on the external leads. This low melting point glass melts at around 350°C and flows over the leads making an air tight seal where the pins come out of the bulb. The glass seal tends to melt and reflow each time the lamp is operated. A more recent lamp design uses a glass rod support formed inside the lamp capsule to hold the filament, as shown in Figure 2.32. In this arrangement, the filament drapes over the support rod and has two distinct sections on either side of the rod. The section that passes over the rod is singly coiled, whereas the sections of the filament free of support on each side are doubly-coiled.

Fig. 2.32.   Tungsten-halogen lamp capsule with filament draped over a glass support rod.

Filaments in tungsten-halogen lamps generally operate at temperatures in excess of 3000°C. At such high temperatures there is substantial production of ultraviolet radiation. The fused quartz bulb is very transparent to short wavelength radiation and allows all of it to come out of the bulb. This can pose some health risks, as UV radiation can cause skin damage and other undesirable effects. To stop the escape of UV radiation, some tungsten-halogen lighting fixtures have a special UV opaque glass mounted just in front of the lamp.

2.5.4 *Modern tungsten-halogen lamps*

Quartz-halogen lamps have been adapted for commercial and domestic applications through a number of innovative design features. Both mains voltage and low voltage lamps are available. Low voltage lamps usually operate at 12 volts using a toroidally-wound lamp transformer that steps down the mains voltage to 12 volts AC. For retail spotlighting and other commercial applications, reflectorized lamps consisting of an outer funnel-shaped parabolic reflector envelope containing a tungsten-halogen lamp capsule are available. These parabolic aluminized reflector (PAR) lamps have a pressed glass front lens that helps in better directing the light out of the lamps. PAR lamps are available in both low voltage and mains voltage versions with a number of different base styles. The first reflector lamps had a specular mirror finish. Later the reflectors were coated with a multi-layer dichroic coating to reflect chosen wavelength bands with high efficiency. Sylvania released the first such lamp in 1960 called the Super Tru-Flector and an improved version in 1962 called True-Beam. The dichroic coating was a 15-layer stack of zinc sulfide (ZnS) and magnesium fluoride (MgF_2). In later years, GE introduced several other types of reflectorized tungsten-halogen lamps, including the very popular multifaceted reflector (MR) lamps in 1971. In these lamps, the inner back surface of the bulb (the reflector) is contoured with many small facets that reflect the light in such a manner as to both narrow down the beam and give it either a soft or a hard edge. MR lamps are available with beam angles from 7° (spot lights) to 60° (flood lamps). Figure 2.33 shows a collection of MR tungsten-halogen lamps.

Here the bulb on the left has a so-called GU10 base — such bases are found on mains voltage bulbs. The other two bulbs are MR 16 and MR 11 types, intended for low voltage operation, with pin bases. The numerals

Fig. 2.33.   Multifaceted reflector (MR) tungsten-halogen lamps.
Courtesy: General Electric Company.

after 'MR' designate the diameter of the bulb in eighth of an inch, measured across the front surface.

Further improvement in energy efficiency has been achieved with
dichroic coatings that recycle some of the heat produced by the lamp filament. This 'hot mirror' technique essentially works by reflecting a significant portion of infrared radiation back to the filament while letting visible
light go out of the lamp. This reduces the heat content of lamp light — a
distinct advantage in several applications, such as retail lighting — but
more importantly, the infrared radiation reflected on to the filament raises
its temperature, allowing it to operate more efficiently. Dichroic tungsten-
halogen lamps were developed during the 1960s and have shown efficacies in excess of 30 lumens per watt.

Tungsten-halogen lamps are also used for heating applications
because, thanks to the halogen-based regenerative cycle, their filaments
can be run at very high temperatures without the risk of a short lamp life.
Figure 2.34 shows a 300 W InfraCare™ medical lamp from Philips which
uses tungsten-halogen lamps as the heat source. It is widely used for
relieving joint and muscular pain in both hospitals and private residences.
Banks of tungsten-halogen lamps are also employed for rapid thermal

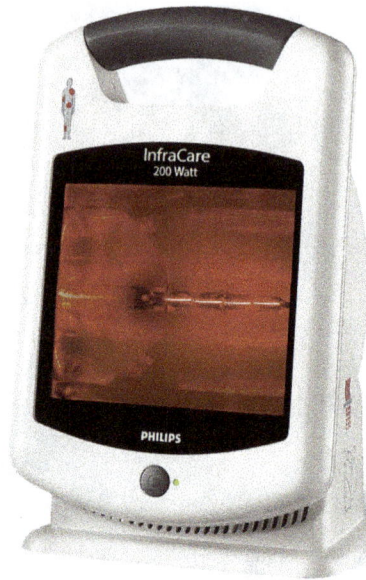

Fig. 2.34.   Philips InfraCare™ heating lamp.

processing of materials, such as semiconductors. Rapid thermal annealers (RTAs) allow material samples to be heated rapidly at rates in excess of 100°C/s, thereby preventing unwanted material diffusions from taking place.

Figure 2.35 shows a commercial halogen lamp-based RTA system for rapid thermal processing of a variety of materials.

Resistance toward the continued use of low efficiency incandescent lamps in recent years has seen renewed efforts at increasing the efficiency of tungsten filament lamps. Although incandescent lamps can never approach the efficiency figures of gas discharge and electroluminescent lamps discussed in the next chapters, clever engineering improvements and the use of modern materials have resulted in lamps that are far superior to those that existed during the first half of the 20th century. One lamp design that has been attracting great interest recently is that of retrofit tungsten-halogen lamps where a halogen capsule is sealed inside a conventional lamp bulb, as seen in Figure 2.36 here. Such lamps are now being increasingly marketed as a higher efficiency alternative to ordinary incandescent lamps, at a comparable price. These lamps are inexpensive, but benefit from the significantly higher efficiency of tungsten-halogen

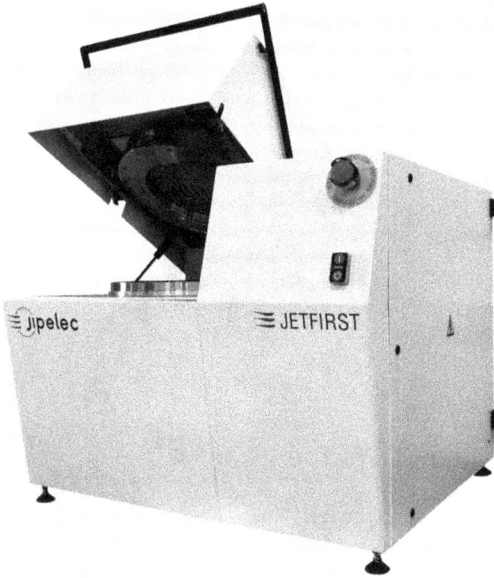

Fig. 2.35. Halogen lamp-based rapid thermal annealing (RTA) system.

Fig. 2.36. Retrofit tungsten-halogen lamp.

Courtesy: General Electric Company.

Fig. 2.37.   Frederick Mosby.

technology. Thus, they represent a superior alternative to plain old tungsten bulbs, and have been sold in large numbers as a higher efficiency
alternative to traditional non-halogen tungsten lamps.

This design was developed by Frederick Mosby of General Electric in
the mid-1960s. At that time, it was known as an A-line bulb and a patent
for it was granted in 1966. Until that time, tungsten-halogen lamps were
still a very recent invention and required special sockets and fixtures for
their use. Mosby, who appears here in Figure 2.37, wanted to take advantage of the halogen lamp technology to develop a more efficient and longer lasting lamp for domestic uses.

He started working on his own to adapt halogen bulbs to a traditional
lamp configuration and developed a working design within a couple of
years of starting on the project. Mosby's original design from his patent
application is shown in Figure 2.38, where a vertical filament is seen with
a midway support, sealed inside a quartz ampoule. The outer glass envelope was made thicker as a precaution against accidental explosion of the
halogen cartridge.

Fred Mosby's lamp was not commercialized by GE at that time
because it could not be produced economically enough to compete with
ordinary incandescent lamps. Much later, with the cost of tungsten-
halogen lamps falling to levels comparable with non-halogen tungsten

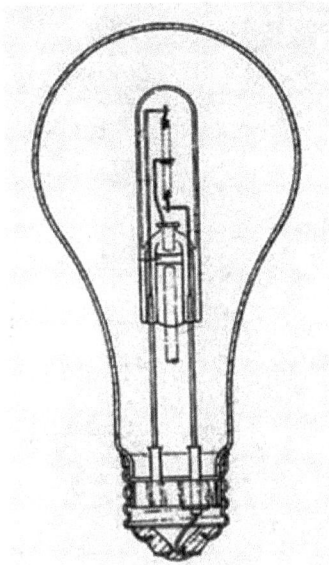

Fig. 2.38. Illustration of a retrofit tungsten-halogen lamp from Mosby's patent.

Courtesy: General Electric Company.

lamps, GE and other companies began to manufacture lamps based on this design, and now these lamps are quite common in both industrial and domestic lighting fixtures.

Lighting manufacturers describe the efficiency measurement or *efficacy* of incandescent lamps as the amount of visible light produced in lumens per watt (sometimes stated as, LPW) of electrical power consumed. A bar of tungsten heated to its melting point has a theoretical maximum efficacy of 52 LPW. Practical studio flood lamps achieve 33 LPW, standard 60-W household lamps with a rated lifetime of 1000 hours achieve 14.5 LPW, or 870 lumens total.

Perhaps the last practical way to enhance the efficiency of traditional tungsten filament bulbs is through selective trapping of radiant heat (infrared radiation) inside the bulb envelope. With the aid of suitable dichroic coatings short to mid-wavelength infrared radiation can be reflected back so as to impinge back on the filament (see Figure 2.39). This makes the filament hotter and increases the efficiency of visible light generation. As most of the input of tungsten lamps is in the infrared region, so this can be a very effective way of boosting the operational efficiency of old-style

Fig. 2.39.   Experimental tungsten filament lamp with a dichroic heat-reflecting filter.

tungsten luminaires. This method has been demonstrated in the laboratory but it is not clear at this time if it can be made economical enough to use with inexpensive tungsten lamps, especially as much higher efficacy LED lamps have nearly supplanted incandescent lamps.

2.6 The Nernst Lamp

Walther Nernst (see Figure 2.40), in Göttingen, Germany, developed an interesting variation of incandescent lamps during the late 19th century. This lamp appeared at a time when the limitations of carbon filament lamps were well known and good metallic filament lamps were yet to make an appearance. Nernst's invention relied on a ceramic resistive electrical element heated to incandescence. This very clever invention, relied on the fact that certain refractory ceramics act as solid electrolytes and become good electrical conductors when heated to very high temperatures (approximately 2500°C). Furthermore, when heated to incandescence, these ceramics emit a brilliant white light. Nernst utilized this knowledge to develop a lamp based on an electrically heated ceramic element. The so-called Nernst lamp contained a ceramic 'glower' in the form of a thin inch-long rod made from a mixture of zirconium, yttrium and erbium oxides. Typically, the composition was zirconium oxide (ZrO_2), yttrium oxide (Y_2O_3) and erbium oxide (Er_2O_3) in the ratio of 90:7:3 by weight.

Fig. 2.40.   Walther Nernst.

The oxides were mixed together with a binder and extruded to form thin rods or filaments which were cut to size and fired to produce the heating elements. As this ceramic material is almost insulating at room temperature so it had to be heated to a high enough temperature to make it conduct electricity. Once it became incandescent, an electrical current could maintain it at its high operating temperature. In some very early variants of the Nernst lamp the glower was heated to working temperature with an oil or gas lamp, but later fully electrical lamps were developed where the glower rod was brought to its operating temperature with secondary filament-based heating elements. When the lamp was first started, current flowed through heaters consisting of fine platinum wires wound around ceramic tubes and over-coated with a thin ceramic layer. In later lamps, expensive platinum was replaced by German silver — an alloy, consisting of 60% copper, 20% nickel and 20% zinc. The heaters served to heat the main incandescent element — the 'glower' — whose temperature rose quickly to above 1000°C, at which point its resistance fell enough to let an appreciable current flow throw it. This took less than 30 s. The glower current was also made to flow through an electromagnet (see Figure 2.41) which became sufficiently magnetized to attract the armature to it. This then cut the current to the heaters and all the current was then diverted to the

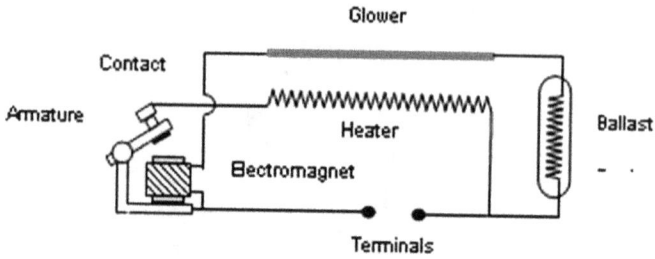

Fig. 2.41.   Circuit for operating a Nernst lamp.

glower. The glower then rapidly attained its operating temperature and became incandescent with a brilliant white light. The operating temperature of the glower was around 3000°C — much higher than that of any incandescent lamp filament of its time. This caused light to be produced with high efficiency so that Nernst lamps had significantly higher efficacies than filament lamps with comparable light output. Also, because of the high glower temperature, the light produced had its spectral peak at a significantly shorter wavelength than for light from filament lamps. The higher color temperature of the light made it look more natural and its day light-like appearance was greatly admired. Nernst lamps were made in wattage ratings from 100 to 2000 W, and millions were produced during the first decade of the 20th century.

Depending on the power rating, the lamps used to have anywhere from 2 to 30 heaters. Every Nernst lamp contained a glass ampoule that looked like a filament bulb and was the only glass part in a Nernst lamp. The actual light producing components — the glower and heaters — were not enclosed as in filament lamps. The glass bulb contained a short length of fine iron wire which was electrically in series with the glower circuit. This wire was placed inside its protective evacuated glass envelope to prevent its oxidation from atmospheric oxygen and water vapor. The iron wire served as ballast, i.e. a means of keeping the current down to the desired operating level. Its need arose because, unlike metals, ceramics are essentially very wide band gap semiconductors that have a negative temperature coefficient of resistivity. The resistance of a glower rod — extremely high at room temperature — would fall quickly at high temperatures, causing a higher and higher current to flow through it. If nothing was done to limit the current, then the lamp would destroy itself due to thermal runaway. The iron wire ballast served to counteract this, as iron has a high positive coefficient of resistivity, so that as the current flowing through it increased, and gradually heated it to higher and higher temperature, its resistance also steadily increased, thus, keeping the current through the glower at most or less a constant value. This clever constant current scheme, using materials with opposite electro-thermal characteristics to maintain a fixed current through Nernst glowers, was one of several inventions by Walther Nernst. He was much admired during his lifetime, and one of his students was Irving Langmuir, who later went on to make extremely valuable contributions in the development of the tungsten filament lamps, as has been already described. The Nernst lamp first appeared in 1897 — before metallic filament incandescent lamps were commercially available. These lamps were widely used for both domestic and street lighting for more than a decade. It was used in the Buckingham Palace and on the clock faces of the Houses of Parliament in London. Figure 2.42 shows a typical Nernst lamp, dating from the early 20th century. The two cylindrical objects at the top are heaters for the glower. There are two glower elements in this lamp, visible as two thin white rods between the heaters. This lamp has a diffuse glass shade for decorative purposes — a Nernst lamp does not require a glass enclosure, unlike filament lamps.

Fig. 2.42. Early 20th century Nernst lamp.

Unlike filament lamps, when the glower of a Nernst lamp failed (usually after hundreds of hours of use), it was simply replaced by another glower element, instead of replacing the entire lamp. The lamps were made by Nernst's company in Germany, and by Westinghouse in the US. Nernst lamps were commercially available until around 1910 when their production stopped due to the growing popularity of tungsten filament lamps. By that time, the latter had become much cheaper, safer, and convenient to use. Special variants of Nernst lamps, however, were made until the 1980s, for use in infrared spectroscopy, as their light contains a very wide band of infrared radiation. The lamps have been supplanted in this use by infrared sources based on silicon carbide heater elements called 'glow bars'. These devices, also known by their trade name 'globar', do not require external heating because silicon carbide has a narrower band gap than the oxides used in ceramics, and, thus, it can conduct electricity even at room temperature. In use, a glow bar lamp's silicon carbide element is heated to around 1200°C by passing an electric current through it. The radiation it produces is a good approximation to that from an ideal black body at the same temperature. This radiation can be filtered through long pass or band pass interference filters to obtain radiation in the 2–20 microns range, which is difficult to obtain from other sources. Figure 2.43 shows a set of silicon carbide globar elements.

Fig. 2.43.   Silicon carbide (SiC) Globar elements.

Silicon carbide elements have been tried for making visible light lamps as well. In 1989, John Milewski and his son Peter Milewski were awarded a patent for lamps based on silicon carbide filaments. They developed technology for growing single crystal filaments (whiskers) of silicon carbide and hafnium carbide that could be made incandescent by passing an electric current through them.

It is interesting to note that some oxides, when heated to incandescence, actually produce much more visible light and relatively little infrared radiation. Examples include calcium oxide (CaO) which was used in the lime lights of the early 19th century. A calcium oxide (lime) cylinder was heated by an oxygen–hydrogen blow torch and gave off an intense white light. This type of illumination was widely used for theater performances, and although it became obsolete with the invention of electrical lighting, the phrase 'being in the lime light' has survived to this day. Other materials that exhibit this 'candoluminescence' effect include the oxides of cerium and thorium, which were widely used in gas mantle lights before the advent of electric lighting. Candoluminescent materials have higher emissivity in the visible region than in the infrared region, when compared with a black body.

2.7 Black Light Lamps

Tungsten incandescent lamps produce almost all of their output in the infrared and visible regions with hardly any emission close to 400 nm. This wavelength is usually taken as the boundary between visible and ultraviolet regions of the electromagnetic spectrum. If, however, the tungsten filament is operated at a very high temperature, even close to its melting point, by passing a large current through it, then the light's spectral distribution changes. In accordance with Plank's radiation law (corrected for this case by taking account of the emissivity of tungsten), the radiation becomes both more intense and the peak of its spectrum moves toward shorter wavelengths. In this way, a small amount of near-ultraviolet radiation is also generated. Bulbs produced specifically for generating small amounts of near-UV radiation through incandescence are called black light lamps. A typical such bulb has the appearance as seen here in Figure 2.44. The inner surface of the bulb is coated with a visible light-blocking material which filters out almost all visible radiation, leaving just the UV radiation to escape.

Fig. 2.44.   Black light lamp.

Incandescent tungsten black light lamps are fairly in-expensive and are easily available from specialist lamp stockists. These are used for applications such as mineral identification through induced-fluorescence, and lighting up UV-fluorescing paints and pigments in decorative and entertainment applications. As incandescent tungsten produces only very small amounts of UV radiation so this type of black light lamp is extremely inefficient. More efficient black light lamps based on gas discharge tubes are also available. As discussed in the next chapter, these consist of low-pressure mercury vapor lamps with appropriate visible light-blocking filters.

3

Electric Discharge Lamps

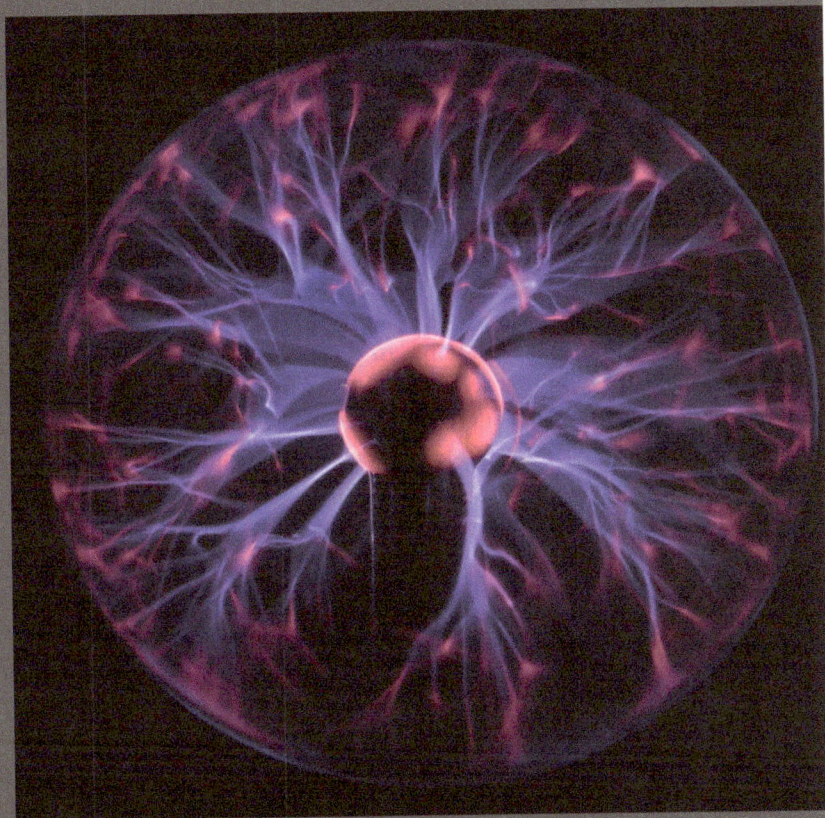

3.1 Introduction

Electrical discharge in a gas is a very effective means of transferring electrical energy to the atoms, ions or molecules that constitute the gas. Through this process, energy can be transferred from an electric field to the constituents of an ionized gas, i.e. gaseous plasma by exciting them to higher energy states. This happens when electrons, accelerated to high velocities, under the electric field established between two electrodes, collide with ions, atoms and molecules, and a transfer of energy takes place. Thus energized, they emit resonance radiation a short time after getting excited. The gas then appears to glow, providing an excellent source of visible and invisible radiation. However, no energy conversion process is 100% efficient in transforming energy from one form to another desired form, and discharge lighting is no exception. Losses here arise from inelastic collisions and collisions with the walls of the plasma container, in addition to electrical losses. Nevertheless, this general technique of impact excitation of gaseous media is made use of in an exceptionally wide class of so-called electric discharge lamps. In fact, most outdoor lighting these days is produced by discharge lamps of one kind or another. Inside buildings too their applications are increasing, such that these lamps are a ubiquitous feature in our day-to-day life. Compared to incandescent lighting, discharge lighting is vastly more efficient with efficiencies of around 40% being extremely common. With the use of incandescent lamps on rapid decline, the appeal of discharge lamps has been grown steadily. All prominent discharge lamps are described in this chapter in roughly a chronological order. A strict chronological description is not possible because the developments of several different types of lamps often happened in parallel. The lamps mentioned here include various noble gas-filled devices, mercury vapor lamps and its different variants, fluorescent lamps in their several commercially-available forms, sodium vapor lamps and a number of specialized lamps, such as xenon, krypton and deuterium arc lamps. The discussion opens with a brief discussion of lightning and auroras, as natural discharge-produced light emissions. Thereafter, we look at the historical development and structural characteristics of various types of discharge lamps.

3.2 Natural Light Generation due to Atmospheric Electrical Activity

Electrical discharges due to dielectric breakdown of the medium between charged objects in a rarefied gaseous atmosphere often result in light emission. This process is seen both in nature and in man-made lighting devices. Lightning during thunderstorms is a familiar example from nature. Lightning takes place when electrified clouds accumulate so much charge that the resulting electric field causes the surrounding air to break down. While this natural phenomenon is observed all over the world, it is significantly more common in the tropical regions where thunderclouds develop more frequently. It is estimated that around 70% of all lightning activity on earth takes place over the tropics, where up-drafts of moist air create an abundance of clouds. Convective airborne movements of these clouds cause friction between cloud masses, and between clouds and the surrounding air. This can result in charge stripping and the formation of electrically-charged clouds called thunderheads. The exact mechanism behind triboelectric charge build-up in clouds is not fully known, but the processes that lead to a lightning discharge are relatively well-understood. When the electric field due to charge build-up exceeds the dielectric breakdown strength of the air surrounding charged clouds, a sudden discharge takes place, as charge neutrality is restored. This process releases energy as sound, light and heat. The discharge can take place between different regions of the same cloud, between two different clouds, or between a cloud and the ground. During the electrical discharge, air becomes highly ionized, forming a plasma conduit, through which a large current can flow. This conduit can be several kilometers in length. Excited atoms and molecules in the plasma release light that is seen as lightning (see Figure 3.1). Rapid release of thermal energy also causes air in the

Fig. 3.1.   Lightning strikes create brilliant flashes of light.

vicinity of the plasma conduit to expand and contract rapidly, resulting in the familiar thunder clap.

Benjamin Franklin was the first person who performed systematic, scientific studies of lightning during the second half of the 18th century. Later, several other investigators worked to better understand this spectacular natural phenomenon. Nowadays, several government funded agencies around the world carry out research on lightning. Sensors on-board several NASA spacecrafts routinely monitor terrestrial lightning activity. The global lightning frequency image shown below (see Figure 3.2) from such an orbiting sensor shows that most lightning on earth occurs over land masses rather than over oceans.

Aurorae in the earth's polar regions (see Figure 3.3), and on several other planets in our solar system, are another type of natural light emission, involving charged particles. These do not involve electrical discharges but instead are caused by high-energy charged particles, streaming out from the sun and slamming into atmospheric gas atoms and molecules at high altitudes. Earth's magnetic field concentrates these charged particles from the so-called solar wind into two doughnut-shaped regions close

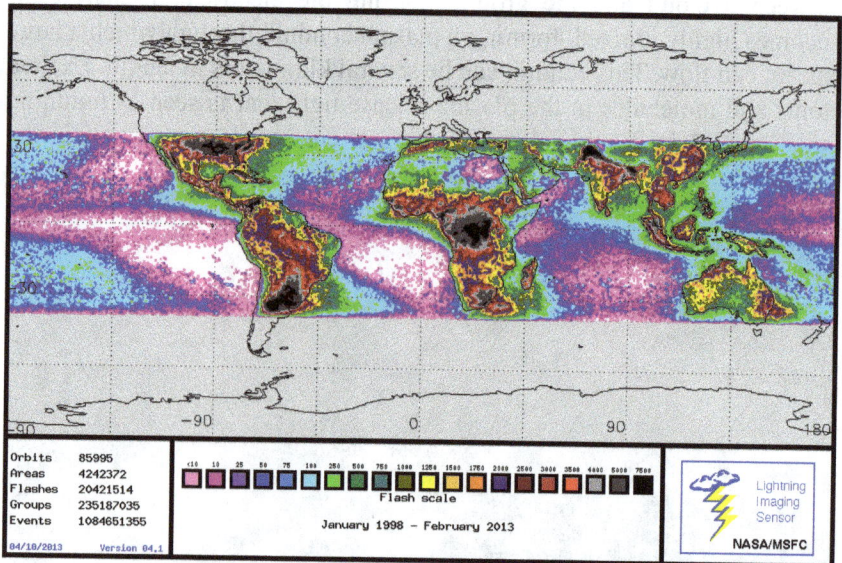

Fig. 3.2. Average global distribution of lightning frequency, as measured by earth-orbiting satellites.

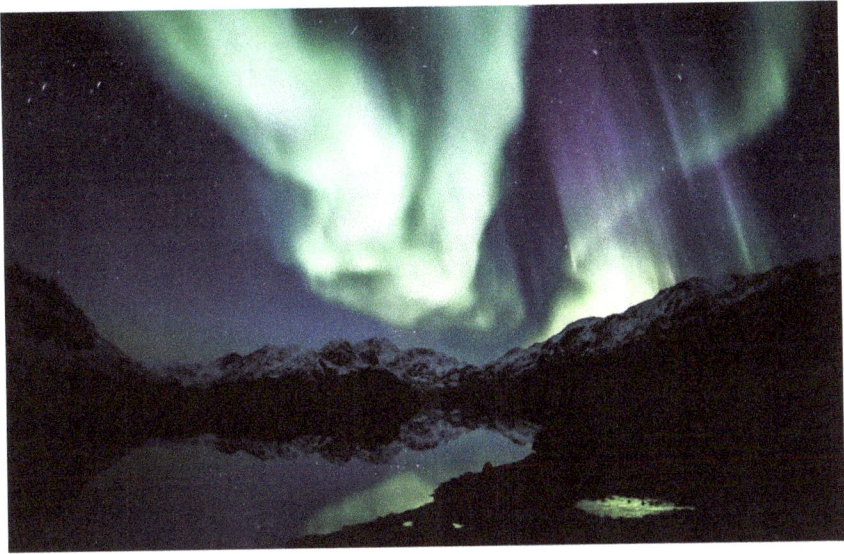

Fig. 3.3.   Aurora Borealis.

to the north and south magnetic poles. The tenuous atmosphere at very high altitudes becomes ionized from impact with these charged particles, resulting in auroral light emission. Due to the concentration of solar wind in the polar regions, auroral displays are primarily seen in, and close to, the Arctic and Antarctic regions, as aurora borealis and aurora australis, respectively. During periods of increased solar activity, the outbound solar wind flux increases considerably from its normal levels. This results in more intense auroral displays and also their spread to lower latitudes. As far back as 1743, Mikhail Lomonossov had suggested that both lightning and aurora were electrical discharges in atmospheric gases. Although this is strictly true for lightning, auroral displays are more subtle, being. The result of excitation of gaseous molecules from energetic charged particles that originate from the sun.

Interaction of the solar wind with planetary magnetospheres produces auroral displays on other planets too. Both lightning and auroras have, in fact, been observed on other planetary bodies, most notably, Jupiter and Saturn. A Hubble telescope image of a persistent aurora on Jupiter is seen here in Figure 3.4. It was taken by the Advanced Camera for Surveys (ACS) imaging instrument on-board the Hubble Space Telescope. The

Fig. 3.4.   Aurora in Jovian atmosphere.

extremely strong magnetic field generated by the, presumably, electrically-conducting interior of Jupiter creates strong concentrations of charged particles in Jupiter's upper atmosphere. These energetic particles interact with Jupiter's escaping atmospheric gases, energizing them and causing the auroral light emission seen here.

This aurora and the one on Saturn, seen in Figure 3.5, also captured by the Hubble Space Telescope, were both seen at ultraviolet wavelengths, and were seen to persist for days.

While lightning and aurorae are created by un-controlled natural electrical discharges, controlled discharges are widely used to generate artificial light. In addition to electrically-induced incandescence, this is another mechanism through which electrical energy can be converted to light energy. A big advantage of discharge-produced light is the much greater efficiency with which this process converts electrical energy to light.

From a historical point of view, lamps based on electrical discharges have been around for significantly longer than even incandescent lamps. However, the very early lamps were used for scientific work rather than as illumination sources. Even to this day, lamps featuring electrically-excited gas discharges are widely used in many scientific and technical applications, such as laser pumping, spectroscopy research, material identification and metrology. Discharge lamps for illumination purposes were

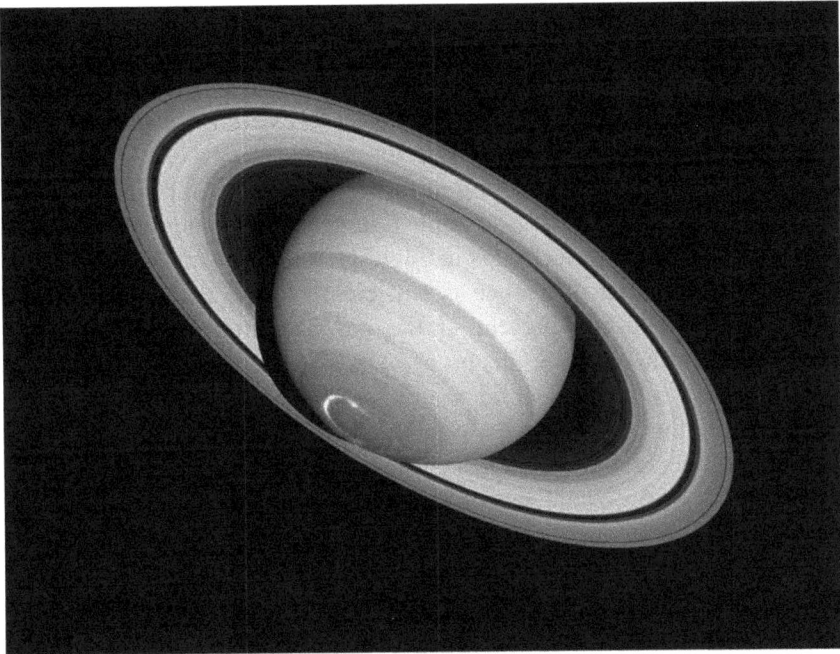

Fig. 3.5.   Aurora in Saturn's polar region.

developed much later, starting from around the 1920s, and are now wide-spread throughout the world. Fluorescent lamps, neon signs, sodium and mercury vapor lamps, all utilize electrical discharge in either low- or high-pressure gases or gas mixtures. Plasma televisions are also based on light produced through local discharges in small cells, forming the glowing pixels on television screens. In this chapter, we look at several types of discharge lamps in roughly the chronological order of their appearance. These include neon lamps, fluorescent tubelights, compact fluorescent lamps (CFLs), mercury vapor lamps, sodium vapor lamps, high-intensity discharge (HID) lamps, and several others.

3.3 Luminous Electrical Discharges

Gases generally do not conduct electric current. However, the atoms or molecules of a gas can be induced to conduct a current by ionizing them with a suitable technique. Once ionized, the gas becomes a plasma and has the ability to transport an electric current. A conducting plasma often emits light due to atomic excitation and de-excitation processes occurring in it. There are several ways to ionize gases and generate plasma. Depending on the gas and the method used to excite it, several different types of discharge lamps can then be constructed.

The oldest and the most straightforward technique for generating an electrical discharge plasma is to apply a large enough potential difference between two separate electrodes in a low-pressure gas ambient. Provided the applied potential difference is sufficiently large, free electrons and charged ions naturally present in the gas are attracted to oppositely charged electrodes. During this process, they get accelerated by the electric field between the electrodes and as they collide with neutral gas atoms, cause them to ionize by kicking electrons out of them. The (very small) initial ionization in the gas is caused by natural background radioactivity and from cosmic ray particles. With high voltage between the electrodes, accelerating electrons free up more electrons from gas atoms and the ionization in the gas quickly builds up to very large levels. The gas, thus, becomes a moderately good conductor of electricity. Depending on the areas of the electrodes, the type and pressure of fill gas and the voltage between the electrodes, currents from a few micro amps to several amperes can then flow through the plasma, and light, characteristic of the gas composition, is emitted. Such plasmas are generally cold, with the discharge operating at barely above ambient temperature. Lamps based on this type of discharge are called cold cathode lamps.

A significant drawback of cold cathode discharges is the necessity to apply very high voltages to initiate and sustain the discharge. These can run into several thousand volts and require specialized high-voltage power supplies. This requirement can be eased by using a heated cathode. Here, the cathode is made of a coiled tungsten filament which is heated to incandescence. It is generally also coated with a material with low thermionic work function. This type of active heated cathode emits a copious supply of electrons which efficiently ionize gas in its vicinity. This then considerably reduces the strength of the electric field needed for a discharge to take place. A hot cathode arrangement is capable of operating at significantly lower voltages, compared to cold cathode devices. Additionally, hot cathode discharges are also more efficient light emitters on account of their higher ion density. Thus, it is not surprising that most illumination–quality discharge lamps utilize hot cathode arrangements. However, these lamps do have a significant downside in that their

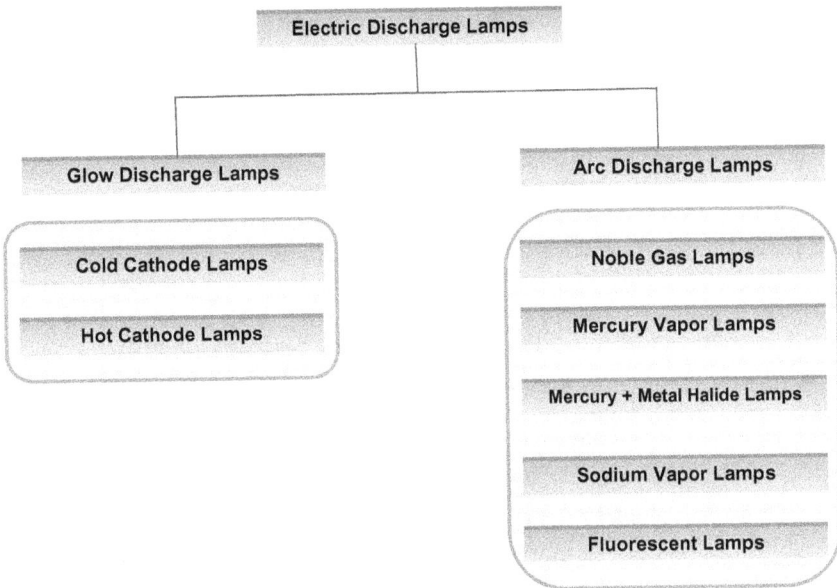

Fig. 3.6.   Types of electric discharge lamps.

lifetimes are much shorter than that of comparable cold cathode lamps. This is due to the use of heated cathodes that gradually evaporate away, thin down, and eventually break up, as in any incandescent tungsten filament lamp. Despite this disadvantage, hot cathode lamps are widespread due to their higher brightness and efficiency. Figure 3.6 shows a rough classification of various practical electric discharge lamps.

Apart from the temperature of the cathode, another important distinguishing characteristic of discharge lamps is the pressure at which they operate. Some lamps operate at pressures that are much lower than atmospheric pressure, but some, such as HID lamps and high-pressure mercury discharge lamps, operate at pressures close to or even well above atmospheric pressure. The operating pressure has great influence on the color content of light emission, as will be seen later in this chapter. We will examine most common electrical discharge lamps in this chapter, beginning with the earliest one that was commercially used for lighting applications — the neon lamp. This lamp is a low-pressure cold cathode device which entered service early in the 20th century and is still being widely used all over the world for applications ranging from electronic indicators and household night lights to artistic displays and outdoor advertising. But before we come to neon lamps, we take a look at the science behind electrical discharge in gases.

3.4 Physics of Electrical Discharges

The production of sparks through electrical means is an observation that goes back to at least the 17th century. In the year 1672, Otto von Guericke described sparking between sulphur balls when they were rubbed with hand. By the end of the 17th century, sparking between electrified objects was well-known, though poorly understood. As the science of electricity, magnetism and eventually, electromagnetism developed during the 18th and 19th centuries, arcs and sparks came to be better understood. During the second half of the 19th century, much work was done to investigate the nature of electric sparks, discharges and arcs. There was active research in this area in many European universities. French and German scientists were particularly involved with such studies. Around 1857, Johann Heinrich Geißler (see Figure 3.7(a)), a German glassblowing technician at the University of Bonn, started working for the physicist Julius Plücker (see Figure 3.7(b). On Plücker's request he constructed several hand-cranked mercury diffusion pumps. These pumps were superior to what was commercially available at that time, and could achieve lower background pressures than other pumps. Plücker, at that time, was

Fig. 3.7.   (a) Johann Heinrich Wilhelm Geißler. (b) Julius Plücker.

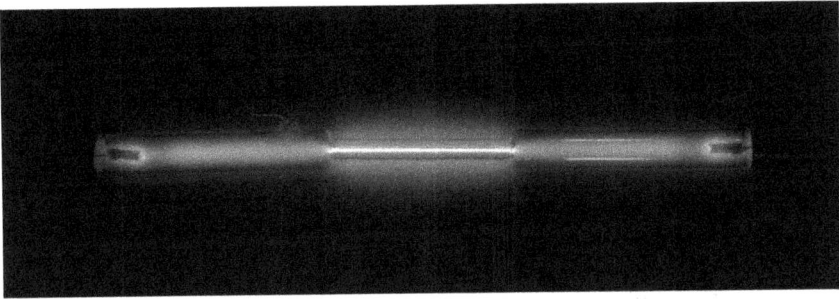

Fig. 3.8.   Geißler tube.

conducting research on electric sparks and other types of discharges. Working with him, Geißler constructed a glass tube with metal electrodes at opposite ends that could be connected to one of his vacuum pumps. He then took out as much air from the tube as possible, and connected the electrodes to a high-voltage source. Nothing happened on the first few tries but on improving the vacuum and applying a higher voltage between the two electrodes he observed bright light emission from the glass tube. This was the first artificially-produced glow discharge. Soon word of his observations spread throughout Germany, and beyond. In later years of the 19th century, Geißler's tubes (see Figure 3.8) became commonplace as fanciful display objects. These tubes were filled with different gases and emitted a range of colors. Still later, when even better vacuum pumps became available, it became possible to evacuate discharge tubes to extremely low-pressures. Such tubes ceased to develop any glow no matter how high a potential difference was applied across them. Instead, their glass, close to the anode electrode, began to glow a yellow-green color. These tubes called Crookes tubes later led to the discovery of the electron and gave birth to electronic vacuum tubes, such as diodes, triodes, tetrodes, and even the cathode ray tubes used as television and radar displays.

Concepts needed to understand light emission from discharge tubes developed only gradually during the late 19th century. In time, it became clear that the excitation and subsequent de-excitation of gaseous atoms or molecules by collisions with accelerating electrons is the mechanism underlying light emission. Thus, the study of electron-atom collisions began to be studied in detail. Already in 1879, Johann Hittorff, through his study of dilute gases, had determined the electrical conductivity of air and different gases. In 1900, this culminated in Paul Drude's theory for the electrical conductivity of gases. His explanation introduced the concept of collision frequency ν_c. The collision frequency received its physical explanation when Ernest Rutherford found the, now famous, Rutherford collisional cross section, σ_c, in terms of which the collision frequency can be written as $\nu_c = n \, \sigma_c \, v$, where n is the number density of gas atoms, and v is the velocity of the moving particle. Equivalently, this allowed the introduction of the collisional free flight distance $\lambda_{ff} = v/\nu_c$ of a particle between two collisions. If v is taken to be the thermal velocity v_{th} then this becomes the mean free path $\lambda_{mfp} = v_{th}/\nu_c = 1/n\nu_c$. Higher collision frequency and larger electron energy work together to enhance light emission from gaseous discharges. Gas pressure, represented by n, in the expressions above, plays a crucial role. For efficient light generation, it has to have an optimum value. At extremely low gas pressure, n has a small value and, consequently, the collision frequency is also small. This leads to infrequent excitation-de-excitation events and, thus, weak light emission. However, at too high-pressures, n can become quite large and while this increases the number of collisions, the mean free path also becomes quite short so that electrons are unable to gain sufficient energy between collisions to effectively excite gas atoms. Thus, there is an intermediate gas pressure where collisional excitation reaches its peak. While a deeper understanding of the light emission process in electrical discharges had to await the development of atomic physics and quantum mechanics, sufficient developments in physics had taken place by the end of the 19th century to qualitatively understand the phenomena observed when electrical discharges took place in Geißler tubes.

To understand the origin of light emission in gas-filled discharge tubes, consider a long glass tube fitted with metal electrodes at its two ends. Assume that the tube is filled with a gas, such as argon, at a pressure of around 0.1 Torr. When the electrodes are connected to a source of DC voltage — around 10 kV — some portion of the interior of the tube begins to glow with a bluish light. In this state, the gas inside the tube has become partially ionized, i.e. has changed into a plasma and has begun to conduct electricity. This is a direct result of the high-electric field that exists in the tube, due to the application of a high-voltage across it. The ionization is initiated by naturally-occurring free electron–ion pairs in the gas inside the tube. These pairs are being constantly formed due to high energy particles from cosmic rays and natural radioactivity traversing through the tube. The free electrons are attracted toward the anode whereas the free ions are attracted toward the cathode. Thus, both experience an acceleration. If the potential difference across the tube is low (say, 500 V) then the acceleration is relatively low and nothing much happens. However, at larger voltages, the electrons gain sufficient energy in the electric field to become capable of ionizing other neutral gas atoms or molecules through collisions. This voltage is called the breakdown or ionization voltage. These collisions produce further free electron-ion pairs which, in turn, get accelerated, gain kinetic energy, and collide with other neutral species to produce even more pairs. Provided the voltage is sufficiently high, this collision cascade rapidly ionizes the gas and turns it into a weakly conducting plasma. This stage of the discharge which is obtained at relatively low voltages and results in only a few micro amperes of current flow is often called Townsend discharge.

At somewhat higher voltages, a luminous discharge is observed. This is also called a glow discharge and appears as alternating bright and dark regions along a long Geißler tube. The structure of a typical glow discharge is seen in Figure 3.9.

Electrons and positively-charged ions in a glow discharge plasma move in opposite directions. Electrons, being much lighter, have a much higher mobility compared to the positive ions and, thus, carry the bulk of the current. On reaching the cathode, positive ions hit it and release electrons from the electrode's surface. These electrons are supplied by the current flowing in the external circuit. Some of the ejected electrons neutralize the ions, while others are repelled by the cathode and accelerate in the opposite direction. Their energies however are low in the immediate vicinity of the cathode. As there is not much ionization and collisions in

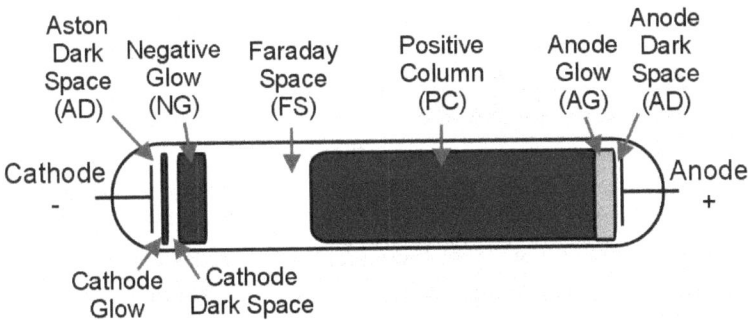

Fig. 3.9.   Various regions in a typical glow discharge.

this region so the space surrounding the cathode is dark — it is called the Aston dark space. This is a region where electrons keep accelerating to higher kinetic energies. Once the electrons gain sufficient energy to excite electrons in neutral species to higher energy levels, they traverse through a narrow glowing region called the cathode glow. Immediately after this region a dark space called the cathode dark space is seen where the electrons have lost energy through collisions with neutral species and are no longer able to excite them. Again, this dark region is a place for electrons to accelerate and gain kinetic energy. This process of gaining kinetic energy and then losing it in collisions keeps repeating, resulting in the formation of several alternating dark and glowing regions. In very short lamps, such as the type NE-2 ampoule neon lamp, it is the negative glow which is primarily visible to the eye, as light emission from the lamp. In longer lamps, such as a neon advertising sign, the light comes from a lengthened positive column which fills the entire tube. Once a glow discharge is established in a tube, it displays a negative resistance behavior, i.e. decreasing the voltage across it actually leads to an increase in current. This necessitates using a 'ballast' in series with the tube, if it is to be operated safely. A ballast limits the current to safe values. Simple neon glow lamps employ a resistor for this purpose whereas most other discharge lamps utilize an inductive ballast.

During the early years of research on electrical discharge in gases, careful investigations were carried out on the dependence of electrical discharge on factors such as the type of gas or gas mixture, gas pressure, electrode spacing and the presence of external ionization sources such as light or radioactive substances. Knowledge gained through such experiments later helped develop the disciplines of spectroscopy and quantum mechanics. For instance, it was found that not all gases offered the same ease of ionization. Thus, the heavier noble gases were easier to excite a glow discharge in, when compared to helium, for instance. This trend was easier to explain later on the basis of atomic structure theory. Electrons in smaller atoms were closer to the nucleus and, thus, more tightly held by it; needing higher energy to remove them and cause the atom to ionize. Given a certain gas, it was also found that the voltage needed to initiate a glow discharge decreased as the gas pressure fell, but at very low-pressures it began to rise with further decrease in pressure. Similarly, the breakdown voltage decreased with reduction in the anode–cathode gap up to a certain distance, but afterwards began to rise again. In 1889, the German physicist, Friedrich Paschen elaborated on these observations by

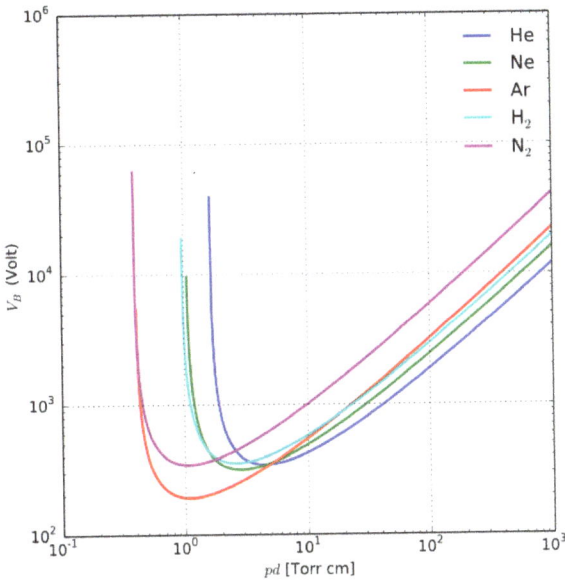

Fig. 3.10.   Paschen plot, showing the dependence of breakdown voltage on pressure times electrode spacing product.

showing that, for any gas or gas mixture, the breakdown voltage was closely dependent on the pressure-electrode distance product (*pd*). He published curves of breakdown voltage variation with changes in the *pd* product for a number of gases (see Figure 3.10). An empirical explanation of these trends is not difficult to provide. At high gas pressures and/or inter-electrode distance, it is difficult to ionize the gas inside a discharge tube. This is because the electric field is low and, thus, below the breakdown field strength for large electrode separations. Even if the separation is small, with high gas pressure there are many gas atoms (or molecules) around. In this case, free electrons do not get a long enough un-hindered path to accelerate and gain sufficient kinetic energy to ionize neutral atoms. Thus, the electrodes need to be close apart (or the applied voltage high enough to exceed the breakdown electric field) for a glow discharge to form. At the same time, the pressure needs to be low enough to result in a sufficiently long mean free path for electrons to cause ionizing collisions. If the pressure gets too low then there are so few gas atoms around that it becomes increasingly improbable for an accelerated electron to collide with one before hitting the anode. Thus, no glow discharge is observed. If the anode-cathode distance is too short then electrons can reach the anode before colliding with any neutral species and, thus, ionization is again suppressed. Generally, for technological applications, a suitable gas mixture, electrode separation and gas pressure are used that results in a sufficiently low breakdown voltage.

The changeover from 19th to 20th century also saw a flurry of activity in spectroscopic study of light emitted by different gases in low-pressure discharge tubes. Spectral line series, such as the Balmer, Ritz, Lyman and Paschen series were discovered during this time. These studies were vital for the development of atomic physics and quantum mechanics during the first few decades of the 20th century. It had become clear early on that the spectral makeup of radiation from different gases and metallic vapors is governed by which electronic transitions are possible in the discharge medium. The spectral profile then controls the perceived color from the emitted light. Thus, the choice of gas or gases used in a discharge lamp is the principal determinant controlling the makeup of radiation emitted by the lamp. Electronic transitions allowed between specific atomic energy levels determine the wavelengths of light emission. In many cases, discharge lamps are filled with a mixture of noble gases and metal vapors in order to enrich the lamps' spectrum with a range of wavelengths. This produces fuller spectral coverage and, thus, light with better color rendering properties. Gas pressure has a secondary effect on the spectrum through the broadening of atomic spectral lines, beyond their natural line widths. Typical atomic transitions produce spectral lines that are only about 16 MHz wide. Due to a variety of dynamical processes inside discharge tubes, actual spectral lines are observed to be significantly wider than this value.

Gas atoms move at high velocities and, thus, photon emission from them shows corresponding Doppler shifts. Thermal motion causes a range of atomic velocities, determined by the Maxwell–Boltzmann distribution, to be present. This causes the spectral lines to appear broadened with the full-width-at-half-maximum (FWHM) in (circular) frequency of emission given by:

$$\Delta\omega_{Doppler} = \frac{2\omega_0}{c}\sqrt{2\ln 2 \frac{kT}{m_0}}$$

Here, ω_0 is the center frequency, T is absolute temperature of the gas, k is Boltzmann constant, c is velocity of light and m_0 is the mass of an atom.

The intensity distribution in such a broadened line is Gaussian, with the maximum intensity I_0 at the center, and the intensity falling away at each side in a Gaussian fashion, given by:

$$I(\omega) = I_0 \exp\left[\frac{-m_0 c^2 (\omega_0 - \omega)^2}{2kT\omega_0^2}\right]$$

Pressure broadening due to the interaction of emitting atoms with each other further broadens spectral lines. Interactions interrupt photon emissions, shortening emission lifetimes and broadening emission line widths. Interactions also broaden the upper and lower energy levels involved in electronic transitions causing additional broadening of the spectral line. At high temperatures and pressures, characteristic of gas discharge plasmas in most modern lamps, there is considerable broadening of emission due to both Doppler and pressure effects. This is usually favorable for achieving richer spectral outputs that are desirable for good quality lighting.

3.5 Neon Discharge Light Sources

Electrical discharge in low-pressure neon gas provided the very first discharge-based light sources. Neon is a relatively abundant noble gas (when compared with krypton and xenon), and produces a bright orange-red glow when used in a cold cathode discharge tube. Several neon-based light sources — ranging from lamps to displays — have been built over the years. We now take a look at all the major neon discharge light sources.

3.5.1 *Neon lamps*

Neon lamps are generally cold cathode devices that were among the first electrical discharge lamps to be developed for illumination applications. Due to the fact that their light is of an intense red-orange color and they require high-voltage power supplies for normal operation, they were soon relegated for use in only decorative lighting applications. There are several variations on cold cathode neon lamps, and although bulb-style lamps were once developed, the only types of neon lamps that are still being produced in significant quantities are small mains-operated lamps for

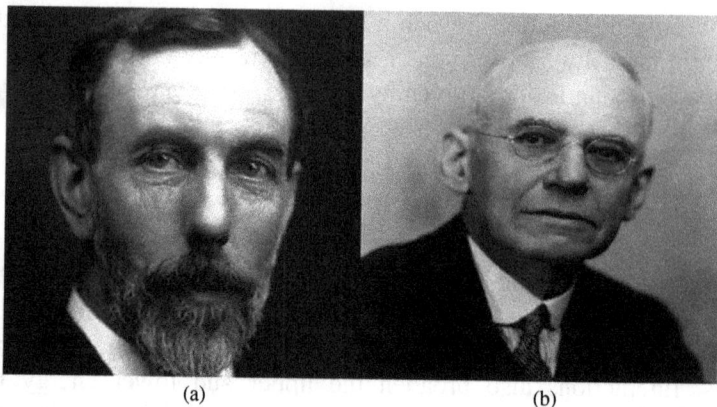

Fig. 3.11.   (a) William Ramsay. (b) Morris Travers.

indication applications and neon signs used, mainly, for advertizing purposes.

The strong orange-red color of neon lamps is the principal characteristic of discharge in neon gas which was discovered in 1898 at University College, London, by William Ramsay (see Figure 3.11(a)) and Morris Travers (see Figure 3.11(b)). A few years prior to this, Ramsay had already discovered argon (1894) and produced a sample of helium (1895) — a gas that had originally been discovered in the sun, through spectroscopic observations. Ramsay — an accomplished chemist — was certain that a similar gaseous element was yet to be discovered, because there was a blank space between helium and argon in the periodic table. At first, he tried to obtain this new purported element from radioactive minerals (from which he had previously obtained a sample of helium) by collecting gases given off when samples of radioactive minerals, such as cleveite and baryte, were heated. Various meteorites were also tried in this quest. When this did not meet with success, he tried a different approach. Working with Travers, he liquefied argon gas and then heated it gently to recover any other gas that was present with argon and had been liquefied with it. Through this fractional distillation process, they obtained a small sample of a gas that boiled off before argon. In order to further establish the identity of this gas, they employed a spectroscopic technique that had previously been used for establishing the identities of helium and argon. An electrical discharge was created in the recovered gas sample, and immediately they noticed its strong orange-red glow. Undoubtedly, they had obtained the missing element which was named 'neon' by Ramsay, after the Greek word for new. Both Ramsay and Travers were awestruck when they first saw neon's glow, and in his memoirs, Travers later wrote that "the blaze of crimson light from the tube told its own story and was a sight to dwell upon and never forget." Thus, the method of discovery of neon itself immediately established the potential of neon as a gas for electric discharge lamps.

Soon after, through the same fractional distillation process on liquid air, the team of Ramsay and Travis isolated two other similar gases — krypton and xenon — which completed the column of noble gases in the periodic table. All of these gases now find applications in gas discharge lamps and various types of lasers, and are, thus, of huge importance for making a variety of light sources. Neon is quite abundant in the universe as a by-product of nucleosynthesis in stars but is relatively rare on earth with its atmospheric concentration only being about 0.0018%. This scarcity somewhat delayed the development of the first generation of neon lamps, as it had to wait for affordable commercial availability of neon in sufficient quantities. During the first decade of the 20th century, Georges Claude (see Figure 3.12), a prolific French inventor, developed an industrial-scale process for liquefying air which enabled low-cost production of not only liquid oxygen and nitrogen but also of the noble gases (excluding helium, which is obtained from natural gas deposits). With his business associate, Paul Delorme, he established the Air Liquide company for producing and selling gases obtained from fractional distillation of air. Commercial availability of noble gases then made it possible to develop discharge lamps based on their spectral emissions. Initially, the noble gases were only considered by-products of the production of liquid oxygen and nitrogen so the business partners devised a plan to make use of these gases in new lighting devices. In this way, what was considered a useless derivative of liquid air's fractional distillation process could also be put to good use. The strong glow in neon's electric discharge was, of course, well known to them. Thus, in December 1910, it was Georges Claude himself who demonstrated an electrically-operated neon lamp at the Paris Motor Show.

The neon indicator lamp has, since the early years of the 20th century, remained virtually un-changed as a small glow discharge lamp. It is widely used for indicating the presence of mains voltage in electrical line testers and mains-powered electrical equipment. Neon indicator lamps

Fig. 3.12.   Georges Claude, with his gas liquefaction apparatus.

are, perhaps, the simplest type of electric discharge lamps possible; consisting of two metal electrodes in a sealed glass ampoule filled with a mixture of 99.5% neon and 0.5% argon — called Penning mixture — at a pressure of around 10 Torr (see Figure 3.13). This mixture is also used in many other gas discharge lamps as it is particularly easy to ionize; requiring a relatively low breakdown voltage. On applying a voltage in the range of 100–300 V, a current of about 0.5 mA flows and a glow discharge forms inside the lamp, with the neon plasma glowing with a bright orange-red color. Such lamps can be operated with both DC and AC voltages. With the former, the glow is seen around the cathode while in the latter case both electrodes are seen surrounded by glowing neon plasma. This means that simple neon lamps can be used to determine if a source is AC or DC and to determine its polarity if it is DC.

Fig. 3.13.   Neon indicator lamp.

Common neon lamps are produced in a cylindrical form factor, with outer diameters of either 3 mm or 5 mm, but other diameters are also not uncommon. The ubiquitous tail seen at the top of the glass envelope (see Figure 3.13) is formed when the tube is sealed after gas filling. These so-called type NE-2 lamps are always operated with a series resistance of around 80 kω, as a ballast, to limit the current flow to safe values. In use, the lamp ampoules are generally contained inside an outer plastic holder which can be screwed onto a front panel. While these types of simple neon indicator lamps are the most common ones encountered in electrical equipment, neon lamps emitting colors other than neon's characteristic orange-red glow are also available from some suppliers. These neon lamp-like tubes contain low-pressure argon or an argon–neon mixture and have internal walls coated with different phosphors (see Figure 3.14). Depending on the gas mixture present in the tube and the phosphor coating on the tube wall, these lamps emit a variety of colors, such as white, pink, blue and green. As the tubes have the same form factor as ordinary neon lamps, so they can be used instead of neon lamps for displaying colors other than neon's orange-red.

While neon lamps, of the type described above, are low luminosity devices, on account of their small size, and only used in indicator and night light type applications, higher intensity lamps were also commercially developed during the first part of the 20th century. Daniel Moore, while working for the General Electric Company (GEC) in 1917, developed a neon lamp with a bulb form factor (see Figure 3.15). This type NE-34 lamp was based on a patent issued to General Electric in 1919.

Fig. 3.14.   Phosphor-coated neon lamps.

Fig. 3.15.   General Electric NE-34 neon lamp.
Courtesy: General Electric Company Ltd.

These lamps were available in different sizes with both bayonet and screw bases. The main innovation in their design, which made their patent protection possible, was their special electrode structure. These had large area metal plates combined with thinner metal wires. This design enabled the glow to take a large area and, thus, resulted in increased light intensity.

Although extremely un-common, glow discharge lamps have also been made from argon. General Electric made an argon-filled derivative of its neon glow lamps for many years as a convenient small source of UV

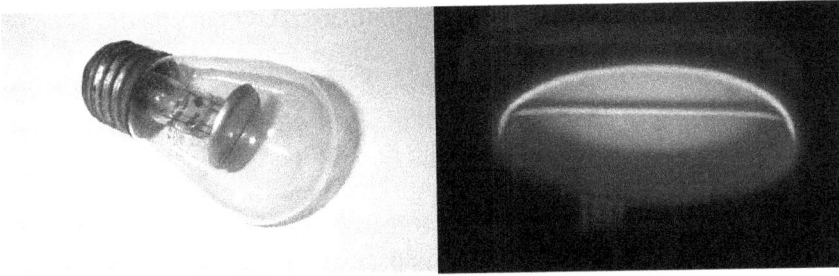

Fig. 3.16. General Electric AR-1 argon lamp (left) and its light emission during lamp operation (right).

Courtesy: General Electric Company.

radiation for hobbyists and mineral prospectors. This type AR-1 lamp (see Figure 3.16) was quite popular during the 1950s and 1960s. The similarity to the neon glow lamp is easy to see from the figure. A patent (US patent: 1,965,587) issued to Ted E. Foulke of the General Electric Vapor Lamp Company, in 1934, covered the essential features of this lamp. This lamp had electrodes made from stamped nickel sheet. The electrodes were coated with metallic barium on their upper surface to enhance electron emission. The glass bulb contained a mixture of 90% argon and 10% nitrogen at around 10 Torr. Operating directly from the 110 V AC mains supply and consuming 1 W of electrical power, this lamp produced a nice-looking blue glow, together with some invisible UV radiation. Compared to mercury vapor UV lamps, the argon glow lamp produced much less UV radiation but also had a relatively short useful lifetime of around 1000 hours. Due to these shortcomings, the argon glow lamp is no longer produced.

3.5.2 *Neon discharge-based displays*

The fact that the neon glow is confined to the cathode, when driven with a DC voltage, was made use of in a numeral display device, called a Nixie tube. This specialized lamp consisted of a set of shaped wire cathodes — each one formed in the shape of a numeral from 0 to 9 — positioned behind a metal mesh anode. The electrode arrangement was sealed inside a neon-filled glass capsule. On applying a suitable DC voltage, of around 170 V, between the anode and one of the ten cathodes, the chosen numeral could be displayed as a glowing character. Some tubes could display other symbols too, such as decimal and percentage signs. These displays were first conceived in the 1930s, but their commercial production did not take place until two decades later. A very early patent for a shaped cathode neon tube display device was awarded in 1938 to the German inventors Hermann Pressler and Hans Richter (US patent: 2,138,197). They described their invention as a 'self-luminous sign'. They also pointed out that the use of shaped-cathodes meant that their display did not require bending glass tubes to elaborate shapes; as is the norm for usual neon sign displays. However, their display could not be made in large sizes and, thus, it did not find success as a replacement for traditional neon signs which were getting into widespread use then for outdoor advertisement purposes. In the same decade, another German engineer, Hans P. Boswau, a resident of Lorain, Ohio, near the shores of Lake Erie, conceived of a much more practical design of shaped-cathode neon display device. He was the first person to come up with the idea of stacking several shaped-cathodes — in the shapes of the numerals from 0 to 9 — one behind the other. Thus, one device could display any of the numerals, as needed. In the year 1934, Boswau filed two patents (US patents: 2,142,106 and 2,268,441) that described his invention. While the patents were granted, Boswau did not go any further to utilize his invention.

The first commercial neon indicator tube was brought out by the National Union Radio Corporation in 1954. The 'Inditron' tube used a stack of shaped-cathodes inside a neon-filled glass tube (see Figure 3.17). Connections for all cathodes were individually brought out of the tube. To display a numeral, the corresponding cathode was connected to the negative terminal of a high-voltage DC power supply while all the other cathodes were connected to the positive terminal. This was a cumbersome arrangement that complicated the drive electronics. These tubes also had a relatively short life. Taken together, these disadvantages doomed the Inditron tubes so that they never found widespread use.

The first commercially successful neon indicator tubes started appearing in 1955, in equipment made by the Burroughs Corporation. Their development was made possible by Burroughs' acquisition of Haydu Brothers — a Plainfield, New Jersey, manufacturer of precision etched metal sheet assemblies that were used for vacuum tube manufacture. Burroughs also hired Saul Kuchinsky, who had earlier worked on the Inditron tube at the National Union Radio Corporation, and put him in charge of developing its own line of indicator tubes. Working at Burroughs Research Center in Paoli, Pennsylvania, Kuchinsky and his team developed the design which is now universally known as the Nixie tube.

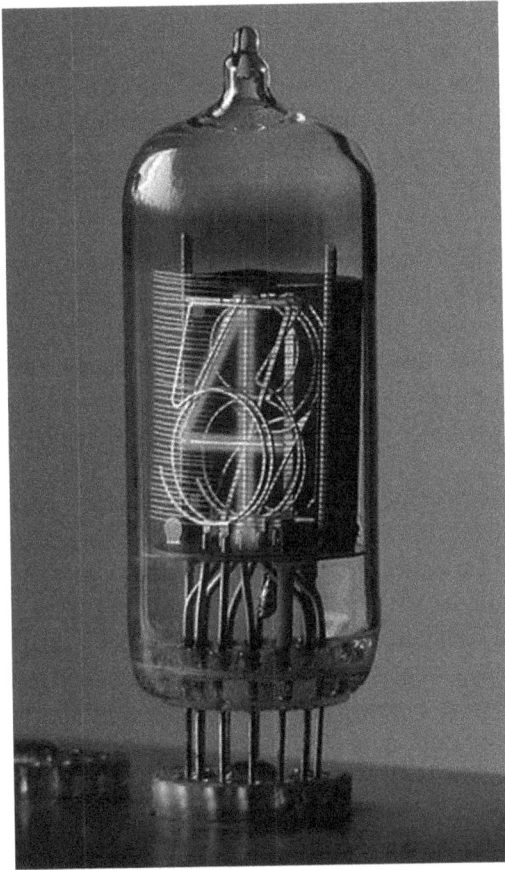

Fig. 3.17.   Inditron numeric display tube.
Courtesy: National Union Radio Corporation.

Kuchinsky not only came up with the name 'Nixie' but also was instrumental in developing the features that made Nixie tubes so successful in
later years. These included the use of mercury vapor fill to greatly extend
the operating life of the tubes, use of precision etched shaped-cathodes
rather than bent-wire cathodes (as was the case with all earlier indicator
tubes) and the use of a mesh anode that surrounded the cathode stack. The
mesh could be permanently connected to the positive terminal of the driving power supply and then only one cathode needed to be connected to the

negative terminal to display the chosen numeral. This greatly simplified the driving requirements for indicator tubes and resulted in the rapid adoption of Nixie tubes in all kinds of electronic equipment. Since August 1955, when the first Nixie tubes were demonstrated at the Wescon show in California, Nixie tubes — like the ZM10120 — became the staple of all kinds of equipment that needed to display numeric information. These tubes were widely used in systems, such as frequency counters, power supplies, voltmeters and oscillators. Special integrated circuits, such as the SN7441 BCD-to-Nixie tube driver were available to drive Nixie tubes. If used properly, these tubes were very long-lived and were aesthetically pleasing too. Nixie tubes were made until the early 1990s when LED-based displays firmly put an end to their use. However, these displays are still on the market for use as replacement parts on legacy equipment.

Use of glow discharge for displaying numbers got a new make-over in the 1970s when segmented alphanumeric displays based on the Nixie tube principle were introduced. These devices differed from the Nixie tubes in having separate straight segments that could be lit in different configurations to display various letters and numerals. Nixie tubes, in comparison, had fixed numeral-shaped electrodes for displaying numbers. The newer displays were more versatile and, thus, became quite popular. Although, like Nixie tubes, these displays have not survived the test of time, they are an important link in the development of more modern image display technologies.

A monochrome video display, based on neon glow discharge plasma, was first developed at the University of Illinois at Urbana-Champaign in 1964 by Donald Bitzer, H. Gene Slottow, and graduate student Robert Willson, for the PLATO Computer System. This display consisted of a two-dimensional array of neon glow discharge cells with row and column electrode connections arranged such that any cell in the display matrix could be individually addressed. Each cell formed an individual picture element (pixel), and by illuminating it a tiny glowing dot could be displayed. In this way, a collection of triggered pixels could be used to display an image (see Figure 3.18). These displays were very popular during the 1970s, and were also used for billboards and public information displays. Their use for displaying digital information declined after the 1970s, but the technology was refined for full-color displays and became the basis of plasma display panels (PDPs). By itself neon can only display its characteristic orange-red color, so in order to display red, green and

Fig. 3.18.   Neon glow discharge-based video display.

blue colors, electric discharge was augmented with the use of different color-producing phosphors.

A PDP consists of a two-dimensional array of tiny chambers, each coated with a red, green or blue phosphor, sandwiched between two plastic sheets. These dielectric sheets are printed with a pattern of parallel conductive lines; with lines on one sheet running orthogonal to lines on the other sheet. These tracks are called address and display electrodes. This assembly is enclosed between a pair of glass sheets and hermetically sealed all around. During the panel sealing process it is filled with a mixture of neon and xenon gases, as well as mercury vapor, at a low pressure. As seen in the schematic cut-away diagram here in Figure 3.19, the chambers or cells are formed by rib partitions and each is coated with a thin red, green or blue phosphor layer. A trio of RGB cells forms a single-color pixel, where each R, G and B sub-pixel is individually addressable. To make any sub-pixel light up, its corresponding row and column tracks are energized such that a glow discharge takes place at their intersection. The gas composition and pressure are chosen so that most radiation in the discharge is emitted in the UV region. The UV photons strike the phosphor and visible red, green or blue light is emitted. This is similar to the way fluorescent neon lamps, described earlier, operate. The intensity of

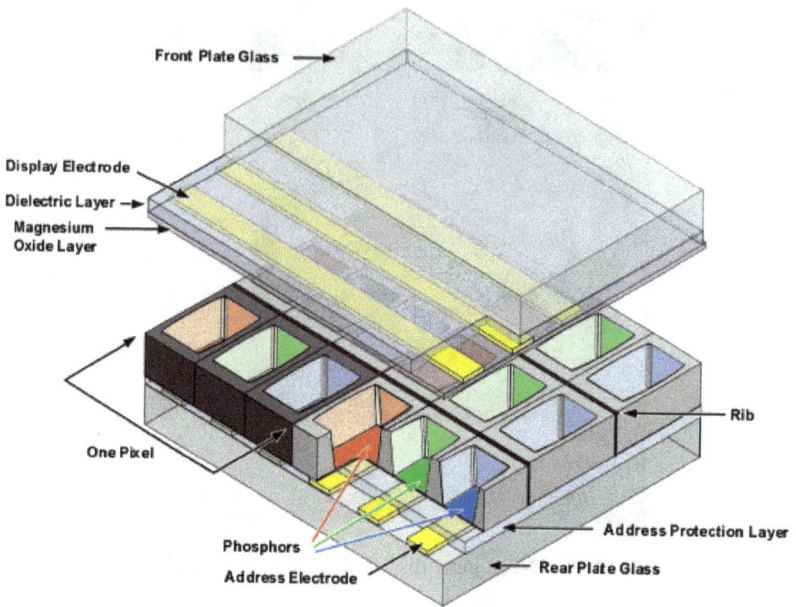

Fig. 3.19.   Structure of a typical plasma display panel, showing RGB phosphor-coated chambers.

Courtesy: The Computer Language Company Inc.

emitted light can be varied by controlling the voltage on the display electrode through a technique called pulse width modulation (PWM). By controlling the intensities of the red, green and blue sub-pixel, any desired color can be displayed. The entire display is scanned at high speed to show video frames at 50 or 60 frames per second.

Plasma displays, like the 150-inch diagonal model from Panasonic shown here in Figure 3.20, were the leading television display technology during the 1990s. PDPs can be made in very large screen sizes and have a number of advantages over other display technologies. Perhaps the most important is the extremely high contrast exhibited by plasma panels. As any pixel can be completely dark when not addressed, PDPs can boast extremely high contrast values. Liquid crystal displays (LCD), in contrast, are not capable of such high contrast images because LCD pixels block light from a rear backlight using liquid crystal light valves. These are not

Fig. 3.20.   Panasonic 150-inch plasma television.
Courtesy: Panasonic Corporation.

100% opaque when blocking light and, thus, some light leaks out, reducing the picture contrast. Self-emissive displays, such as PDPs, on the other hand, can achieve much higher contrasts. PDPs are also 'fast' because their pixel cells can be switched between lit and un-lit state very quickly. For this reason, they do not show any motion blur which once plagued LCD displays. PDP screens don't rely on polarization of light to switch pixels on and off and, thus, can be viewed from any angle. These advantages once made PDP-based televisions extremely popular but falling prices of LCD televisions and their improving qualities tilted the balance toward LCD screens, gradually making PDPs obsolete.

3.5.3 *Neon signs*

By far, the most widespread application of neon lamps today is for illuminated signage (see Figure 3.21). Neon signs — as these are called — form the bulk of all neon-based lighting systems produced in the world today.

Neon signs were extremely popular during the middle part of 20th century and remained popular until well into the 1980s when they began to be replaced by directly-lit billboards and signage based on LEDs. However, there is still a very substantial market for neon signs all around

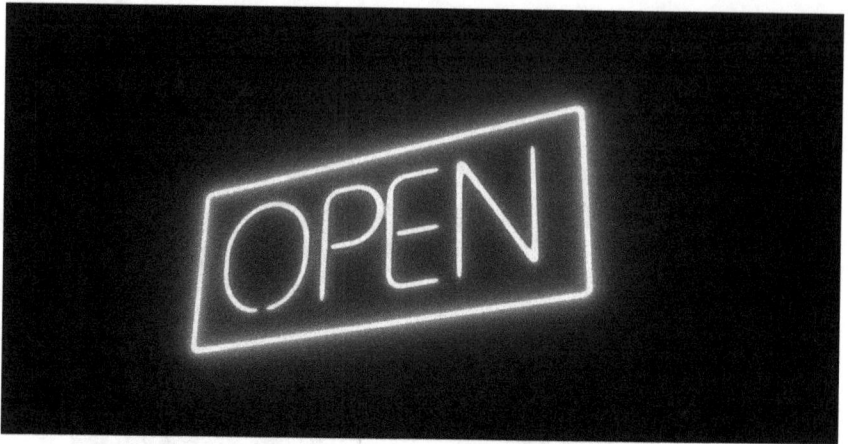

Fig. 3.21.   Modern commercial neon sign.

the world, and many manufacturers produce customized neon signs for indoor and outdoor advertising applications.

Tubular neon signs already had their genesis when Georges Claude demonstrated his neon discharge tubes at the Paris Motor Show from December 3 to 18, 1910. His display consisted of two tubes, each 12 meter long, filled with low-pressure neon and fitted with metal electrodes at both ends. During operation, the tubes emitted a strong red light which took everyone by surprise. At this demonstration, the lights were used to illuminate the area near the entrance of the exhibition at the Grand Palais in Paris. Due to their strong orange-red colored light, it was clear from the beginning that neon lights could not be used for general illumination purposes. Georges Claude was initially disappointed that due to the color of the neon lights they were not useful for space illumination. He was, however, soon persuaded by his close friend Jacques Fonsèque to use the lamps as advertising signs. George Claude took his friend's advice and started experiments on bending his neon discharge tubes into shapes so as to produce electrically-lit signs in desired shapes. These went into production by 1912, with the first one reportedly purchased by a Parisian barber. Later, a sign designed to advertise an Italian alcoholic beverage, a Vermouth Cinzano, in Paris became the first sign to actively promote a commercial product. After he obtained a patent on his invention in the year 1915, titled: Systems of Illuminating by Luminescent Tubes (US

Fig. 3.22.   The very first commercial neon sign.

patent: 1,125,476), Georges Claude further expanded his neon sign business — establishing neon sign making companies in both France and the United States. His French company was called Claude Neon while the American company was called Claude Neon Lighting. Through these companies he started selling neon advertising signs to shop owners and the interested public. By 1919, even the entrance to the Paris Opera boasted a neon sign. However, his real break came in 1923 when he sold two signs to a Packard car dealership in Los Angeles. Earle C. Anthony, the Packard car dealer in California, purchased the two signs reading 'Packard' for $1,250 apiece (see Figure 3.22). By all accounts, this proved to be a valuable transaction for both Georges Claude and Earle C. Anthony, as the signs attracted great attention.

Georges Claude's venture became the premier neon discharge-based lighting company in North America for the next two decades. Neon lighting quickly became a popular fixture in outdoor advertising, and it rapidly spread to other cities, such as Las Vegas in Nevada, and then to all over the country. Neon signs were widespread in all US cities, of any significant size, during the 1950s and 60s, as can be seen in a photograph of Astoria, Oregon, as it looked during the early 1950s (see Figure 3.23). These were (and still remain) particularly prominent in places such as the Times Square in New York City and the Strip in Las Vegas, Nevada. They were often referred to as 'liquid fire' and became somewhat of a cultural icon during that era. Manufacturers produced a dazzling array of colors

Fig. 3.23.   Neon signs in downtown Astoria, OR (early 1950s).

through the use of gases other than neon, and various gas mixtures, together with the use of colored glass tubes. A major role in extending the color palette was also played by tubes with phosphor coatings on their inner walls, similar to the phosphor-coated noble gas lamps described earlier. Jaques Risler, in France, was awarded a patent in 1926 for this development. Although, usually called neon tubes or neon lamps, lamps making use of luminescent phosphor materials do not contain neon. Instead, such tubes are filled with argon and mercury. On applying high voltage, argon plasma forms and begins to conduct electricity which raises the temperature of the tube causing the mercury to vaporize. Mercury atoms then become excited due to collisions with energetic argon ions, and while de-exciting give off UV photons. These are down-converted by the phosphor coating to visible light whose color depends on the composition of the phosphor. Several suitable phosphors have been available since Risler's time and, thus, a wide range of colors have been produced through this technique. Neon and, generally, signage based on electrical discharge is still popular worldwide, despite fierce competition from more modern lighting technologies. A major development in recent years has been the use of compact AC-AC inverters in place of bulky transformers for generating the high voltages needed for neon signs.

Several museums now display neon advertising signage from the early days of neon lighting, as it evokes a certain nostalgia for an era when this form of lighting first made its appearance.

Neon advertising signs are made from both clear and colored soda-lime glass tubes of diameters ranging from 8 to 20 mm and a wall thickness of 1 mm. Tube sections, 4–6 feet long, are worked individually through various gas flame-based bending procedures to develop the required shapes. A number of different types of specialized, fixed and hand-held, propane torches are used to bend the tubular sections in accordance with the shape and lettering design specifications. For longer designs, shaped tubes are joined to each other using a cross-head burner that melts tube ends so that they can be fused to each other. As much as possible, a long continuous tubular structure is constructed by joining a number of tube segments together. This ensures that only one electrical circuit is needed for that single structure. Once the basic shaped tube is ready, two small glass tube sections already fitted with metal electrodes, coated with a thin film of barium carbonate ($BaCO_3$), are fused to the two ends of the long tube. Using a T-shaped tube, previously joined to the main tube, the assembly is connected to a gas manifold. Most, but not all, air is removed from the structure, and then an out-gassing step is performed. This procedure, called 'bombardment', is necessary to ensure a long life for the tube, and consists of driving out gases and impurities adsorbed on the inner walls of the tube. This out-gassing process was first described in Georges Claude's original neon lamp patent of 1915 and it is remarkable that this step is still in use. It consists of attaching the external wires at the two ends of the tube to a high-voltage transformer and operating at a voltage significantly above the normal operating voltage of the assembly. This causes a strong current to flow through the tube which heats up the inner walls of the tube to temperatures in excess of 200°C. This then causes the inner surface of the walls to release adsorbed gases and other contaminants which are removed due to the high vacuum. The heat also causes the $BaCO_3$ coating on the electrode surfaces to decompose into barium oxide. This oxide has a low work function, so it is an efficient emitter of electrons when heated. Although the electrodes in a neon tube are not directly heated, they warm up considerably during normal operation and emit a good number of thermionic electrons. These electrons facilitate the formation of plasma in the tube and help lower the breakdown voltage needed for normal operation. After bombardment, the tube is filled with the desired gas or gas mixture (not necessarily just neon gas). The gas fill port is then sealed off with a burner, the tube assembly is cleaned and its back side coated with black paint to enhance the coloration when the tube is lit. Finally, the electrode connections are attached to wires that lead to an appropriate high-voltage transformer. Properly constructed neon signs easily last in excess of 30 years, and being quite efficient, do not consume a significant amount of electric power during their working lifetime.

3.6 Mercury Vapor Lamps

The earliest electric discharge lamps to be developed for mainstream illumination applications were lamps based on discharges taking place in mercury vapor. Due to the spectroscopic emission properties of electric discharges in mercury vapor, lighting systems based on this general concept can be enormously versatile. Thus, a number of different types of mercury vapor lamps have been developed over more than 100 years, to serve in varied application areas. There are low-pressure, medium pressure and high-pressure lamps. Then we have cold and hot cathode lamps, as well as arc discharge lamps and metal halide lamps. All these types will be discussed here. A particularly attractive feature of mercury vapor lamps is their high brightness and, thus, these lamps are used in several applications, such as large area outdoor lighting and projection displays, where high brightness from the light emitter is the principal requirement. Also, being discharge lamps, mercury vapor lamps are also highly energy efficient. Fluorescent tube lamps and their derivatives are also basically mercury vapor lamps, but those lamps will be discussed separately in their own section.

3.6.1 *Early mercury vapor lamps*

Light emission from electric sparks and discharges in mercury vapor was first observed in the first half of the 19th century. By that time both voltaic batteries and mercury manometers were common and several experimentalists, including Charles Wheatstone, observed the bluish light that emanates from electric discharge in the atmosphere of a low-pressure mercury vapor. It was noted that the spectrum of this light contains emission lines in both UV and visible regions. Around 1821, Humphrey Davy showed that it is possible to form an arc between mercury and a carbon electrode.

During the second half of the 19th century, several crude mercury vapor lamps were developed by English and German inventors. Particularly notable among these was the arc lamp developed by Professor John Thomas Way in early 1860. This device was, essentially, an open carbon arc lamp, containing an amount of mercury. Operation of the arc generated heat which evaporated some mercury and, thus, the lamp's arc operated in a mercury vapor atmosphere. Unlike simple carbon arc lamps, the light in this lamp was generated from both the incandescent carbon electrode tips and the mercury plasma. The latter added a strong blue and green component to the light from this lamp. This was the first purpose-built lighting device that employed discharge in mercury vapor. As the lamp was of an open construction so mercury vapor was continuously lost from it to the atmosphere. Way used his lamp to illuminate London's Hungerford suspension bridge which crosses the Thames River between the Waterloo and Westminster bridges. By early September 1860, the entire span of this bridge was illuminated by Way's mercury-containing carbon arc lamps. Although these lamps were brighter than ordinary carbon arc lamps, they did not seem to have seen much use, for reasons that are not known. One speculation is that concerns over the toxic effects of mercury emanating from Way's lamps prevented their widespread use in enclosed spaces, and later their use was generally discouraged.

Several other mercury vapor lamps were developed after Way's demonstration, but there is general agreement that the first practical mercury vapor lamp to find commercial success was the one developed in 1901 by the American engineer, Peter Cooper Hewitt (see Figure 3.24). On September 17, 1901, Hewitt was granted US patent 682,692 for his invention. His patent was simply titled: Method of manufacturing electric lamps.

Cooper Hewitt lamps soon became quite widespread. Their strong light was very popular, but, at the same time, the quality of light was perceived to be very poor as these lamps produced a harsh bluish light. This proved to be a hindrance for the initial adoption of Cooper Hewitt lamps. The first-generation Cooper Hewitt lamps (see Figure 3.25) also had an awkward design, in that the lamp contained a large amount of mercury in a highly evacuated glass envelope and had to be tilted to start it. Tilting brought mercury, originally all contained at one end in the bulbous cathode reservoir, into contact with the anode. This established an arc between the anode and the surface of mercury. Some mercury evaporated and then ionized. The arc operated in this medium, producing a spectrum typical of

Fig. 3.24.   Peter Cooper Hewitt.

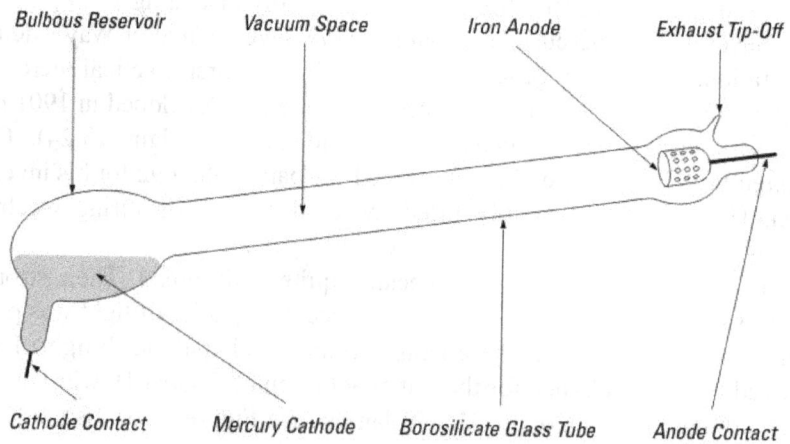

Fig. 3.25.   Diagram of Hewitt's first mercury vapor lamp.

an arc discharge in low-pressure mercury vapor. Such spectrum contains mainly emissions in the UV, blue and green regions; not particularly suitable for ordinary lighting applications. Needless to say, these lamps failed to attract the popularity that its inventor had hoped for. Hewitt then developed another design for his lamp with a U-shaped tube and no need to tilt it to strike the arc. This second-generation lamp also had somewhat better spectral properties, and during the first decade of the 20th century it became quite popular for general space illumination in factories and printing presses. With finance and backing from the industrialist George Westinghouse, Hewitt established a company, named, Cooper Hewitt Lamp Company, to produce and market his lamps (see Figure 3.26).

Despite improvements, Cooper Hewitt lamps produced light that hardly contained any red wavelengths; its major output being in the blue and green regions. Thus, it suffered from poor color rendering properties which eventually limited its use to industrial space lighting applications. Its one greatly appreciated quality, however, was its great energy efficiency when compared to the very inefficient incandescent lamps of that period. The inventor himself stated in 1902 that "When it is considered that this light, when obtained with mercury gas, has an efficiency at least eight times as great as that obtained by an ordinary incandescent lamp, it

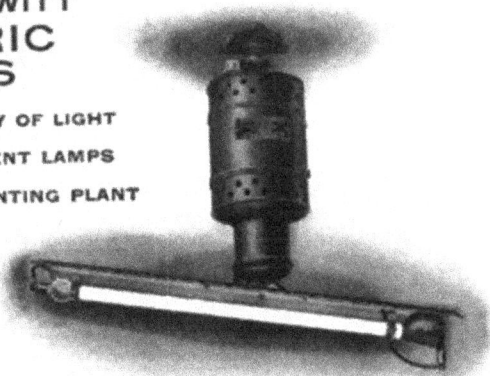

Fig. 3.26.   Advertisement for Cooper Hewitt electric lamps.

will be appreciated that it has its use in places where lack of red is not important, for the economy of operation will much more than compensate for the somewhat unnatural color given to illuminated objects". It is notable that despite this shortcoming, the Cooper Hewitt Lamp Company promoted its lamps as producing 'best quality of light'.

In an effort to improve the color balance, Hewitt came up with an ingenious solution for his lamps. He wired a number of incandescent filament lamps as ballast, in series with his mercury lamp (see Figure 3.27). The filament lamps both served as ballast and their light supplied some of the missing red component. This type of 'blended' lamp was not quite as efficient as just a mercury vapor lamp, but provided much superior light, compared to just the mercury arc lamp. The Cooper Hewitt Lamp Company remained a profitable business for many years until, in the year 1919, it was finally bought by GE. Production then relocated from the

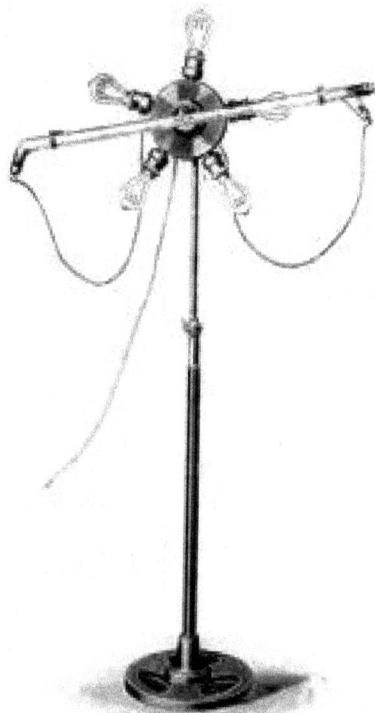

Fig. 3.27.   'Blended' Cooper Hewitt lamp with incandescent lamp ballasts.

original factory at 220 West 29th Street in Manhattan to GE's premises in Hoboken, New Jersey.

An improvement of Hewitt's lamp design was brought in 1907 by R. Küch and T. Retschinsky of W.C. Heraeus Company in Hanau, Germany when they developed quartz envelopes for such lamps. Küch and Retschinsky described their development in an issue of the journal *Annalen der Physik*, in their paper, titled: Temperaturmessungen im Quecksilberlichtbogen der Quarzlampe.

Their lamp differed from all previous lamps in having a quartz rather than a glass envelope. Otherwise, it was fairly similar to the Cooper Hewitt lamp. In their paper they described their technique for assembling the lamp, including a method to form glass-to-metal seals. It was crude by today's standards but did result in successful, albeit short-lived mercury vapor lamps. Use of quartz (or synthetic silica, also called fused silica) was a major innovation because quartz is much more resistant to high temperatures and chemical attacks than is ordinary glass. Quartz later enabled new high-temperature and high-pressure mercury lamps with much improved characteristics compared to their forerunners. The Küch and Retschinsky lamp (see Figure 3.28) did not enjoy much commercial success because of poor construction (the glass-to-metal seals were a particular problem as they often leaked). Also, the quartz allowed most of the short wavelength UV radiation to come out of the lamp causing skin burns and other problems. It took several more years before high quality mercury vapor lamps for mainstream lighting applications came to the market.

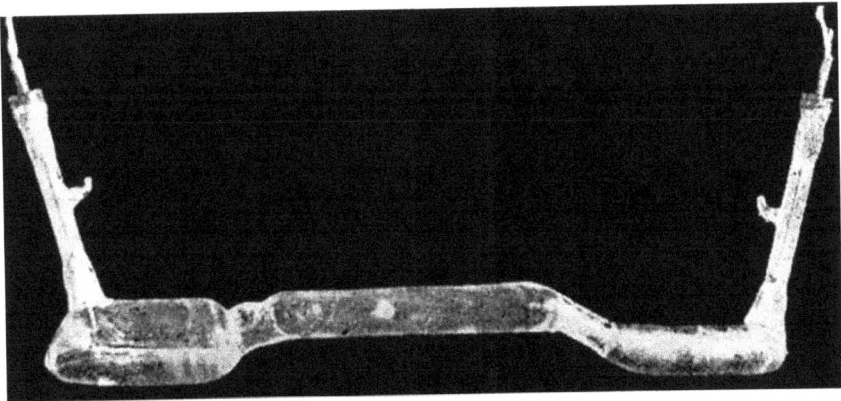

Fig. 3.28. Küch & Retschinsky lamp.

3.6.2   *UV-emitting mercury vapor lamps*

While a great deal of attention went into developing visible light-emitting mercury vapor discharge lamps, devices for the purpose of specifically emitting UV radiation were also developed in the early years of the 20th century. These lamps are still widely used and have basically retained their form since those early years. Their prevalence derives from their simple construction and low cost. Even to this day, these lamps are far cheaper sources of mid and deep UV radiation than more modern products, such as LEDs. Such lamps find applications as germicidal lamps for surface cleaning and water treatment, as well as for resin cross-linking in paint and printing industries. Other common applications include tanning booths, forgery detection, and use in ozone-generation for air purifiers. There are several types of mercury vapor UV lamps available. The simplest type is based on a U-shaped glass tube containing argon at low-pressure (around 0.005 Torr) and a few drops of mercury (see Figure 3.29). This basic design dates from the days of Peter Cooper Hewitt. The lamp is operated from as little as 5 V using a DC-AC inverter which generates the high voltage necessary to operate the lamp. This type of lamp mainly emits radiation at 254 and 365 nm with a small amount of radiation also emitted at 404 nm. Versions of this lamp are also made with a

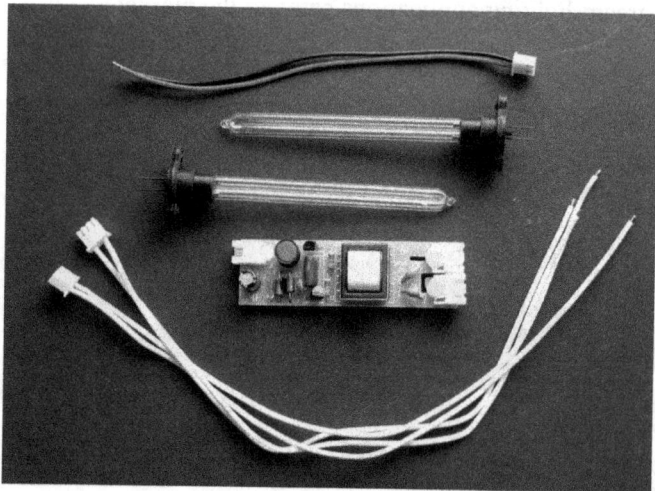

Fig. 3.29.   Cold cathode mercury vapor lamps with high voltage inverter.

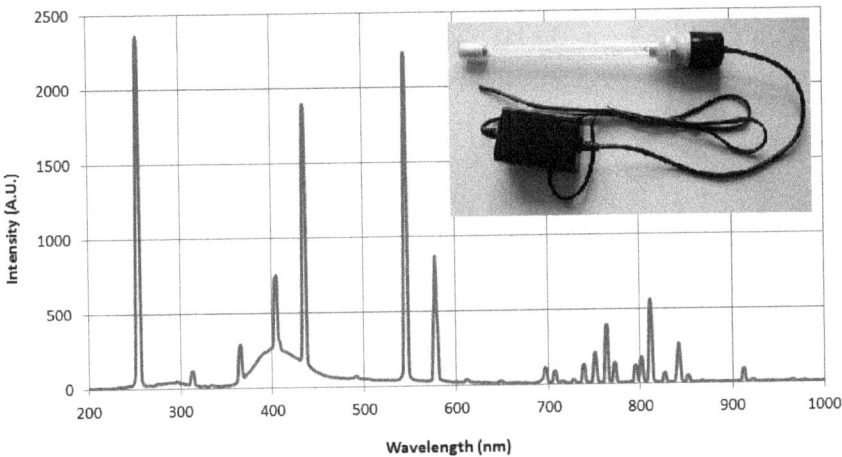

Fig. 3.30.   Hot cathode mercury vapor lamp and its emission spectrum.

quartz or fused silica tube which allows more of the 254 nm radiation to come out due to its higher UV transparency.

More powerful (higher brightness) deep UV lamps utilize a hot cathode construction with a thermionic heated cathode for higher efficiency. These lamps have the main UV discharge tube confined inside an outer quartz tube. Figure 3.30 shows such a lamp with its mains power supply and emission spectrum. The unfiltered light from this kind of lamp contains both UV and visible radiation.

3.6.3 *High-pressure mercury vapor lamps*

Mercury vapor lamp development really took off once it was realized that lamps operating at higher pressures have much improved efficiency, as well as lighting quality. Higher pressure lamps have better spectral balance in that they also contain the red component which is missing in low-pressure mercury vapor lamps. The reason for this lies in the self-absorption of UV radiation together with pressure-induced broadening of mercury's emission lines at higher pressures. Already, at atmospheric pressure the light is quite white, and its quality improves further at above-atmospheric pressures. At low pressures, a mercury discharge mainly emits at UV and near-UV wavelengths. Prominent spectral lines in this region are at 189 nm, 254 nm, 365 nm and 404 nm. As the mercury vapor pressure is increased to a few atmospheres, the principal emission lines begin to broaden which enhances the spectral coverage of the emitted light. At the same time, a continuum radiation with characteristic continuous spectrum begins to develop. This continuum radiation is weak but provides coverage of orange and red wavelengths. At very high pressures of a few hundred atmospheres, the broad spectral lines become broader still but decrease in overall intensity whereas the continuum radiation gains in strength. This makes the light look much whiter and it loses its bluish-green hue. These changes in the spectral power distribution of light from a mercury vapor discharge lamp, as its pressure is changed, are seen in Figure 3.31.

For mercury vapor lamps, the total amount of visible radiation produced per watt of electrical energy consumption, i.e. the luminous efficacy in lumens/W, also has a remarkable dependence on mercury vapor pressure, as seen here in Figure 3.32. At very low-pressures, the emission is weak because there aren't many mercury atoms available. With increasing pressure, from A to B, as the mercury atoms become more numerous, the emission increases in intensity. In this region, the UV emission itself decreases in intensity. As the pressure rises above 0.1 Torr, the density of mercury atoms increases; reducing the mean free path of accelerating electrons. As the electrons begin to have more frequent direction randomizing collisions with mercury atoms, their mean energy is reduced so they

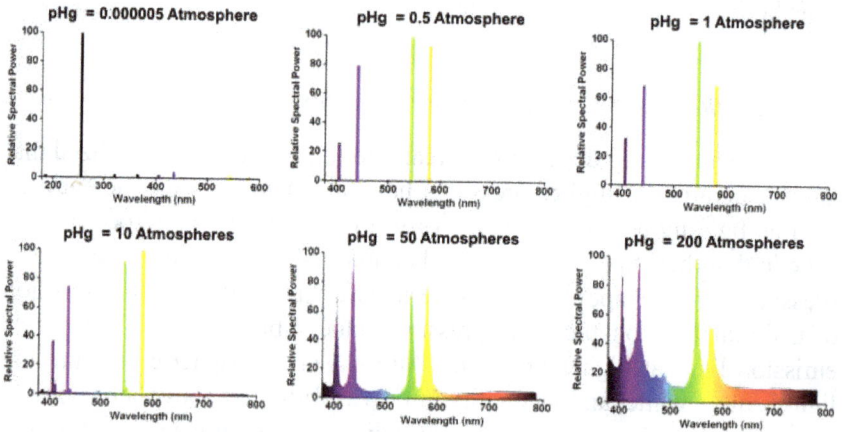

Fig. 3.31.   Emission spectra of mercury vapor lamp in the visible region, at different mercury partial pressures.

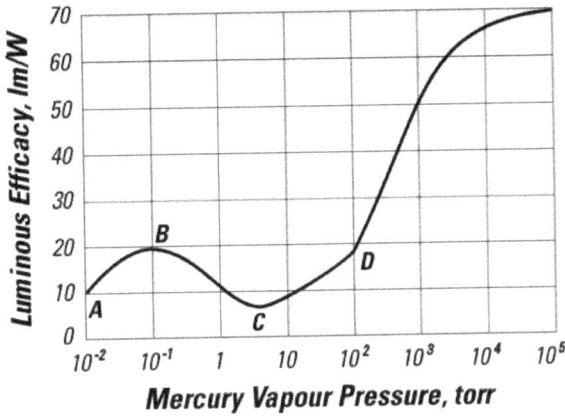

Fig. 3.32. Luminous efficacy of mercury vapor lamps as a function of mercury partial pressure.

are not able to effectively excite mercury atoms. This reduces the emission intensity. At the same time, in the region of pressure between points B and C, the mercury atoms gain enough energy from low energy elastic collisions with electrons that their temperature increases substantially. The plasma, thus, heats up to temperatures around 500°C. In this pressure region, the discharge dissipates most of its energy as heat and visible light production efficiency remains quite poor.

Soon after the development of the mercury lamp housed in a quartz envelope, researchers in Europe, including Küch and Retschinsky, started investigating lamp performance at higher pressures. They found something very interesting. While, with increasing pressure the light output from mercury vapor lamps decreased in the region from B to C, after reaching a minimum at around 5 Torr, it started to increase unexpectedly. This happens when the discharge plasma detaches from the tube wall and shrinks axially toward the center of the tube. The effective resistance of the plasma channel increases as its cross-section decreases. This causes higher dissipation in the plasma channel and its temperature increases rapidly to thousands of degrees Celsius. A higher temperature plasma is both brighter and a more efficient producer of light, as the axially-confined plasma does not lose any significant amount of heat to the tube wall. Thus, both light output and efficacy increase with increasing pressure. At the same time, the continuum radiation also increases in intensity, causing the light to become redder, producing a more pleasant hue. This transition in lamp behavior is so well-defined that point C where the plasma detachment starts to take place is considered the boundary between low-pressure mercury vapor lamps (pressure <5 Torr) and higher-pressure lamps (pressure >5 Torr). With further increase in pressure, these general trends continue and even accelerate beyond point D (200 Torr), eventually beginning to slow down only after the pressure increases to 10,000 Torr. Beyond that pressure further increase in brightness and lamp efficacy slows down, so as to slowly approach a limiting efficacy of around 70 lumens/W, at pressures of the order of 100,000 Torr (130 atm). Thus, we see that it is very beneficial to operate mercury vapor lamps at high pressures. For this reason, most low-cost visible light-emitting mercury vapor lamps are built to operate at pressures close to 1 atm (760 Torr). Some specialized very high brightness lamps even operate at pressures as high as 200 atm, as will be seen later.

After 1910, for many years, most research work on higher pressure mercury vapor lamps was carried out in Europe, and European companies took the lead in developing more advanced visible light-emitting mercury vapor discharge lamps. In 1923, Hirst Research Laboratories of the GEC in East Lane, Wembley, England, started research on developing an improved mercury vapor discharge lamp than had existed until then. Up to that point, while the potential of the mercury arc discharge lamp was well-proven, mass-scale deployment of such lamps had not proven feasible. GEC's aim was to open new markets for itself in high brightness lamps that could replace high-wattage tungsten lamps. After intense research for nearly a decade, GEC came out with its 'Osira' lamp in 1933 (see Figure 3.33). This lamp was extensively marketed, and by all accounts was an outstanding success; making the mercury vapor lamp the main choice for high brightness illumination applications for many years.

Its arc tube, made from aluminosilicate glass, was 35 mm in diameter. This specialized type of glass contains 20% aluminum oxide (Al_2O_3) as well as calcium oxide, magnesium oxide and boric oxide in relatively small amounts, but with only very small amounts of soda or potash. It is able to withstand high temperatures and thermal shock and is, thus, very suitable for use as the arc tube of high-temperature gas discharge lamps. Another essential property of aluminosilicate glass — its high coefficient of thermal expansion — was essential to its success in this role. This is because for use in an arc tube, metal electrodes need to be sealed through glass tubes. As metals have a significantly higher expansion coefficient

Fig. 3.33.   Advertisement for GEC Osira lamp.
Courtesy: General Electric Company Ltd.

compared to most glasses, so the expansion mismatch at high temperatures could lead to metal-to-glass seal failure. Aluminosilicate glass developed at the GEC glassworks at Lemington was, thus, the key to the Osira
lamps, as it allowed molybdenum electrode wires to be sealed through it
with high reliability. The arc tube was filled with argon at 5 Torr and just
a small amount of mercury, which completely vaporized at the operating
temperature of the lamp. This produced an un-saturated mercury vapor
atmosphere inside the arc tube. Not having excess mercury (and, thus, a
just-saturated mercury atmosphere) presented an operating advantage,
because the mercury did not condense or evaporate as the temperature
changed; keeping a constant concentration of mercury in the tube. This

Fig. 3.34.   GEC Osira lamp, showing the aluminosilicate arc tube.
Courtesy: General Electric Company Ltd.

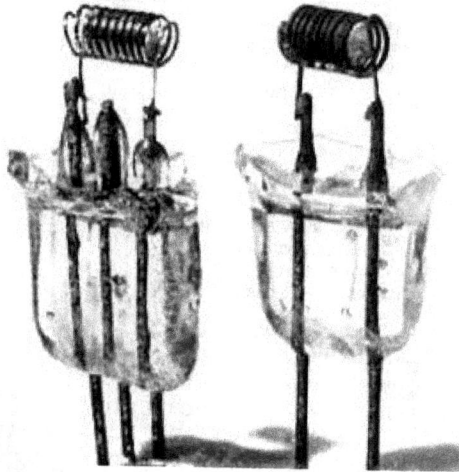

Fig. 3.35.   High electron emission efficiency thermionic electrodes used in Osira lamps.
Courtesy: General Electric Company Ltd.

stabilized the luminous output of the tube against changes in temperature. The arc tube itself was enclosed inside an outer glass jacket, with the space between them filled with nitrogen (see Figure 3.34). This arrangement provided thermal insulation to the arc tube, making it immune to external temperature variations.

Due to a cross-licensing agreement with GE in the US, the aluminosilicate glass technology was quickly adopted by US manufacturers and by 1934 they had also started producing similar lamps.

The high electron emissivity electrodes used in the Osira lamp consisted of a pelleted mixture of barium, calcium and strontium silicates held in a tungsten wire coil (see Figure 3.35). During operation the cathode coil was heated by an electric current. This caused the silicates to decompose and the reactive low work function metals coated the cathode, increasing its thermionic electron emission efficiency many folds over that of bare tungsten.

The very first, type MA Osira lamps took up to 9 min to warm up to full operating brightness. The discharge first operated in argon atmosphere and filled the tube with a weak diffuse blue glow. The mercury vapor pressure initially was very low, so most radiation was emitted at UV wavelengths. The discharge gradually heated the tube, so that more mercury evaporated; increasing its pressure and the light gradually became white with a concomitant increase in brightness. After about 9 minutes, the lamp was operating in a steady state. Later, lamps were built with a heat-reflecting platinum coating at both ends of the arc tube to accelerate the warm up process, shortening the run-up time to 6 minutes.

The first generation, type MA 400 W Osira lamps were used on June 22, 1932, to illuminate the area immediately outside the Hirst Research Laboratories in East Lane, London. This was the first ever street lighting in the world with mercury lamps. While this was an experimental trial, a proper installation was done later, and on March 2, 1933, lights were switched on along Watford Road, Wembley. This demonstration was so successful that many government agencies, both British and from other countries, expressed interest in adopting the new lamp (see Figure 3.36) for public area lighting in their own cities.

Fig. 3.36.   Packaging of commercial Osira lamps.

Courtesy: General Electric Company Ltd.

While Osira lamps were distinctly much superior to earlier mercury vapor lamps and were a commercial success, they still had some prominent peculiarities. The relatively low softening temperature of the aluminosilicate glass used in this lamp limited the power loading of the electric arc to a maximum of 100 W per centimeter of arc length. This, in turn, limited the maximum brightness obtainable from an Osira lamp. Due to convective effects in the gas inside the arc tube, the arc had a tendency to bow upward at its center, if the lamp was operated in a horizontal position. This could cause the arc to come into contact with the inner wall of the arc tube and eventually melt it. Thus, the Osira lamp had to be operated in a vertical position. Vertical burning made it difficult to make suitable lamp housings that could collect light from an upright lamp and direct it downwards. This problem was, eventually, solved by the use of an electromagnet that kept the arc straight when the lamp was operated horizontally.

Mercury vapor discharge lamps continued developing over the next several decades. Mostly, efforts were directed toward further increasing the operating pressure of mercury lamps, because, as described earlier, this increases both efficacy and color rendering capability of mercury discharge lamps. To accomplish this, materials and designs were appropriately modified. Philips, based in The Netherlands, initially took the lead in developing higher pressure mercury vapor lamps, introducing a high pressure, water-cooled, mercury capillary lamp in 1935, a quartz arc tube mercury vapor lamp in 1936 (Philora HP300; see Figure 3.37) and a fluorescent material-coated mercury lamp in 1937. Due to the fact that the mercury vapor lamps are more efficient at higher wattages (and operating pressures), they have been used as high brightness luminaires; generally used for illuminating large areas. For this reason, they do not, usually, come in wattage ratings below 150 W. The focus of development, later, shifted toward developing even higher pressure mercury lamps.

Quartz has been the material of choice for making high-pressure mercury lamps. Its refractory nature, chemical inertness, mechanical strength, and high transparency, make it a material that is hard to beat for this application. However, it is also one of the most difficult materials to work with. Sealing metal conductors through quartz is especially difficult, as metals have large thermal expansion coefficients, in contrast to quartz's very small coefficient of thermal expansion. This disparity in temperature-induced expansion tends to crack metal-to-quartz seals. While working in Germany, during the 1930s, Denis Gabor — a Hungarian-British

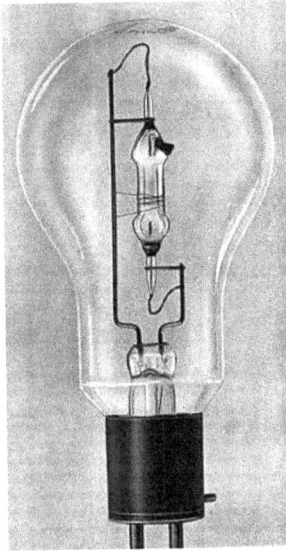

Fig. 3.37.   Philips Philora HP300 lamp.
Courtesy: Philips N.V.

scientist — developed a special technique for constructing quartz-metal seals. This involved sealing through quartz, extremely thin (~20 microns thick), narrow strips of molybdenum foils with sharply polished edges. In this geometry molybdenum strips expand very little with increasing temperature, and the integrity of the seal is maintained all the way to the softening temperature of quartz. Use of quartz and molybdenum foil seals made it possible to develop high-pressure mercury vapor lamps during the 1930s. Today, the mercury vapor lamp is still widely used for outdoor lighting in public places where its high efficacy, bright light, and long life is highly valued. Modern forms of mercury vapor lamps, for general illumination, employ a fluorescent coating — both to block the escape of any UV light and to convert it to visible radiation.

As seen in Figure 3.38, modern mercury lamps are similar in external form to incandescent light bulbs, but have a much more involved construction. The outer envelope is made from borosilicate glass whereas the inner arc tube is made of quartz and filled with nitrogen gas at about 50 Torr, with a few milligrams of mercury. The space between the arc tube and the outer jacket is filled with nitrogen for thermal insulation, and also to prevent oxidation of the metallic conductors. Tungsten electrodes pass

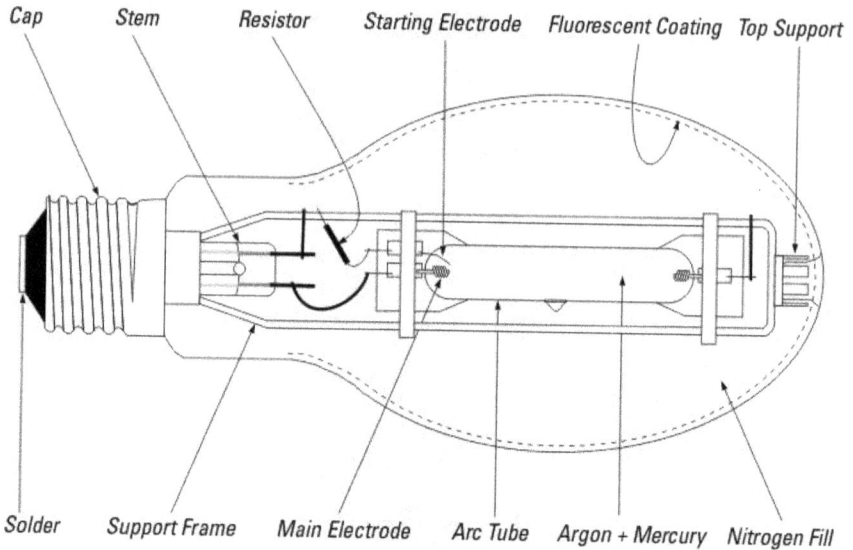

Fig. 3.38.   Construction of a modern visible light mercury vapor discharge lamp.

through the top and bottom of the arc tube. There is also a discharge-starting electrode at the bottom, in series with an internal resistor. As the lamp ignites, a small arc develops between the starting electrode and the main electrode closest to it. This arc produces heat that vaporizes some of the mercury and also becomes a source of electrons, ions and photons. This then facilitates the main discharge to start between the two main electrodes. As the external resistance along this path is lower (because of the presence of the resistor in the starting circuit), so the discharge gradually changes over to the longer arc, while the starting arc gets automatically extinguished. High wattage lamps often have two starting electrodes, one at each end, to help in initiating the main discharge. As the lamp warms up, the mercury vapor pressure inside the arc tube rises and the spectrum of light emitted by it changes. UV radiation decreases in intensity while visible light gets more intense. When fully operational, the mercury gets completely vaporized inside the arc tube and its pressure rises to a few times the atmospheric pressure. Most of the light is then emitted in the visible region, but a significant amount is still produced as UV radiation. All visible light-emitting mercury lamps now have a fluorescent coating (see Figure 3.39) that changes the UV radiation to visible

Fig. 3.39.   General purpose fluorescent mercury vapor discharge lamp.

Courtesy: Philips N.V.

yellow and red light. This improves the color balance of mercury lamps; enhancing their appeal as good sources of bright white light.

3.6.4 *Ultra high-pressure mercury vapor lamps*

Certain applications, such as projector lamps, require the highest brightness light sources. All major types of projectors, such as cine film, LCD and digital light projection (DLP), make use of compact lamps with extremely high light outputs. For this purpose, ultra-high-pressure (UHP) mercury vapor lamps are the usual choice. Several manufacturers, including Philips, Ushio and Osram/Sylvania make UHP mercury lamps. More recently UHP has come to mean ultra-high-performance (especially by Philips, who introduced this type in 1995). Hanns Fischer at Philips, in The Netherlands, was responsible for its development. These lamps operate at internal pressures as high as 200 atm to achieve high efficacies and desirable spectral distributions. UHP lamps incorporate several

Fig. 3.40.   Ultra-high-pressure mercury vapor discharge lamp, showing spiral antenna for lamp start-up.

Courtesy: Philips N.V.

innovations not found in medium pressure mercury lamps. For instance, their discharge initiation mechanism is completely different. Instead of making use of a starting electrode, UHP lamps utilize a spiral 'antenna', which is wound around a section of the arc tube (see Figure 3.40). On initial application of electric power, the antenna induces an electric field inside the tube that helps ionize the argon gas filling in the arc tube, initiating the discharge. Some lamps also have a so-called UV enhancer — a small chamber that itself is a miniature discharge lamp. Activated by the spiral antenna, the UV enhancer generates short wavelength (mainly 254 nm) UV radiation that is capable of ejecting electron from tungsten electrodes through the photoelectric effect. These electrons then help in ionizing the lamp gas and initiating the discharge. A dichroic heat reflecting coating that recycles heat for enhancing plasma generation and raising its temperature is also included. The entire lamp tube is surrounded by an elliptical reflector, such that the center of the arc is located at one of the foci. This assembly is then enclosed in a suitable housing which makes it easy to install and change the lamp, as needed. UHP lamps also feature very small gaps between the anode and the cathode. Typical electrode separation in these short-arc lamps is about 1 mm. Such close separations are essential for obtaining a small light source volume of about 1 mm^3.

This is a key requirement for extracting the maximum possible amount of light from the lamp, using the integrated elliptical reflector. The plasma within this small volume attains temperatures of around 7000°C, which places substantial heat load on the arc tube. Any further increase in plasma temperature can push the temperature of the quartz tube close to its recrystallization temperature. Recrystallization vitrifies the quartz and can cause premature arc tube failure. On the other hand, the inner wall of the quartz arc tube has to have a minimum temperature significantly above the boiling point of mercury (which is higher in the high-pressure arc tube than 357°C — the value at 1 atm pressure), so that all the mercury remains in the vapor state. Thus, proper thermal management is essential for the successful operation of UHP lamps. If this is properly managed then such lamps can last for up to 10,000 hours of operation. When the lamps do fail then it is due to some structural failure such as electrode wear or tungsten-to-quartz seal failure. This is inevitable, because modern UHP lamps operate very close to ultimate material endurance limits. When first introduced, UHP lamps were only available at 100 W power rating, but now lamps with power ratings up to 300 W are available from multiple manufacturers. In the future, even higher power UHP lamps are likely to become available.

3.7 Metal Halide Lamps

Another modern mercury vapor lamp technology is one that relies on using other metals, in addition to mercury, to enhance the spectral coverage of emitted light. Instead of relying on elaborate construction techniques, as in UHP lamps (which necessarily make lamps expensive), the metal halide lamp technology simply incorporates metal salts in the arc tube to obtain the wide-spectrum benefits of high-pressure lamps, while operating at much lower pressures. This simplifies lamp design — making it possible to produce relatively in-expensive, high brightness, full-spectrum lamps, for area illumination applications. Metal halides, instead of the corresponding metals, are used because the latter would require very high temperatures to vaporize throughout the arc tube, without some condensing on the tube's inner surface. Indeed, M. Wolke had tried adding cadmium and zinc to a mercury vapor lamp in 1912, but this attempt turned out to be unsuccessful due to a low lamp cold-spot temperature (600°C), which led to insufficient zinc and cadmium vapor pressures. Halide salts, on the other hand, are much easier to vaporize and keep in that state as long as the lamp keeps operating. The efficacy of mercury halide lamps can reach in excess of 100 lumens/W, which is much more than the typical 40 to 50 lumens/W efficacy of ordinary mercury vapor lamps. Furthermore, modern varieties of metal halide lamps feature outstanding full spectrum coverage; resulting in color rendering index (CRI) values of up to 96.

The halide lamp owes its genesis to a 1910 invention by Charles Proteus Steinmetz — a Prussia-born American electrical engineer of much repute. Steinmetz worked on a wide range of problems in electrical engineering including lamp design, and was nicknamed the 'Wizard of General Electric'. He applied for, and was granted, a patent (US patent: 1,025,932) on his invention of a mercury discharge Lamp, containing metal halide salts, in 1912. The patent, seen in Figure 3.41, described a U-shaped discharge tube with mercury electrodes at each end and metal halides placed on the top of the mercury pools. This was done in an attempt to improve the intensity and chromaticity of light from very early mercury vapor lamps. As an invention, it did not work well because the arc was unstable and the light continuously fluctuated in both intensity and color. Decades later, in another attempt in 1942, K. G. Schnetzler made a mercury-thallium lamp having an efficacy of 70 lm/w, almost twice as high as its mercury-only counterpart. He had been granted a patent for this invention a year earlier (US patent: 2,240,353). The desired thallium vapor pressure was reached by operating the arc tube at three-times its normal power loading. This, however, greatly reduced the lifetime of the lamp to only around 20 hours of operation. Clearly, this was not a successful design. Much later, the first really successful metal halide lamp was developed by Gilbert H. Reiling (see Figure 3.42) of GE in 1961. After some altercation with the US patent office, related to the novelty of the invention, Reiling was granted a patent (US patent: 3,234,421) for his invention in 1966. GE quickly commercialized this development into a practical product, and the first metal halide lamps appeared at the New York World's Fair in 1965 as GE's MultivaporTM lamp. Almost simultaneously, Philips in The Netherlands also brought out their own metal halide lamps, as seen here in Figure 3.43. Then other companies in Europe and Japan also followed suit with varied compositions, in order to meet different lighting needs and also to circumvent competitors' patents.

Modern metal halide lamps were further developed in the 1960s and 70s by several different manufacturers. Owing to a number of significant advantages, metal halide lamps have been growing in popularity for many

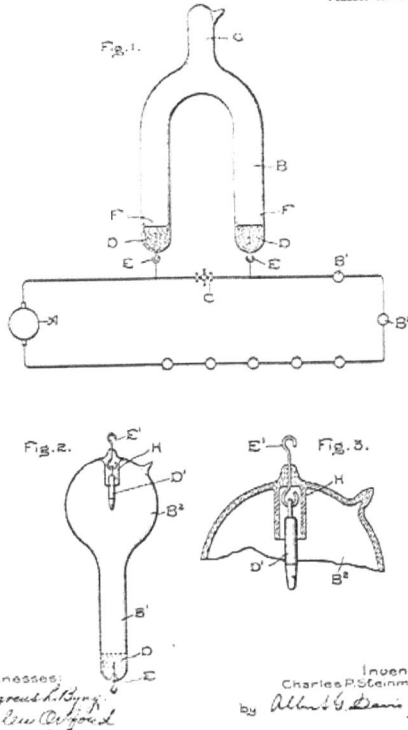

Fig. 3.41.   Charles Steinmetz's patent on halide-containing mercury vapor discharge lamp. *Courtesy*: General Electric Corporation.

years, and for the past couple of decades, these lamps have been increasingly replacing the more traditional medium and high-pressure mercury-only lamps for lighting streets, parks, stadiums and factory floors. Because of more light being emitted in the visible rather than the UV region, these lamps have higher efficacy than ordinary mercury vapor lamps. They also produce very bright light and have a spectral distribution close to that of natural daylight. The last-mentioned feature also makes these lamps suitable for indoor plant cultivation, and, thus, these lamps are much favored by horticulturists and aquarists.

Fig. 3.42.   Gilbert Reiling.
Courtesy: General Electric Corporation.

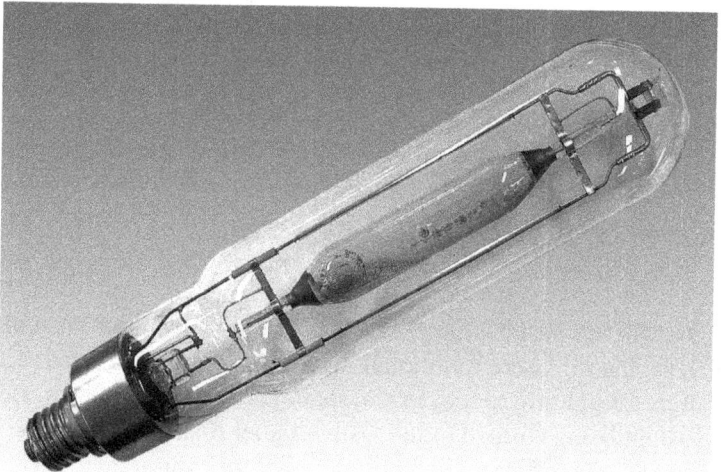

Fig. 3.43.   Philips metal halide lamp.
Courtesy: Philips N.V.

As described above, metal halide lamps are basically mercury vapor lamps that, in addition to argon and mercury, also contain a suitable metal halide — sodium iodide (NaI) being very common — inside the arc tube. Besides this, scandium iodide (ScI_3) and silver iodide (AgI) are also fairly common in such lamps. Other metals, whose halides are used, include thallium, tin, indium, dysprosium, gallium, thulium and holmium. While iodides are the most common halides, bromides are sometimes also used. At the operating temperature of the arc tube, the metal halides melt, evaporate, and dissociate into metal and halogen atoms. The metal atoms, now a component of the plasma in the arc tube, are excited by collisions with ions and electrons in the plasma and emit their characteristic atomic transition wavelengths. This process takes around 10 min to get fully activated from a cold state. As the lamp gradually starts up, and the various halides evaporate, its light changes through different colors, until it settles to its particular shade of white. Simultaneous emissions from mercury and other metal atoms from added halides enrich the spectrum of the lamp (see Figure 3.44), and with proper choice of metal halides a desirable lamp spectrum can be obtained.

Metal halide lamps come in many power ratings, from as low as 75 W to several kilowatts, and, generally, utilize a construction which is very similar to that of modern halide-less mercury vapor lamps. Thus, they feature very similar quartz arc tubes, starting electrodes and

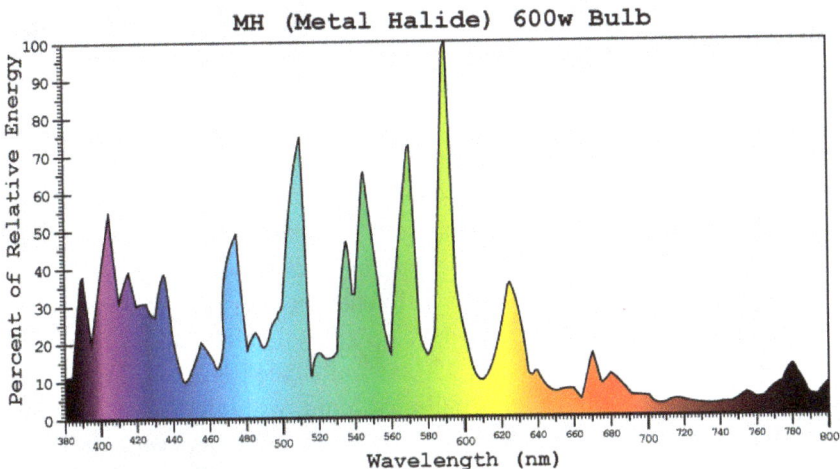

Fig. 3.44.   Emission spectrum from a multiple metal halide lamp.

Courtesy: Philips N.V.

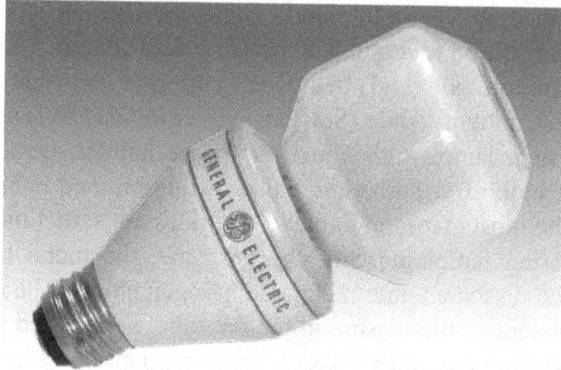

Fig. 3.45.   GE Halarc metal halide lamp.
Courtesy: General Electric Corporation.

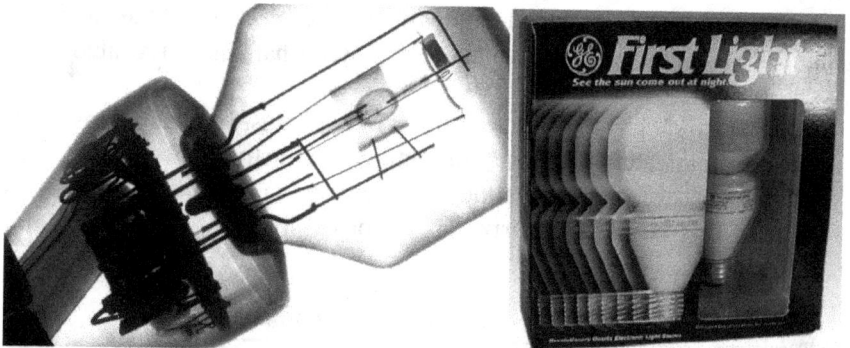

Fig. 3.46.   X-ray image of a metal halide lamp with integrated drive electronics.
Courtesy: General Electric Corporation.

metal-to-quartz seals, as well as a double jacket design with a borosilicate glass outer envelope. These lamps have served well for large area and professional lighting applications, but have not found their way into homes where lamps have to meet a different set of desirable criteria. In the 1970s, designers at GE did try to develop a metal halide lamp suitable for domestic use. Called Halarc Miser Maxi-Light (see Figure 3.45), this was a revolutionary design of a self-ballasted lamp, with all drive electronics contained in the lamp base. An X-ray image showing the inside details of this lamp appears at the left in Figure 3.46.

Several of its features were the result of the work of GE's Elmer Fridrich on low wattage metal halide lamps. It utilized a sodium and scandium iodide mixture within an arc tube filled with an argon-nitrogen mixture. It also incorporated a filament lamp — partly to act as an internal ballast and, more importantly, to provide some light immediately after the lamp was switched on. The filament also helped give the light a warm appearance. GE had great expectations from this lamp and invested millions of dollars in its development. However, it did not prove a commercial success because of such factors as difficulty starting the lamp once power was briefly interrupted (hot re-strike problem), a pronounced flicker and restriction in its use in only a vertical position, with base in the downward position. These, coupled with its very high price, caused it to meet its demise by the end of 1984, when it was withdrawn from the market.

All quartz-based arc tubes suffer from metal halide attack to various degrees. To combat the corrosive effects of metal halides, the British Thorn Lighting Company developed a sintered polycrystalline alumina (PCA) ceramic arc-tube in the early 1980s. Ceramics have advantages over quartz in being more resistant to the corrosive effects of halide salts, and also having higher resistance to thermal shock. PCA tubes can be made translucent, if properly processed during the sintering operation. Translucent ceramics, such as PCA, transmit at least 92% of the light that is incident on them. They also serve to reduce glare by diffusing the light, as well as blocking UV radiation from escaping the arc tube. Thorn made their own ceramic tubes using their patented 'Stellox' PCA tubing. For lamp developers at Thorn, while the ceramic tube itself did not pose any problems, its seal with metal conductors at both ends proved very tricky to get right. When trying to make a leak-tight seal with metal electrodes, ceramics are even more difficult to work with than is quartz. Certain metals, such as niobium can make good metal-to-ceramic seals, but these metals are also readily attacked by metal halide vapors. This leaves metals like tungsten or molybdenum as the only practical choices to serve as arc tube electrodes. Thorn solved this problem by making use of cermets as seals, capping the ends of the ceramic arc-tube. A cermet is a composite of a conventional ceramic with one or more metals and can be engineered to have the desirable properties of both ceramics (high chemical resistance, high temperature resistance) and metals (high plasticity and good electrical conductivity). Thorn employed a molybdenum-containing electrically-conducting cermet as tube end seals that also acted as electrical contacts. The cermet seals were resistant to halide vapors, and served to connect tungsten electrodes inside the tube to molybdenum stems outside the tube, without any continuous conductor passing all the way through the seal.

Fig. 3.47.   Metal halide lamp with ceramic arc tube.
Courtesy: Thorn Lighting.

This appeared to be an ingenious solution, and in 1981 Thorn Lighting exhibited the world's first metal halide lamp (see Figure 3.47), sporting a ceramic arc tube, at the Hannover World Light Fair in Germany. An additional feature of this lamp was the presence of tin iodide (SnI_2) in the arc tube, which resulted in a very pleasant color of the emitted light due to broad and balanced spectral coverage. While this 'TSH' lamp looked quite promising, Thorn desisted from commercializing it because it required too high a voltage to operate with existing ballasts, and the cermet seals proved less robust than had been initially hoped for.

Successful ceramic arc-tube-based metal halide lamps only appeared in 1994, with Philips coming out with its CDM family of halide lamps. The very first offering was a 35 W lamp, filled with an argon/krypton mixture and containing a mixture of sodium iodide, together with a range of rare-earth iodides. The most important innovation was in the design of the ceramic arc tube which had long tubular extensions at both ends through which molybdenum conductors passed through (see Figure 3.48). In this way, the seals were located some distance away from the main heat source at the center of the arc tube. Thus, being kept at a much lower temperature, heat-related seal failure was completely eliminated. This design, often referred to as ceramic metal halide (CMH), is now standard for all metal halide lamps that contain a ceramic arc tube. Thanks to the ideal match of ceramic for this application, CMH lamps can last for up to 24,000 hours of operation.

The tungsten electrodes used in modern halide lamps usually contain a small alloyed-in content of thorium metal. This increases the thermionic

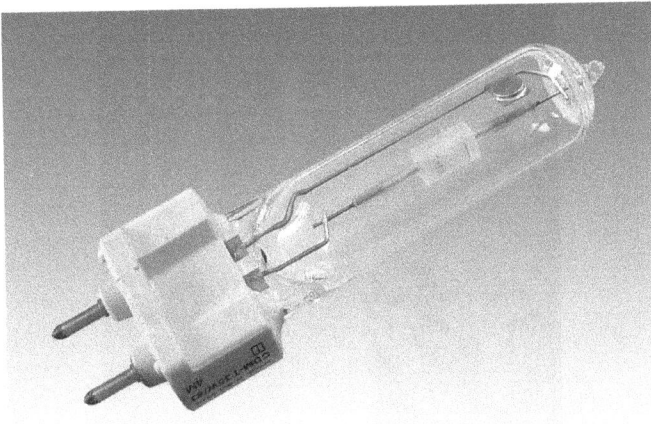

Fig. 3.48.   Seal failure-resistant ceramic arc tube in a metal halide lamp.
Courtesy: Philips N.V.

electron emission efficiency of the electrodes and helps in initial ionization during lamp start-up. This happens because thorium is a radioactive
metal that emits energetic alpha and beta particles that cause ionization of
gases present in its vicinity. The tungsten electrodes are welded to molybdenum foil conductors that are sealed to the quartz or ceramic arc tube.
Even with this type of 'active' electrode, metal halide lamps still need
to use a starting arc. The arc is generated by a small trigger electrode,
connected to an external resistor. This external resistor is visible in
Figure 3.49. Once the main long arc has been initiated, the external resistor causes the initiating short arc to extinguish, as this route then offers a
higher resistance to current flow than the main long arc discharge.

The ends of the arc tube usually have an external zirconium oxide or
zirconium silicate infrared-reflective coating that serves to reflect infrared
radiation back to the electrodes to keep them hot and emitting thermionic
electrons with high efficiency. In contrast to mercury vapor lamps, metal
halide lamps do not use any fluorescent coating, but rely on metal and
other atoms to emit pressure-broadened wavelength bands that make up
their white light output. Some UV radiation is inevitably produced, and
usually the outer glass envelope is made from a specially-doped UV-stop
silica to prevent this light from coming out. Alternatively, the outer borosilicate jacket has a UV-absorbing coating on its inner surface to absorb
any residual UV radiation coming from the arc tube. During normal
operation, metal halide lamps have to be kept continuously lit. If the

Fig. 3.49.   Structure of a modern metal halide lamp.

power supply to the lamp is interrupted then, due to the high-pressure inside the hot arc tube, the arc is rapidly extinguished and is prevented from re-striking. A cool-down period of 5–10 min is required, before the lamp can be restarted. Modern metal halide lamps sometimes come with special ignitors that allow a lamp to be re-started immediately following a power interruption. Many lamps, such as the one shown here in Figure 3.50, meant as direct replacements for tungsten halogen and mercury vapor lamps, also come with built-in ballasts, so that they can be used with existing fixtures.

Many failure mechanisms are possible with metal halide lamps that ultimately cause it to fail. These include: deterioration of the arc tube,

Fig. 3.50.   Metal halide lamp with built-in ignitor.

sputtering away of the lamp electrodes, and cracking of the metal-to-glass seals. As a lamp nears its end-of-life, it becomes susceptible to a commonly observed phenomenon called 'discharge cycling'. This manifests as repeated flashing of the lamp at low intensities. During normal operation, as the tube heats up and its internal pressure rises, it requires a higher voltage to maintain the arc. The ballast is designed to provide the maximum voltage that the tube requires at its highest rated operating pressure. With age, gradual deterioration of arc tube, and electrode materials (such as increase in the distance between the electrodes) leads to increase in the voltage required to sustain the arc above what the ballast can supply. Thus, as the lamp heats up and its pressure rises, the voltage required to sustain the arc increases beyond the capability of the ballast, and the arc goes out. The lamp then cools down and the arc tube pressure falls. The arc tries to establish again as the striking voltage is lower at lower pressure. However, as the pressure rises, again the sustaining voltage exceeds that available from the ballast and the arc fails again. This happens

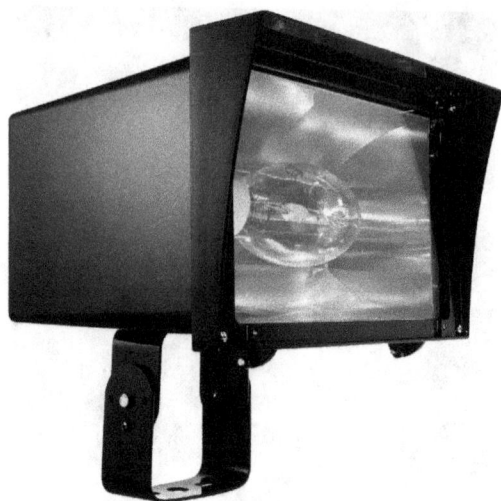

Fig. 3.51.   Metal halide lamp housed in a protective metal enclosure.

repeatedly and gives the appearance of the lamp flickering. At this point, the lamp needs to be replaced. In some cases, failure can also be catastrophic; causing the lamp to explode, but that is, thankfully, very rare. However, to guard against such an outcome, which can cause heated glass fragments to be flung away at high speeds, metal halide lamps are usually installed inside protective metal enclosures with a thick safety glass or plastic front to contain any glass shards, in case of a catastrophic bulb explosion (see Figure 3.51).

Metal halide lamps are now so well-engineered that they last for many thousands of hours. A modern mercury halide lamp can last for more than 20,000 hours, providing bright, spectrally-rich light over large areas. These lamps come in several different formats, including both single-ended and double-ended designs, and with a variety of base styles, such as the Edison screw base shown here. Construction sites and sports stadiums the world over use banks of high wattage metal halide lamps for large area illumination. Many public buildings, including the Statue of Liberty, in New York Harbor, feature architectural lighting with high-power metal halide lamps. In 2010, the city of Chicago, after a careful assessment, chose CMH lamps for city-wide street lighting — putting its faith in this remarkable lighting technology. Continuing research is serving to improve these lamps even more, so as to open new application areas for halide

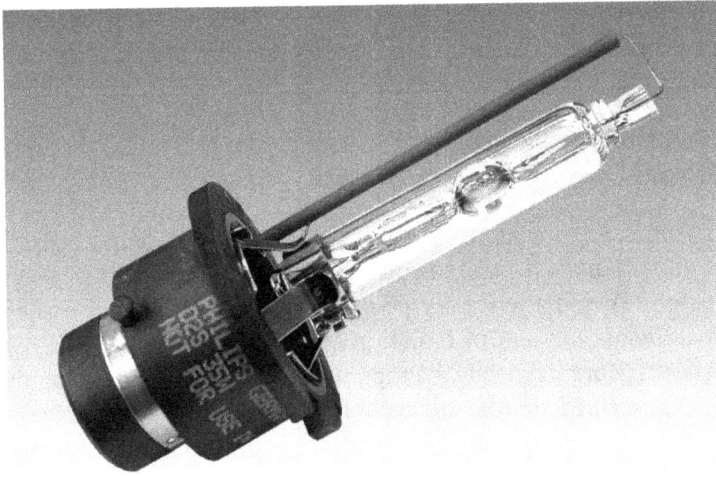

Fig. 3.52.   Philips D2S Ultinon metal halide auto lamp. Philips N.V.

lamps. For instance, further developments have resulted in smaller halide lamps for use as automobile headlights, such as the Philips D2S Ultinon auto lamp shown here in Figure 3.52. This was the result of a European project called VEDILIS (Vehicle Discharge Light System), started in 1991 and involving a number of European lighting manufacturers, such as Philips, Osram and Valeo.

These lamps contain xenon instead of argon as the fill gas, and have a much shorter start-up time, compared to argon-filled lamps. European auto manufacturers first started using them, and later these lamps were also adopted by US manufacturers. The intense bluish-white light from metal halide car headlights are now a frequent sight on our roads. It is interesting to note that even smaller metal halide bulbs have been developed for use in handheld flashlights. However, they require a large battery and power supply for operation, so their use is limited to military and search and rescue applications. Being much cheaper than UHP mercury vapor lamps, metal halide lamps are also used in low-cost projectors, which attests to the high-quality light such lamps are capable of producing. Due to the ability to control color with the use of particular metal halides, single color halide lamps have also been developed. These include a thallium iodide-based green lamp from Tesla and an indium iodide-based blue lamp from GEC. Due to the opportunity for such spectral customization, metal halide lamps continue to gain popularity, and are expected to remain a firm favorite for various illumination applications for many years to come.

3.8 Short-arc UV-emitting Mercury Lamps

Many technical applications in fields such as semiconductor manufacturing, materials analysis and medical diagnosis require high flux short wavelength UV radiation. Low power UV lamps, described earlier, produce insufficient amount of radiation and are, thus, not suitable for such applications. More intense UV lamps are short-arc, high-pressure, devices which, as described shortly, are capable of operation with high brightness and efficacy. These lamps find applications in analytical instruments, photolithography, medical research etc.

High-power mercury vapor UV lamps have a distinct envelope style, consisting of a straight purified fused silica tube with a bulbous center (see Figure 3.53). Synthetic fused silica is made by reducing silicon tetrachloride with hydrogen, and is the purest form of glass. This material is extremely transparent to infrared, visible, and UV radiation, in the range of 180 nm to 4000 nm, and is impermeable to most gasses at high temperature and pressure. It also has low expansion coefficient and high

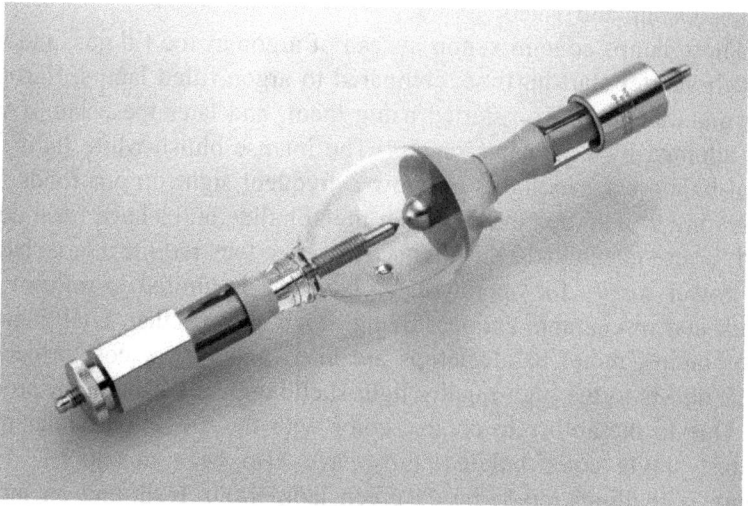

Fig. 3.53.   Short arc mercury vapor UV lamp.

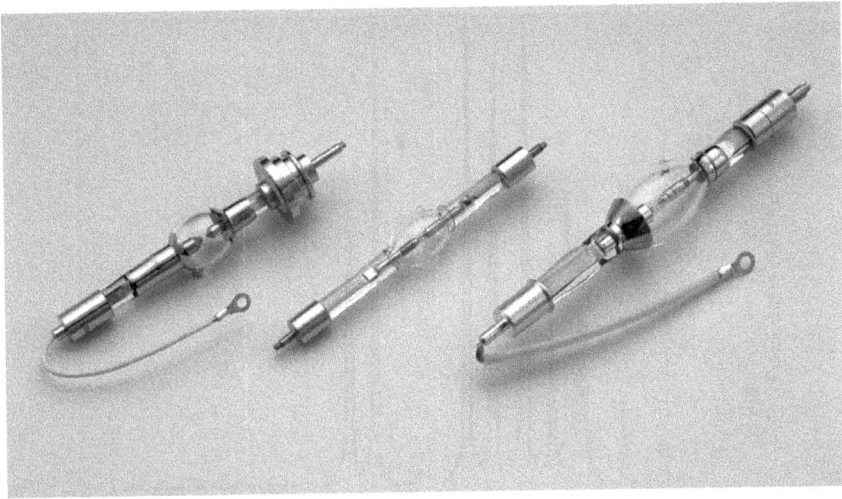

Fig. 3.54.   Different styles of short arc mercury vapor UV lamps.

mechanical durability, that is needed for operation at extreme conditions of temperature and pressure. The electrodes are made from forged high purity tungsten which, due to its high melting point, low vapor pressure, and reasonable cost, is the only metal suitable for this application. Short arc mercury lamps are cold cathode devices that feature conical electrodes with a sharply-pointed thoriated cathode. This electrode emits a copious supply of electrons when operated at high voltages. The anode-cathode gap ranges from 0.25 mm to 5 mm, depending on the lamp's power rating (see Figure 3.54). Both electrodes are highly polished, with the cathode shaped to a sharp point to enhance the field emission of electrons. The anode is not as sharp but is carefully radiused to increase the stability of the arc. The anode is also relatively massive in order to tolerate intense electron bombardment from the plasma. Heat generated at the anode can be removed by externally cooling the anode extension that comes out of the lamp. Most common short arc mercury discharge UV lamps operate at around 30 atm pressure, and consume about 120 W of electrical power. Super high-pressure arc lamps are also available, with operating pressures of several hundred atmospheres.

Short-arc high-pressure mercury vapor UV lamps are designed to produce mainly UV radiation below 400 nm. The emission spectrum contains many mercury resonance lines in the region of 250 to 600 nm (see Figure 3.55).

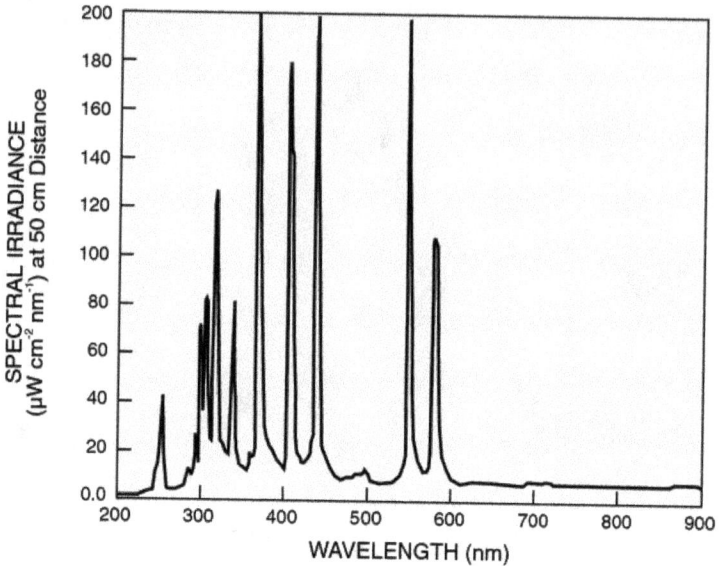

Fig. 3.55.   Emission spectrum of a typical short arc mercury vapor UV lamp.

Specific sub-regions can be selected with the aid of appropriate filters, to bring out only those wavelengths that are needed for particular applications. Thus, *i*-line lithography, for example, is performed with 365 nm radiation which is selected with *i*-line filters.

Short-arc mercury vapor UV lamps usually approach their end-of-life through gradual wear of their electrodes, which is apparent as a steady decrease of radiation intensity over time. There is also some risk associated with over-heating which may result in bulb explosion. For this reason, these lamps are cooled with dry nitrogen gas during operation and cool-down. They are also housed in an enclosed chamber, for containment during a possible lamp explosion event.

3.9 Xenon Arc Lamps

A large class of gas discharge arc lamps are ones that contain xenon as the primary fill gas. These lamps are highly valued in applications where the brightness of light and/or rich spectral coverage are the principal requirements. Xenon plasma emits many strong spectral lines in a broad region, from deep in the UV to the mid-infrared portion of the electromagnetic spectrum. Such broad spectral coverage, coupled with the sheer intensity of emission, make xenon arc lamps suitable for a variety of applications where it will be difficult to use some other lamp technology.

There are two main types of xenon lamps: devices meant for continuous wave (CW) operation and those used in pulsed mode (also called, flash lamps). CW xenon lamps can employ either short-arc or long-arc construction. While there are several variants of xenon arc lamps, they all have certain features in common. All xenon lamps require special power supplies for their operation due to their negative resistance characteristics and the need for pulse triggering (for flash lamps). Almost invariably, they all use fused silica or fused quartz envelopes with thoriated tungsten electrodes. Most lamps contain a pure xenon gas fill, but some are also manufactured with mercury, in addition to xenon.

Xenon arc lamps were developed in Germany during the 1940s. Some years after the end of the Second World War, Osram started marketing the first xenon arc lamps in 1951. These lamps were mainly targeted as replacements for carbon arc lamps in cine projection systems. Thereafter, xenon arc lamps were further developed by several manufacturers, and now-a-days these lamps are available from more than a dozen companies around the world.

The chief reason for the great utility of xenon lamps is the fact that they emit a profusion of rather closely spaced spectral lines from the deep UV to the mid-infrared. The high density of spectral lines results in the spectrum appearing more or less continuous from the UV to the infrared region. Thus, in the visible region, xenon lamps appear to emit a very white light. This makes xenon lamps very valuable as sources of continuous spectrum radiation, for both illumination and instrumentation applications. One can construct continuously-tunable light sources, starting from a xenon lamp and following it with a suitable monochromator. Such a source can provide continuous spectral coverage, thanks to the unique spectral make-up of light from xenon lamps.

A typical short-arc xenon discharge lamp (see Figure 3.56) consists of a transparent fused silica or fused quartz envelope fitted with thoriated tungsten electrodes. In general appearance, such a lamp can look almost identical to UV-emitting short arc mercury vapor lamps. The main difference is, of course, in the fill gas, which is pure xenon in this case. Fused silica or quartz is the purest form of glass, consisting of only silica (SiO_2) in amorphous (non-crystalline) form. It can be made either by melting (fusing) naturally-occurring white quartz sand (which is almost pure

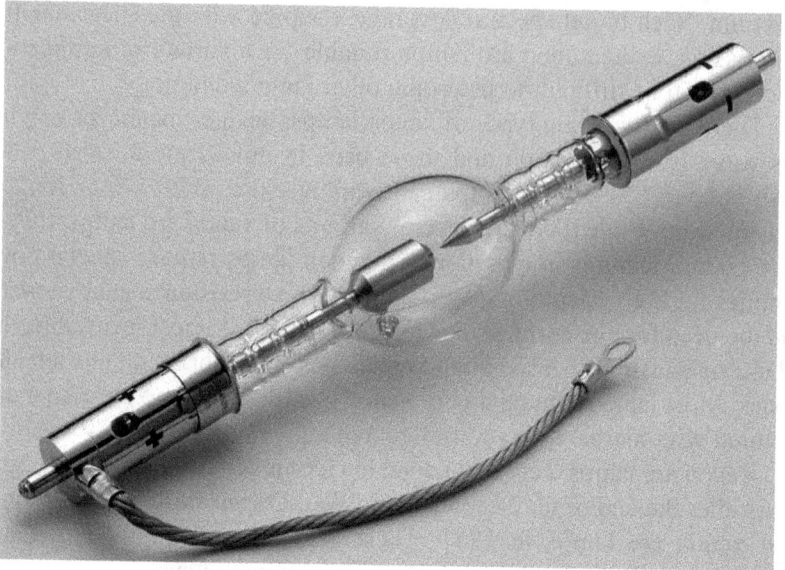

Fig. 3.56.   Short arc xenon discharge lamp.

natural silica), in which case the product is called fused quartz, or by melting synthetic silica, in which case it is called fused silica. Either material contains only trace amounts of impurities and, thus, offers extremely high light transmission.

Generally, fused silica, being synthetic, is superior in this respect to fused quartz. Ordinary glass, in contrast, contains very large amounts of elements other than silicon and oxygen; added to reduce the melting point of glass and to make it more 'workable' during glass shaping processes employed in the industry. The additional elements in commercial glass absorb many wavelengths; reducing its transparency. The transparency is especially reduced in the UV and infrared regions; making ordinary glass unsuitable for use in those regions. Fused silica and fused quartz, due to their high purity, are highly transparent for both short and long wavelengths in the range from approximately 180 nm to 2 microns (see Figure 3.57).

This range almost matches the range over which xenon plasma emits. Thus, making them perfect partners in lamp tube applications (see Figure 3.58). Various grades of Suprasil™ synthetic fused silica, made by the German company, Heraeus GmbH, are particularly favored for this application.

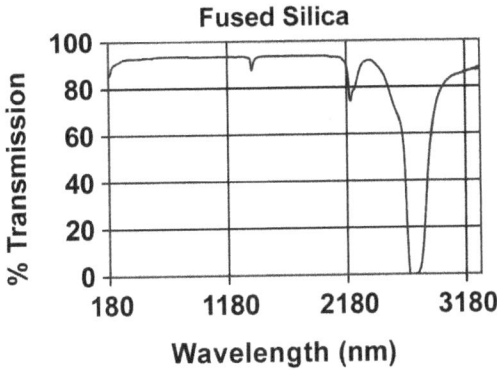

Fig. 3.57.   Transmission spectrum of fused silica.

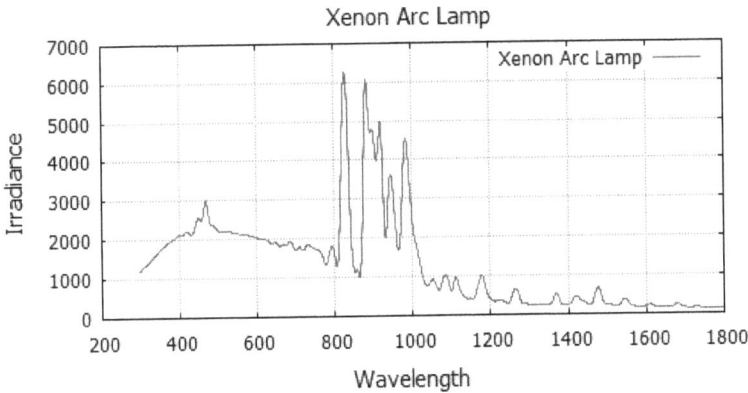

Fig. 3.58.   Emission spectrum of xenon lamp with fused silica envelope.

There are many applications for CW xenon arc lamps. Although these lamps are very infrequently used for direct space illumination, due to their brilliance and color balance, they are almost universally utilized in large cine projectors where lamps rated at 800 W to 15 kW are common. Due to explosion hazard, xenon projection lamps are delivered in special protective enclosures. The enclosure is removed *after* the lamp has been installed inside the projector. When removing a used lamp, it is first encased into its enclosure before the assembly is removed from the projector. The latter step is even more important because aged lamps pose a higher explosion risk. Xenon projection lamps of even higher power ratings are used for large screen IMAX projection systems. These lamps can have power ratings as high as 20 kW, and are water-cooled to dissipate the large amounts of heat they produce during operation. Such lamps are built with water cooling ports that carry cold purified water to the electrodes where heat generated during arc operation is absorbed. The heated returning water is passed through a heat exchanger and fed back to the system. Many scientific broadband light sources are also based on xenon arc lamps, for their un-paralleled spectral coverage. These include solar simulators, because with appropriate spectral filters to mimic atmospheric absorption, light from xenon arc lamps can be made to resemble natural sunlight quite closely. Photographic studios and printing facilities, as well as document scanners and copiers, often employ xenon arc lamps for their daylight-like lighting quality; providing exceptional color rendering. Floodlights and searchlights used for both civilian and military applications also utilize xenon arc lamps. Xenon lamps have also been developed for automotive use, but these are, in principle, metal halide lamps where a xenon arc is only used to start the lamp so that light is generated immediately on switching the lamp on. This is a safety requirement for automobile headlights as metal halide lamps take some time to achieve full brightness. Thus, a xenon arc has been incorporated in automotive metal halide lamps to make them generate some light as soon as the lamps are turned on.

3.10 Deuterium Arc Lamps

Deuterium lamps are somewhat similar to xenon lamps in that they too generate radiation with a broad spectral coverage ranging from deep UV (down to 160 nm) to the infrared. As UV light sources, however, these lamps are superior to xenon lamps because of the presence of a broad UV continuum, without any sharp spectral lines imposed on it (see Figure 3.59). This continuum arises due to molecular excitation of deuterium gas fill in the discharge tube. Atomic excitations, in contrast, produce sharp line emissions. Furthermore, deuterium arc lamp emissions in the visible and infrared regions are relatively weak and can be easily filtered out, making them near-deal for generating UV radiation.

The molecular continuum as well as the Fulcher-α band are generated by rotational and vibrational transitions of the deuterium molecule. In more complicated molecules, similar features are observed at much lower energies that correspond to the infrared region of the electromagnetic spectrum. However, the low atomic masses in deuterium molecules shift

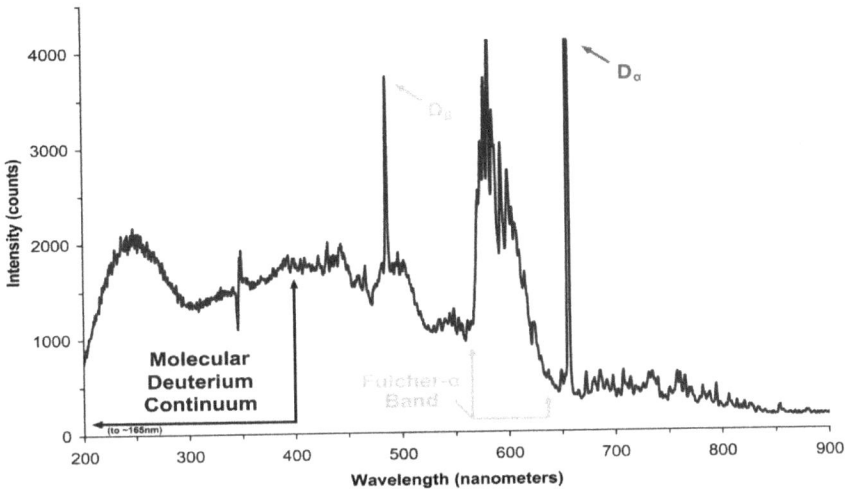

Fig. 3.59.   Emission spectrum of a deuterium arc lamp.

molecular transitions to much higher energies and, thus, shorter wavelengths. Hydrogen-filled arc lamps are also available but their radiation intensities are several times lower than that of deuterium-filled lamps, and, thus, deuterium lamps are used in preference to hydrogen lamps, despite being significantly more expensive.

Several specialist lamp manufacturers, including Hamamatsu of Japan and Photron of Australia, make deuterium arc lamps (see Figure 3.60). These lamps are filled with deuterium, at low-pressure and contain a distinct box-shaped enclosure, made from sheet nickel, that encases the lamp electrodes. The cathode is formed out of thoriated coiled tungsten filament which is initially heated to start the lamp with a generous supply of thermionically-emitted electrons. The start-up time is typically 30s. Once the arc is struck due to high voltage between the cathode and the anode, the cathode heating is stopped. The lamp then takes anywhere from 10 to 30 min to stabilize to a constant power output. As seen in the diagram in Figure 3.61, the lamp anode itself sits separated from the cathode by a diaphragm with a small aperture. This arrangement serves to focus the electron flow and does not let the discharge spread to other locations inside the tube. The box is positioned upright inside a transparent tube made of some suitable material. Three wires come out of the tube. One is the anode connection whereas the other two connect to the cathode. Deuterium lamps emit only from one side and in one direction, unlike

Fig. 3.60.   Commercial deuterium arc lamp.

Courtesy: Hamamatsu Corporation.

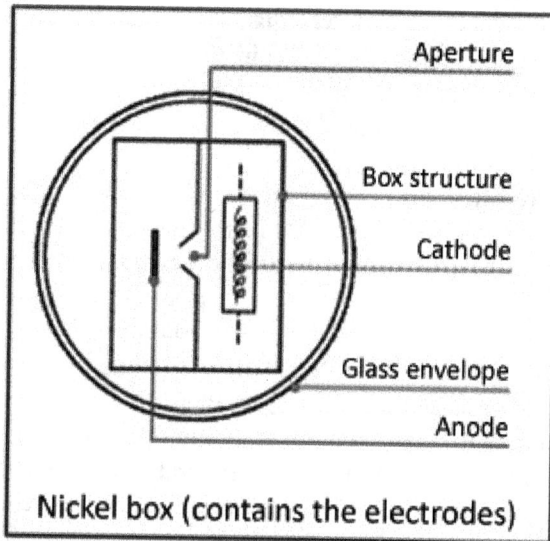

Fig. 3.61.   Structure of a deuterium arc lamp.

other arc lamps, which radiate in all directions. Radiation collection is, therefore, easy and efficient. The window material determines the UV spectral coverage available. Fused quartz windows allow radiation down to 180 nm to come out of the lamp, whereas lamps fitted with magnesium fluoride (MgF_2) windows allow radiation down to 160 nm to be extracted. Special deuterium arc lamps are available for vacuum ultraviolet (VUV) radiation, having wavelengths as short as 110 nm. These lamps are fitted with flexible light conduits that allow such short wavelengths to come out of the lamp without being absorbed by ambient air.

Deuterium arc lamps are often available with tungsten halogen lamps, in systems designed to provide extra-broad spectral coverage for scientific applications such as spectroscopy, water purity analysis, and industrial quality control. Such a system is shown in Figure 3.62. Due to demanding requirements on the constancy of light intensity, most deuterium lamp-based light sources have to demonstrate intensity fluctuations of less than 0.03 ppm/s and long-time intensity drift of less than 0.5 ppm/s. Systems from reputable manufacturers generally satisfy these requirements.

Although somewhat expensive, due to their high intensity and broad spectral coverage, deuterium lamps are widely used in scientific and technical applications, such as UV-visible spectrometers and high-performance

Fig. 3.62.   Compact laboratory deuterium lamp light source.
Courtesy: Ocean Insight Inc.

liquid chromatography (HPLC) systems. Several manufacturers supply deuterium lamp tubes, as these need to be replaced frequently, owing to their rather short lifetimes of around 2000 hours. If used properly, with their designated power supply, deuterium arc lamps provide continuum UV radiation that is hard to match with any other small source.

3.11 Krypton Arc Lamps

Krypton arc lamps are somewhat similar to xenon arc lamps. They too generate a broad spectrum, spanning the region from UV to the infrared, and due to this reason, their light appears whitish in appearance (see Figure 3.63); similar to light from xenon lamps. However, being relatively inefficient, compared to xenon arc lamps, krypton lamps are not used for applications that require visible light generation. A major difference between the two is that, unlike xenon lamps, krypton lamps are almost always long-arc lamps, with anode-cathode separations of several centimeters. Apart from this, another difference — one which makes krypton lamps useful as a separate lamp type — is the different spectrum it generates. The main application of krypton arc lamps is in laser systems, where these lamps are routinely used as optical pump sources for solid-state

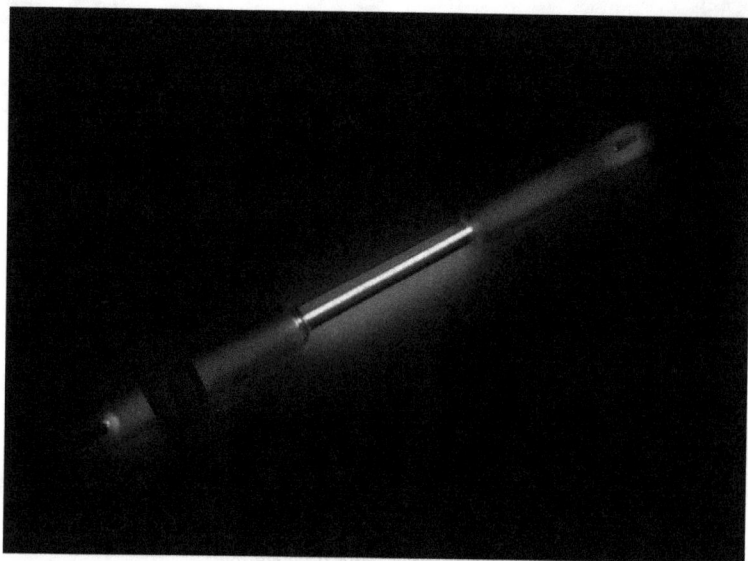

Fig. 3.63.   Light emission from a krypton arc lamp.

lasers; in particular neodymium-doped yttrium aluminum garnet (Nd:YAG) lasers. In this application, the krypton arc lamp offers optimum spectral matching because the emission spectrum of krypton lamps is similar to the absorption spectrum of Nd:YAG crystals. Their intense near-infra-red radiation, between 750 and 900 nm, is perfectly suited for this application. Additionally, for this purpose, krypton discharge lamps also offer high brightness, high-power capability, long operating life, and low cost per hour of operation. Thus, the vast majority of krypton arc lamps are now produced for laser pumping application alone. Several companies manufacture krypton arc lamps. These include, Heraeus, Amglo Kemlite Laboratories, Osram and Perkin Elmer.

Like xenon lamps, krypton lamps also come in both CW and pulsed (flash lamp) varieties for pumping CW and pulsed Nd:YAG lasers, respectively. Krypton laser pump lamps are usually manufactured as long narrow tubular devices of the form seen here in Figure 3.64.

Like xenon lamps, these lamps also have distinct electrode pairs: the cathode being highly pointed while the anode has a gently rounded shape. The electrodes, made from thoriated tungsten, are positioned at the ends of a 4 mm diameter quartz tube that can be 15–30 cm in length (see Figure 3.65). The electrodes are sealed to the discharge tube with a graded seal construction that consists of layers of quartz and tungsten interleaved with each other and fused to a hermetic finish. Further details of lamp construction can be seen in Figure 3.65. The quartz tube is quite thin with typical thickness of only 0.5 mm so that heat can be easily transferred out

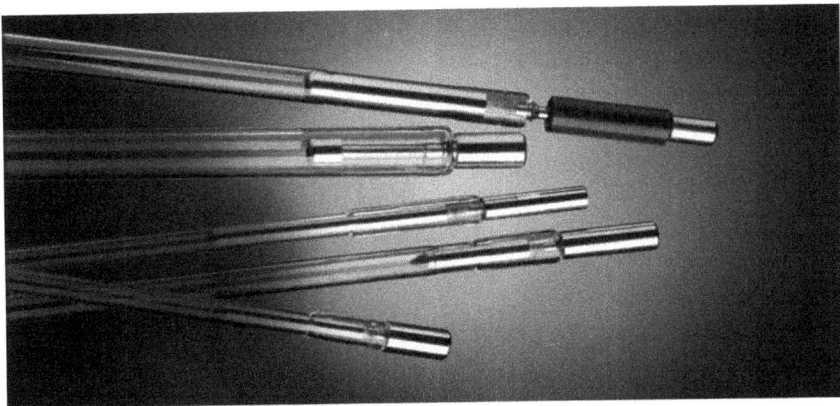

Fig. 3.64.   Tubular krypton arc lamps for laser pumping.

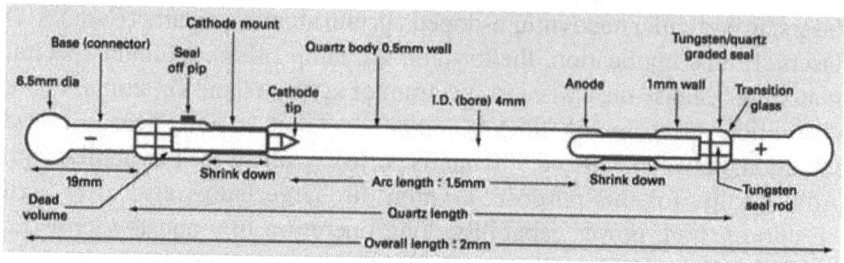

Fig. 3.65.   Structure of a krypton arc lamp.

to a cooling fluid (usually a stream of deionized water). Extreme operating conditions, however, shorten lamp life such that even with proper use, krypton arc lamps have a relatively short lifetime of 500–1000 hours.

Like xenon lamps, krypton lamps too require special power supplies to operate. Pulsed operation is achieved by charging capacitor banks and discharging them into flash lamps through an appropriate capacitor-inductor-based pulse forming network (PFN). Hundreds of amperes are typically discharged in a few milliseconds. To prevent the tube from exploding during such rapid current flow events, they are filled with krypton at less than 1 atm pressure so that during operation their pressure does not rise above safe levels. In usual practice, the series connection of capacitor bank and the PFN is directly connected to the lamp electrodes. The capacitors are charged to a voltage slightly less than the breakdown voltage of the lamp tube. The impedance of the lamp prevents discharge of the capacitor bank until the lamp is ionized by a separate trigger pulse. The capacitors then discharge through the PFN and the lamp, generating a pulse of current and a burst of light. The capacitors can be charged and discharged repeatedly at high repetition rates to generate a steady stream of laser pulses. In recent years, solid-state switch mode power supplies have taken over from older capacitor bank/PFN systems that often-employed thyratron triggering tubes. Modern switch mode converters can generate customized high current pulses under microprocessor control. This facility enables the generation of short laser pulses with desired temporal profiles, which enables such applications as highly advanced laser-based materials processing.

CW krypton arc lamps are used for pumping CW Nd:YAG and Nd:Glass lasers. These lamps operate at currents in the 10–50 Amps range. CW krypton lamp tubes are filled with krypton at much higher pressures — typically 5–10 atm. Just as with high-pressure xenon arc lamp tubes, high-pressure krypton lamps too must be installed and used with great care, as mishandling or improper operation can lead to catastrophic explosions. Because of the fact that CW krypton lamps contain gas at very high-pressure but are operated at relatively low voltages, in the range of 100–200 V, they are difficult to start. Sophisticated power supplies are, therefore, required to provide the multi-level voltage waveforms for successful lamp operation. The power supply first applies a short duration (approximately one microsecond long) high voltage (20–30 kV) trigger pulse to the lamp, which causes the fill gas to ionize. Next, a boost circuit drives a low current through the lamp at a moderate voltage for a few milliseconds, after which the main power supply can take over, and run the lamp at its normal operating current and voltage (typically 20 amps at 150 V).

A secondary application of krypton lamps is in spectroscopy. Electrical discharge in krypton produces very distinct and stable spectral lines which are often used as reference markers in spectroscopy. For this purpose, hot cathode krypton discharge lamps are used. A typical such lamp manufactured by Philips in 1984 is shown in Figure 3.66. It is made of a krypton-filled quartz arc tube, enclosed inside an outer soda-lime glass jacket, with the space in between evacuated. Both electrodes are

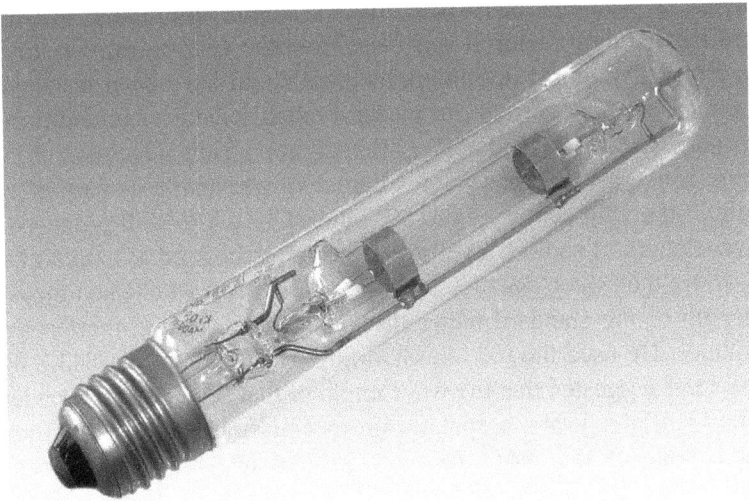

Fig. 3.66. Philips hot cathode krypton arc lamp.

Courtesy: Philips N.V.

heated tungsten filaments with the space inside them filled with a mixture of barium, calcium and strontium oxides to boost the thermionic electron emission efficiency. Initially, both filaments are heated with a DC current, in order to emit electrons and ionize the gas inside the arc tube. After applying 400 V to these electrodes to start the arc discharge, only the cathode filament is kept heated. Lamps like these are still available for spectroscopic applications.

An important historic use of the krypton discharge lamp has been in defining the meter as the international standard unit of length. This is yet another application that was made possible due to the sharp spectral lines that can be obtained from a krypton discharge. The meter was originally defined as one-ten-millionth of the length of the meridian between the equator and the geographic North Pole, passing through Paris. On the instructions of Napoleon Bonaparte, a survey was carried out by Jean-Baptiste Delambre and Pierre Méchain to measure this distance. They traveled between the extreme north French town of Dunkirk and the northern Spanish city of Barcelona in the 1790s, in order to measure the meridional distance between these two sites, with Paris on the way. Through their measurements and calculations, they arrived at the arc length through Paris, joining the equator to the North Pole. This allowed them to figure out the physical length of a meter; defined as above. A platinum bar was then constructed of that length, and kept as the definition of the standard meter. It was clear from the very outset that this was not a physically robust definition, as there could have been inaccuracies in the measurement carried out by the French scientists. In addition, the meter bar, being a physical object, was not universally accessible. In spite of these concerns, no other viable alternative was available until after the middle of the 19th century. In 1859, James Clerk Maxwell proposed that the wavelength of sodium's yellow spectral line be used as a natural standard. More than three decades later, in 1892, Albert Michelson measured the length of the standard meter bar kept in Paris using interferometric techniques. He used the red emission line from cadmium at 643.8 nm to do this, and suggested that the wavelength of that light be used as a length standard. In later years, a specific green emission wavelength from the ^{198}Hg isotope of mercury came into prominence for this purpose. An electrode-less RF discharge-based ^{198}Hg lamp was devised by Jacob Wiens, Luis Alvarez and William Meggers to serve as a source for interferometric determination of the meter. This lamp came to be known as the Meggers lamp. Although, the ^{198}Hg lamp was widely proclaimed as the most suitable light source for an interferometrically-defined meter, it suffered from reproduction problems due to the difficulty in maintaining a uniform operating temperature and pressure. For that reason, it never became part of an official definition of the standard meter, but the quest for a suitable spectroscopic source continued. After much worldwide deliberation, the International Bureau of Weights and Measures decided on a monochromatic emission from a krypton discharge lamp to serve for the definition of the standard meter. On October 14, 1960 the Bureau adopted the 605.6 nm emission wavelength of the orange emission line, from isotopically-pure ^{86}Kr, as the basis for defining the meter. Note that this is one of 33 known isotopes of krypton. Figure 3.67(a) shows one of the ^{86}Kr lamps that were constructed for this purpose. This lamp was operated in a liquid nitrogen bath and the meter was defined as exactly 1,650,763.73 wavelengths in vacuum of the orange line at 605.6 nm, arising out of the unperturbed $2P_{10}$-$5d_5$ atomic transition in ^{86}Kr atoms. Figure 3.67(b) shows the cryostat housing which is used to enclose ^{86}Kr discharge lamps (of a different design), to maintain a fixed operating temperature. The light emission port appears in the lower front portion of the cryostat. This definition for the meter, based on optical interferometry, remained in effect until 1983 when it was replaced by another definition based on the speed of light and the unit of time.

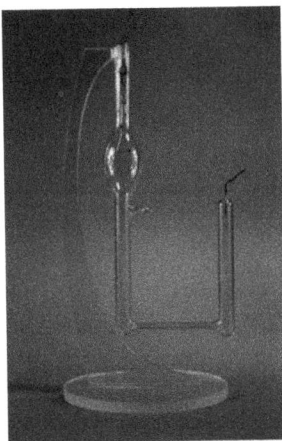

(a)   (b)

Fig. 3.67. (a) ^{86}Kr arc lamp used as an interferometric length standard. (b) Cryogenic lamp housing.

3.12 Sodium Vapor Lamps

Sodium vapor lamps are widely used for outdoor lighting of public spaces such as roads, bridges and parks, as well as grounds outside large buildings. They owe their popularity to low operating costs, maintenance-free operation and desirable lighting characteristics. There are two distinct types of sodium lamps: one operating at low-pressures and the other at high-pressures. They differ considerably in their construction details, spectral coverage, and efficacy. The low-pressure sodium lamp was developed first, followed by its high-pressure variant, some three decades later.

The yellow light of sodium lamps is a familiar sight the world over, especially in European cities, where their use has been widespread since the 1930s. Their most outstanding attribute is their extremely high efficacy. Low-pressure sodium vapor lamps can convert electric power to visible light at efficacies in the range of 150–200 lumens per watt. This is much higher than that of other types of lamps, except those based on light-emitting diode technology. The reason for the exceptionally high efficacy lies chiefly in the emission being confined to a very narrow range of wavelengths close to the peak of human eye's sensitivity curve. The lamp does not waste energy in producing emissions at wavelengths for which the eye is insensitive, or not sensitive enough. Because the generated light is close to the maximum of human photopic sensitivity so it appears particularly bright; thus, increasing the perceived amount of light in lumens. The remarkable efficacy of sodium lamps, combined with their long lifetime of around 20,000 hours of operation, and superb lumen maintenance, has made them an almost universal choice for large space illumination where their single-color character is not an outright disadvantage.

The light from a sodium vapor lamp is spectrally the same as that emitted from a handful of table salt (sodium chloride) thrown into a fire. In each case, the light is generated by electronic transitions from a set of two energy levels, very close together in energy, to the ground state of sodium. This pair is formed as the ^3P level of atomic sodium splits into two closely-spaced levels by spin-orbit interaction. This is seen in Figure 3.68 where the dominant transitions from $3P_{1/2}$ and $3P_{3/2}$ levels to the $3S_{1/2}$ ground state are seen, together with other much less probable transitions. Among all transitions producing visible light, these transitions are dominant, and produce photons with energies of 2.103 eV and 2.105 eV. These correspond to wavelengths of 589.0 nm and 589.6 nm, respectively. This closely-spaced sodium doublet appears as the characteristic bright

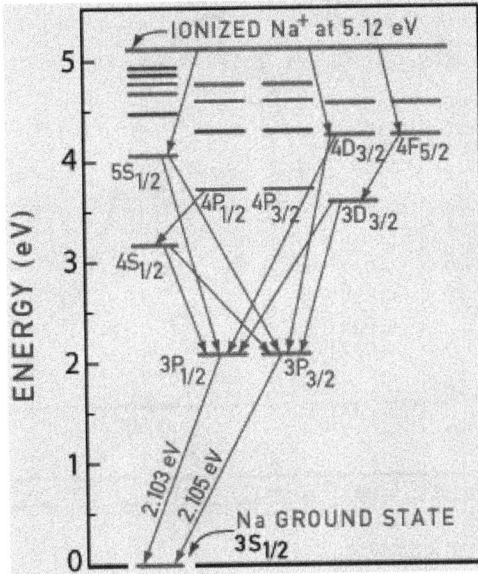

Fig. 3.68. Energy levels and possible radiative transitions for sodium atom.

yellow emission from excited sodium ions, whether it is excited thermally or electrically. These spectral lines are separated by only 0.6 nm in wavelength, which makes them appear like a single line, unless a high-resolution spectrometer is used to analyze the light. Observed with a high-resolution spectrometer, emissions at these two wavelengths appear to be of the same intensity because both $3P_{1/2}$ and $3P_{3/2}$ levels have about the same occupation probability. Note that these lines may also appear as a single narrow band, centered at 589.3 nm, due to collisional broadening if the discharge takes place under high-pressure and/or high temperature conditions.

Figure 3.69 shows the narrow, line-like sodium lamp emission superimposed on the human photopic response, showing its proximity to the maximum of human eye's sensitivity.

3.12.1 *Early development of low-pressure sodium vapor lamps*

It is difficult to say who was the first person to consider electrical discharge in sodium vapor for producing a light source. While it is very easy to obtain sodium's characteristic emission by thermal excitation of a salt,

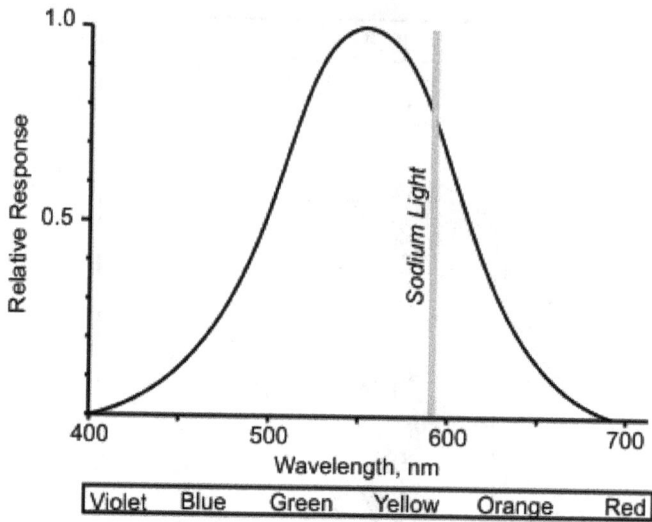

Fig. 3.69. Human eye's photopic sensitivity curve and the relative position of sodium emission line.

Fig. 3.70. Arthur H. Compton.

generating the same emission through an electrical discharge is far from simple. The physicist and Nobel laureate Arthur H. Compton, while working as a research engineer at the Westinghouse Lamp Company, in Pittsburgh, Pennsylvania, is generally credited as the first person who attempted to make a sodium vapor-based light source. Figure 3.70 shows

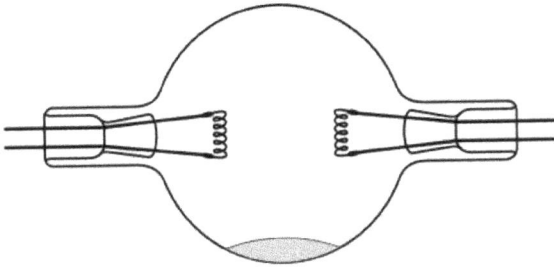

Fig. 3.71.   Compton's sodium vapor lamp design.

a picture of Arthur Compton. In the year 1919, he took out a patent for a discharge source based on sodium vapor contained in a glass globe.

Compton's design of his sodium lamp appears in Figure 3.71. It consisted of a spherical borate glass bulb containing a small amount of sodium metal and two filamentary tungsten electrodes sealed at opposite ends.

The lamp had to be first heated thermally to melt and vaporize the sodium metal. Then, the filament electrodes were heated to incandescence, in order to provide a good supply of electrons into the sodium vapor. A discharge was then created between the two electrodes, and this generated a bright yellow emission. The spherical design of the lamp was necessitated because once the lamp was turned off and the sodium vapor cooled, the metal had the tendency of migrating to the coolest part of the bulb where it solidified. With a tubular design, it was found that the sodium would migrate to the outer ends of the tube, and there the sodium would destroy the electrodes over time as well as not get hot enough to vaporize. The spherical design avoided the undesirable migration of sodium and, thus, was used with the first ever demonstration of a sodium vapor lamp.

Another problem with this primitive lamp design was the inability of the glass to resist attack from highly corrosive sodium vapor. Sodium rapidly attacks glass, forming dark-colored sodium silicate which both stains glass and, over time, leads to its fracture. This was the main problem that stopped the commercial development of sodium vapor lamps after Compton's demonstration. There were no significant developments for the next 10 years. Further development of sodium vapor lamps started only when in 1931 Marcello Pirani, a German physicist of Italian origin, working for Osram in Berlin, demonstrated a tubular lamp design with an outer glass jacket. The outer envelope was needed to provide thermal

Fig. 3.72.   Marcello Pirani.

insulation in order to keep the sodium in its vaporized state, as the lamp operated. Pirani's lamps also contained krypton as a discharge gas. Most importantly, the Pirani lamp had a sodium-resistant inner gas discharge tube (arc tube). Figure 3.72 shows a picture of Marcello Pirani.

3.12.2 *Commercial low-pressure sodium vapor lamps*

Pirani's development was followed a year later, in 1932, by the first commercial availability of sodium vapor lamps. Philips in Eindhoven, The Netherlands, became the first company in that year to offer commercial sodium vapor lamps. These lamps had a removable outer glass jacket, with vacuum between it and the inner glass discharge tube. Thirty 'Philora' lamps were installed on lamp posts between the Dutch cities of Beek and Geleen, as the first large scale demonstration of sodium vapor lighting. Figure 3.73 shows a Philora lamp, from which its general construction can be ascertained.

The Philora lamp had a central tungsten filament cathode, coated with alkali-earth oxides for enhanced electron emission. It also had two ring-shaped anodes, two centimeters above and below the centrally-located cathode. Thus, the lamps operated with two separate arcs, filling the length of the arc tube. These lamps were operated at 15 V DC with a current of 5.5 amperes. The Philora lamps were very successful, despite their

Fig. 3.73. Philips Philora sodium vapor lamp.
Courtesy: Philips N.V.

low efficacy and lifetimes, when compared to their modern versions. These lamps proved the commercial feasibility of lighting up roadways with sodium vapor lamps, and started a trend that soon saw most European highways being equipped with sodium vapor lamps.

It did not take long for the development to cross the ocean. Across the Atlantic, General Electric soon developed its own version of sodium vapor lamp. The GE, NA-9 lamp, rated at 180 W and 10,000 lumens, was released in 1936. Before their mass production in that year, GE's sodium lamps were tested with experimental installations in Revere, Massachusetts, in January 1934. Eleven lamps were used to illuminate a highway underpass. Other experimental lamps were installed at the Boston-Worcester Turnpike at Newton, Massachusetts, Hartford-Meriden Road at Wallingford, Connecticut, and at Balltown Road near Schenectady, New York.

The NA-9 lamp had two tungsten filament electrodes at either end of an inner glass bulb which was surrounded by an outer Dewar jacket for thermal insulation. The filament cathodes were surrounded by open-ended molybdenum boxes that served as anodes. Each anode was connected to

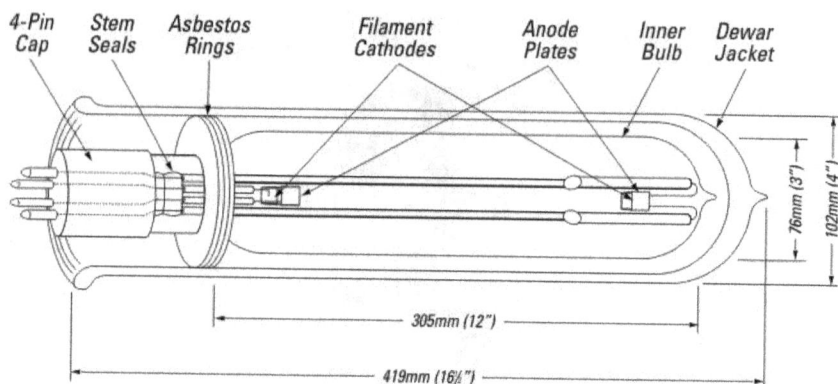

Fig. 3.74.   Structure of a GE NA-9 lamp.

Courtesy: General Electric Corporation.

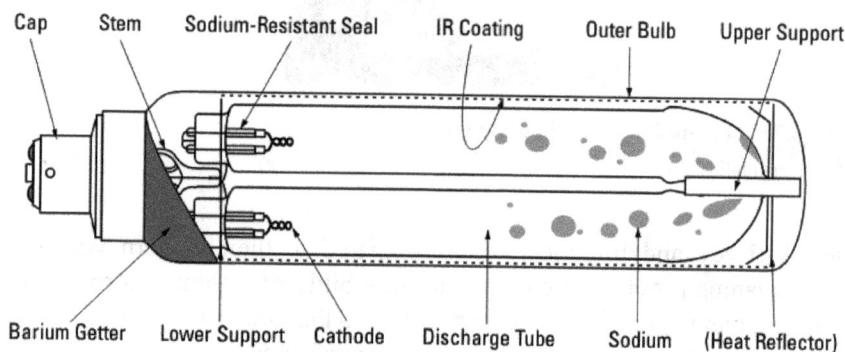

Fig. 3.75.   Diagram of a modern sodium vapor lamp.

one side of the filament. The filaments were initially heated with a 10 amp current at 2–3 V, to generate a copious supply of electrons. A mechanical relay then applied a high open circuit voltage across the two anodes, to strike an arc in the neon fill gas. After a few minutes of discharge operation, the tube warmed up enough for the sodium metal to evaporate, and the lamp attained its full brightness. During operation, the lamp had 28 V across it and drew a current of 6.6 amp. Figure 3.74 is a schematic diagram of a GE NA-9 lamp. Note the asbestos rings toward the base of the lamp. These rings supported the arc tube and provided further thermal insulation to minimize heat transfer between the two glass bulbs.

The low-pressure sodium lamp has had further improvements over the past decades. Figure 3.75 shows a modern version of this lamp. The first thing to note is that the discharge tubes are now usually U-shaped. This reduces the overall length of the lamp and also helps in conserving heat.

The outer glass jacket has been integrated with the inner discharge tube such that these are no longer separate objects. The vacuum in the intervening space serves to thermally insulate the inner tube, as before. There is also a thin (~0.32 microns) tin-doped indium oxide coating on the inner surface of the outer bulb, as shown above. This coating serves to reflect infrared radiation back into the inner bulb, to further reduce any radiative heat loss, and, thus, enhances the efficacy. Better thermal blanketing also improves the reliability of the lamp in cold weather. A barium metal getter is used to improve the atmosphere inside the bulb assembly, by adsorbing any contaminants and outgassing products. More recent designs have also used a zirconium-cobalt pellet getter for the same purpose. With the lamp at room temperature, one can see shiny deposits of sodium metal on the inner surface of the discharge tube. Often there is no dedicated sodium container in the lamp but some arc tubes are made with little bumps on the sides to serve as sodium reservoirs. As the tube starts to function, sodium evaporates from the reservoirs, and when the tube is switched off the metal returns to the reservoirs, which appear as cooler locations on the tube wall. This can be seen in Figure 3.76 which shows a British-made 35 W 1993 Philips SOX-plus low-pressure sodium vapor

Fig. 3.76.   Philips SOX-plus low-pressure sodium vapor lamp. Philips N.V.

lamp. Modern lamps, like these, are filled with a Penning mixture consisting of about 99% neon and 1% argon gas. The argon serves to reduce the potential required to strike an arc in the gas mixture, as has been described before.

When first started, current flows through the tungsten thermionic emitters which heat up to white hot incandescence. The profuse emission of electrons from the cathodes helps ionize the fill gas. Soon after, a brief high voltage pulse, from an igniter built into the ballast, is applied to create a discharge between the two electrodes. The current is then limited to its steady value by the inductive ballast. The initial discharge in the Penning mixture shows as a weak pink glow due to the discharge in neon, as is seen in Figure 3.77. Thus, when first started, all sodium vapor lamps begin to glow with a pink color. The gas discharge heats up the discharge tube, and as its temperature exceeds 98°C — the melting point of metallic sodium — the metal deposits begin to melt and evaporate, gradually filling up the discharge tube with sodium vapor. The color of the discharge gradually changes to yellow, with full brilliance reached in around 8–12 min after starting the lamp. The normal operating temperature of the lamp (inner discharge tube) is about 250°C. The yellow glow seen around both filaments in the figure is because of the early availability of sodium vapor in the regions around the heated filaments, due to their heat generation.

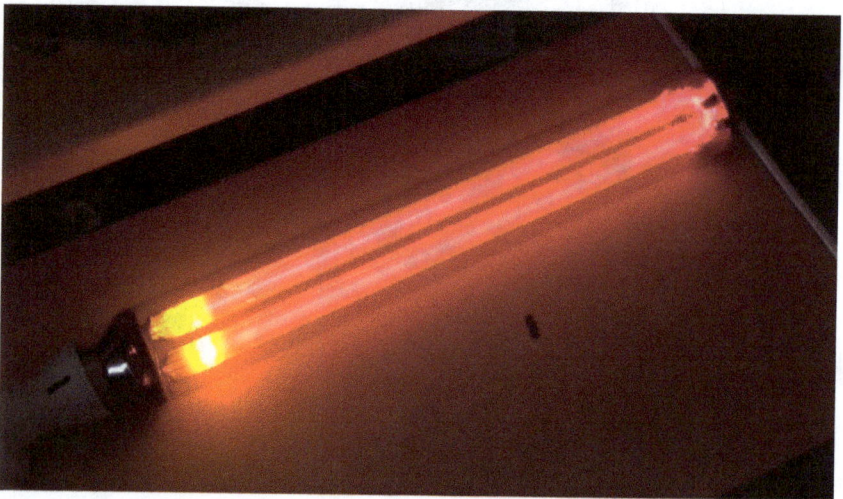

Fig. 3.77.   Start-up of a sodium vapor lamp, showing the pink glow from the neon fill gas.

The starting voltage and current required for sodium vapor lamps are nearly independent of the lamps' temperature over a wide range. This makes it easy to start and operate them under different weather conditions. This is in contrast to mercury vapor lamps, and is a significant advantage for the widespread use of sodium lamps in all geographical locations. Moreover, the lamps are immune to short periods of power loss, lasting for a few seconds. The lamps reignite automatically, and quickly regain their full brightness after episodes of brown-outs or short power dips, as the discharge is re-established in the still hot sodium vapor. These features, combined with their simple operating circuit and almost constant brightness over their entire lifetime, reduce any maintenance requirements; making them especially suitable for installation on lamp posts and other hard-to-reach locations. The only care that has to be taken in their use is to make sure that the lamps are operated within ±20° of the horizontal. This ensures that sodium does not migrate to the cooler parts of the lamp, and cause accidental short circuits.

3.12.3 *High-pressure sodium vapor lamps*

After the introduction of the low-pressure sodium lamps in the early 1930s, it was quickly realized that a high-pressure version of the lamp can provide better spectral coverage than the almost monochromatic spectrum generated by the standard low-pressure lamp. The approach is similar to that used with high-pressure mercury vapor lamps for enhanced spectral coverage. After several years of development, high-pressure sodium vapor lamps became commercially available from General Electric in 1964. These lamps feature a much broader spectrum than their low-pressure counterparts, at the expense of somewhat reduced efficacy. Compared to low-pressure lamps, the high-pressure variety is also physically smaller, and has a longer average life of around 24,000 hours. These advantages have made high-pressure sodium vapor lamps extremely popular.

While the feasibility and advantages of high-pressure sodium lamps were recognized quite early, the practical implementation had to wait for several years, until a suitable container that could withstand sodium vapor at high temperature and pressure could be developed. Borate glass, while suitable for low-pressure sodium vapor lamps, is not resistant to attack from sodium vapor at elevated temperature and pressure, and, thus, cannot be used for containing the discharge in a high-pressure lamp. This

Fig. 3.78.   Robert Coble.

problem was solved by Robert Coble — a young ceramist working for General Electric, during the late 1950s. Working with Joseph Burke's group, and under his guidance, at GE's Schenectady lab, Coble investigated aluminum oxide ceramics as a possible envelope material for high-pressure sodium lamps. After a few years' work, he succeeded in developing a sintered aluminum oxide ceramic, containing a small amount of magnesium oxide that made satisfactory discharge tubes for this purpose. Robert Coble is seen here in Figure 3.78.

By 1960, Coble had left GE to join MIT as an assistant professor of ceramics in the Metallurgy Department, but his invention was patented by GE in 1962, under the trade name 'Lucalox' (derived from, transLUCent ALuminum OXide). This material, mentioned earlier in connection with high-pressure mercury vapor lamps, also enabled the manufacture of the first generation of commercial high-pressure sodium vapor lamps. Lucalox is a PCA-magnesia ceramic. It is translucent, with greater than 90% transparency to visible light when formed into tubes with 1 mm wall thickness. These tubes are remarkably resistant to attack from sodium vapor, at temperatures exceeding 1000°C. Tests have shown no degradation from sodium attack, even after operating for 10,000 hours as lamp discharge tube material, whereas borosilicate and quartz discharge tubes get severely attacked within an hour, under the same conditions. Lucalox discharge tubes were later used by William Louden, Kurt Schmidt, and

Fig. 3.79. William Louden, Kurt Schmidt and Elmer Homonnay (left to right) at GE's Nela Park facility.

Courtesy: General Electric Corporation.

Elmer Homonnay (see Figure 3.79) at GE's Nela Park facility to make the first commercial GE high-pressure sodium lamps.

Compared to their low-pressure version, high-pressure sodium lamps (often also known by the abbreviation, HPS) are sufficiently distinct to merit their classification as a separate lamp type. Like high-pressure mercury discharge lamps, high-pressure sodium vapor lamps are classified as a HID lamp. The lamp consists of a narrow, semi-transparent, inner Lucalox discharge tube and an outer protective glass jacket which also provides thermal insulation. The outer jacket is generally oval-shaped and smaller than that of low-pressure lamps. The discharge tube itself is filled with xenon instead of the neon-argon Penning mixture used in low-pressure lamps. Xenon has a very low thermal conductivity, and also a low ionization potential. Both properties are made good use of in HPS lamps, as the low thermal conductivity of xenon prevents excessive heat loss from the discharge tube, and the low ionization potential enables operation at low arc strike voltage. Another point of difference from

low-pressure sodium vapor lamps is in the use of a sodium-mercury amalgam, instead of pure metallic sodium. The amalgam is stored in special
reservoirs, and this removes the requirement for operating the lamp in
horizontal or near-horizontal positions.

The HPS lamp operates at higher temperature and pressure than its
low-pressure variant. The arc tube has sufficient wall thickness to resist
the high outward pressure at operating temperature and pressure. Voltage
and power requirements for high-pressure lamps are similar to that of the
low-pressure variety, and, thus, similar ballasts and starting circuits are
used. The discharge is started in the xenon gas fill and gradually warms
up the lamp so that the amalgam melts. Initially, the lamp lights up with a
bluish-white-colored emission, as the discharge strikes in the xenon fill
gas. Mercury, with its lower vapor pressure, evaporates first, and then
sodium melts and evaporates to fill the discharge tube with a mixture of
mercury and sodium vapors. The light from a high-pressure sodium vapor
lamp contains emissions from both mercury and sodium atoms.
Furthermore, the emission lines are considerably broadened due to the
high temperature and pressure employed in the discharge tube. The result
is light with much more spectral content than that from low-pressure
sodium lamps, as seen in Figure 3.80. Pressure-broadened yellow emission from sodium provides spectral coverage at long wavelengths,
whereas broadened mercury line emissions provide some coverage at

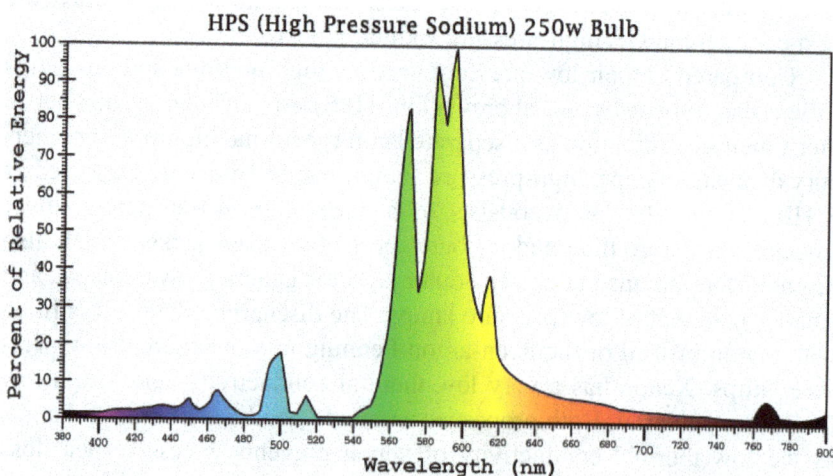

Fig. 3.80.   Emission spectrum from a high-pressure sodium vapor lamp.

shorter wavelengths. The light appears a golden yellow color, providing a much better color rendition than the pure yellow illumination of the low-pressure lamp. As a result, high-pressure sodium vapor lamps have become popular replacements for their low-pressure variants, in applications where color discrimination is an essential requirement.

An interesting phenomenon observed in high-pressure sodium vapor lamps is that of spectral self-reversal, which tends to provide additional broadening over that from thermal and pressure-induced effects. Self-reversal changes the line shape of spectral lines by reducing intensity at the line center while increasing it at both ends. This happens when sodium spectral emission is absorbed by cooler sodium vapor away from the center of the arc tube. Due to spectral resonance, photons with wavelength at the line center have a higher absorption probability and are thus preferentially absorbed, leaving the radiation enriched in photons at shorter and longer wavelengths. This results in an apparent broadening, which further contributes to the rich spectrum displayed by high-pressure sodium lamps.

Other advantages of HPS lamps include their longer lifetime of 28,000–30,000 hours, and a shorter re-strike time than for either mercury vapor or metal halide lamps. In case of a brown-out, i.e. the power failing for up to a few seconds, the gas composition inside a high-pressure sodium arc tube does not change appreciably. This enables the arc to strike instantaneously as soon as the power is restored. These advantages, together with their smaller physical size, efficacies of well over 100 lumens per watt, and the ability to work at temperatures down to −40°C, have led to their widespread adoption throughout the world. GE now markets an entire family of high-pressure sodium vapor lamps called GE Lucalox lamps. A typical lamp is shown here in Figure 3.81. High-pressure sodium lamps are also produced by a number of other manufacturers, including Osram and Philips.

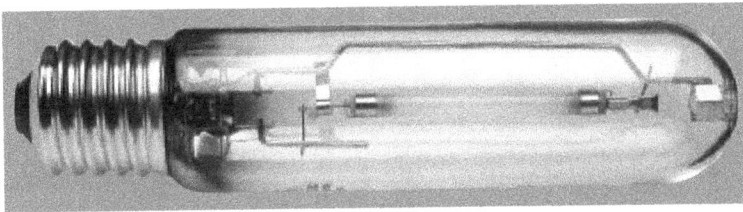

Fig. 3.81.   GE Lucalox lamp.

Courtesy: General Electric Corporation.

3.12.4 *Characteristics of sodium vapor lamp light*

While low-pressure sodium vapor lamps are a great invention, given their outstanding luminous efficacy, lifetime and lumens maintenance, their light is nearly monochromatic, making them unsuitable for general illumination applications. As such these have only found widespread application in the lighting of roadways and open public spaces, where their energy efficiency and low maintenance requirements are highly advantageous. The monochromatic character of sodium vapor light means that objects appear in black-and-yellow tones. Yellows appear bright while all the other colors appear very dark. This can be tolerated for the illumination of highways and large public spaces, but otherwise the low-pressure sodium vapor lamp is the worst source for distinguishing between colors of illuminated objects. For illuminating roads, however, these lamps are in fact better than other alternatives. This is because of the high contrasts produced by monochromatic sodium lamp illumination, as well as the fact that their light penetrates mist and fog very well (see Figure 3.82). In some European countries, such as Britain, road signs and car number

Fig. 3.82.   Outdoor illumination with sodium vapor lamps.

plates utilize a black-on-yellow scheme to take full advantage of the greater visual acuity produced by this type of illumination. In several countries, including Japan, where very long road tunnels are common, sodium vapor lamps are the preferred illumination source inside tunnels because there is evidence to suggest that their illumination produces less fatigue in drivers than that from other types of light sources, such as mercury vapor lamps. Yet another benefit of these lamps is the absence of glare. As the light comes from a glow discharge in sodium vapor instead of an arc discharge so the luminosity arises from the entire extended volume of sodium vapor. This reduces the surface brightness of the source, making it glare-free, and reducing shadows. For wider spectral coverage, high-pressure sodium vapor lamps are used. These find widespread use in manufacturing facilities, high bay areas, and other similar commercial buildings. Before the emergence of LEDs, high-pressure sodium lamps were also the preferred choice for indoor horticultural farms.

3.13 Fluorescent Lamps

Lamps that combine radiation generation through electric discharge with wavelength conversion using luminescent materials are some of the most common lighting devices in use today. From tubelights to CFLs, lamps based on wavelength conversion are in widespread use in our homes, offices and public spaces. Fluorescent lamps have a long genesis; appearing in primitive forms during the early years of the 19th century, before being seriously developed during the 1930s and 40s. Luminescent materials, usually called phosphors, that absorb light of one color and emit light of another color have been known for nearly 200 years. Initially, such materials were used for various decorative applications; often in combination with radioactive minerals as a source of luminescence excitation. Development of gas discharge lamps later provided another source for energizing luminescent materials. As we saw earlier, UV radiation emission from discharges in mercury vapor was an early observation. UV radiation is, of course, invisible but using suitable luminescent coatings it can be converted to visible light. This thought must have occurred to several people, and, thus, there is no single person responsible for the initial development of fluorescent lighting devices. As early as 1859, the French physicist Edmond Becquerel — father of Henri Becquerel, the discoverer of radioactivity — tried to use Geißler tubes filled with fluorescent materials in order to get a practical light source. His trials were not successful, as the devices did not produce a good amount of light. Later in 1896, a year after the discovery of X-rays by Wilhelm Conrad Röntgen, Thomas Edison made an X-ray lamp, internally coated with calcium tungstate which radiated a bluish white light. This source was three times as efficient as carbon filament lamps, but it proved to be un-practical because of the severe dangers posed by X-ray exposure.

The first commercial application of fluorescent coatings in lighting was for expanding the color palette available for neon advertising signs. The number of colors that could be obtained with various noble gas mixtures was limited, so developers turned to using tubes containing mercury vapor and coated with various fluorescent materials. This expanded the number of colors available for making lighted signs. The first phosphors used in early colored neon lamps were long-known naturally occurring minerals, such as zinc silicate (willemite) for green, cadmium silicate for pink, and calcium tungstate for blue. As mentioned earlier, Jacques Risler received a French patent for fluorescent neon tubes in 1926. While only a few colors were initially available to sign designers, after the Second World War (1939–1945) phosphor materials were researched intensively for use in color televisions. About two dozen colors were available to neon sign designers in the 1960s, and today there are nearly 100 available. Thus, fluorescent coatings saw their first widespread application in televisions, rather than in light generation for space lighting. In the present time, of course, fluorescent coatings are chiefly used for visible light production through the conversion of short wavelength UV radiation from gas discharges to longer wavelength visible light. There are three main types of fluorescent lamps: medium and high-pressure phosphor-coated mercury vapor lamps, fluorescent linear tubelights and CFLs. The first of these is generally considered just a mercury vapor lamp, whereas the other two are what people usually mean when they mention fluorescent lamps. However, fluorescent mercury vapor lamps came first and, thus, this is where we start our discussion.

Low-pressure, and even medium pressure, mercury vapor lamps generate large quantities of UV radiation. Visible radiation can be obtained by coating the inside surface of such lamps with an appropriate phosphor. UV radiation can then be down-converted to longer wavelength visible radiation. Obviously, suitable phosphors need to be developed for this purpose. From a historical point of view, the first application of phosphors with mercury vapor lamps was not so much to generate light as it was to supplement the bluish light of early mercury vapor lamps with a red color component. This way, the spectral makeup of light could be balanced to create emission that had a more neutral white appearance. A few years after the introduction of the Osira lamp series, so-called color-corrected lamps with fluorescent coatings began to appear. These were based on phosphors that had been known for some time, but also utilized some newly developed phosphor compositions. In 1934, cadmium sulphide was found to be a suitable fluorescent material, although it provided only a mild color correction. The introduction of true color-corrected mercury lamps was only made possible with the development of manganese-activated magnesium germanate and fluoro-germanate in 1950. This greatly improved the color quality, and also had a beneficial effect on the lamp's efficacy. Three years later, tin-activated orthophosphate was introduced, which further improved the color balance. In later years, the outstanding efficiency of rare-earth doped phosphors came to be recognized, and they began to see widespread use in high quality fluorescent mercury vapor lamps. In the year 1967, the hugely successful europium-activated vanadate and phospho-vanadate phosphors, originally developed for use in color televisions, were introduced and these phosphors are still in use today.

3.13.1 *Early fluorescent lamps*

The first purpose-built fluorescent lamp was developed in 1934 by GEC in Britain. It was a long tubular device filled with mercury vapor at low-pressure, and had heated electrodes. The arc generated relatively long wavelength UV radiation (in the UVA band), which was down-converted to visible light by a fluorescent coating, covering the inner wall of the tube. This development was motivated by a report authored in 1926 by Friedrich Meyer, Hans Spanner and Edmund Germer, from the Osram Company in Berlin, Germany. They had set out to develop a small, easy-to-use mercury vapor discharge lamp for medical and air purification applications. Their paper described a small form-factor low-pressure mercury vapor lamp, fitted with heated oxide-coated electrodes. Most importantly, its design was such that, unlike all previous mercury vapor lamps, this lamp could operate from the AC mains supply. Although, Mayer, Spanner and Germer did not commercially develop their lamp, they did obtain a patent on their invention (US patent: 2,182,732). This patent was later purchased by GE for $180,000, for use as a foundational patent for GE's other subsequent patents on related lamps. Following some of the

Fig. 3.83.   Eugene Lemmer, John Aicher, Richard Thayer and George Inman (left to right) at GE's Nela Park facility.

Courtesy: General Electric Corporation.

design principles set forth in this patent, and adding a fluorescent coating to convert UV radiation to visible light, the fluorescent tube lamp was principally developed at GE's Nela Park, Ohio, lamp-making facility. The development team comprised of Eugene Lemmers, John Aicher, Richard Thayer and George Inman. They appear from left to right, in Figure 3.83. George Inman was the team's leader. This 1934 development served only as a demonstration of the feasibility of this concept and never made it to the market. An improved tubular fluorescent lamp, also developed by the same team at GE, was demonstrated before the Illuminating Engineering Society in Cincinnati, in September 1935. That lamp had a vastly improved phosphor, comprising of a combination of zinc-beryllium silicate and magnesium tungstate, to give off a decent shade of white light. This phosphor was the work of a phosphor research and development team at GE, consisting of Clifton Found and Willard Roberts at Nela Park, together with C. A. Nickel and G. R. Fonda at GE's Schenectady, New York, facility. This lamp can be considered as the progenitor of all fluorescent lamps that came later.

By April 1938, General Electric had introduced fluorescent lamps in the North American market — a development which was soon repeated in Europe as well. The first-generation GE-Mazda F14T12 Fluorescent Tubes, rated at 14 W, were 15 inches long and an-inch-and-a-half in diameter. The lamp employed a two-component phosphor, consisting of a combination of pale yellow-fluorescing zinc beryllium silicate and blue-fluorescing calcium tungstate. By adjusting the ratio of these two phosphors, lamps with different color temperatures could be offered. Due to their low cost and relatively high luminous efficacy of 30–35 lumens/W, these lamps quickly became very popular. By 1939, Westinghouse was also selling fluorescent tubelights. Further improvements followed in the 1940s, when in 1942 A. H. McKeag, from GEC, discovered calcium and strontium-activated halophosphate phosphors. These phosphors had no toxicity problems and were cheaper to manufacture, compared to rare-earth-activated phosphors. Their use enabled the attainment of higher luminous efficacy and even better color quality in new fluorescent tubelights that were introduced from 1946 onward. In later years, linear fluorescent lamps (see Figure 3.84) continued to be improved and their

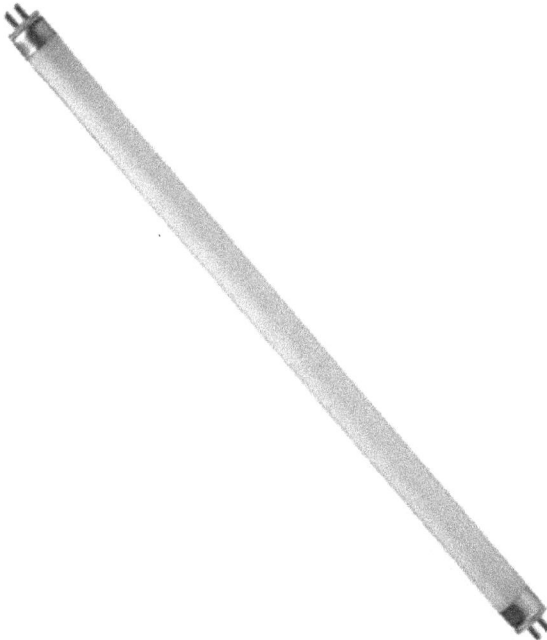

Fig. 3.84.   Linear fluorescent lamp.

popularity peaked during the 1960s. However, during later decades, these lamps gave up a lot of their market share due to a number of factors, such as environmental concerns over their use of hazardous mercury, competition from many other newly-developed lighting technologies and, indeed, the development of yet other types of fluorescent lamps.

3.13.2 *Modern fluorescent lamps*

A tubular fluorescent lamp (see Figure 3.85), informally called a tube-light, consists of a long soda-lime glass tube, that may be anywhere from 9 mm to 40 mm in diameter and 30 cm to more than a meter in length. The most common tubelights are 4 feet in length, and are rated at 40 W. Fitted to each end is a heated electrode that consists of a coiled tungsten wire, coated with low work-function materials, such as calcium, barium and strontium oxides. This coating is applied as a mixture of carbonates which then decompose into the respective oxides when the filament is operated during tube construction. The coil, which is similar to the coiled tungsten filament used in low-wattage incandescent lamps, is usually surrounded by a short cylindrical metal shield. This part, if present, serves to contain evaporated filament material and prevent it from reaching the inner wall of the lamp tube and blackening it. The filament and shield assembly are held by a glass stem which is fused to aluminum metal end caps. The tube is, generally, filled with either argon or a mixture of argon and nitrogen, at low-pressure. It also contains a small amount of mercury (about 5 mg) that contributes mercury vapor for the tube's main discharge. Thus, essentially, this lamp is a hot filament low-pressure mercury vapor lamp that produces long wavelength UV radiation. This radiation is converted to visible light by a fluorescent coating on the inner walls of the tube. In order to apply this coating, finely-milled phosphor is mixed with solvents, nitrocellulose binder and a low melting point glass compound. This mixture is fed through a slowly rotating tube whose inner walls get coated with the 'phosphor lacquer'. Next the tube is passed through an oven which heats it, first to remove all solvent and nitrocellulose, and then to melt the glass compound, forming an adherent coating of fine phosphor dust inside the tube. As it operates directly with AC mains so there is no designated anode and cathode. The electrodes at either end alternate in their role as anode and cathode, as the AC waveform alternates in polarity. The discharge, therefore, switches its direction many times every second which gives a lit tube its uniform appearance. Nearly 50% of the input

Aluminium Cap   **Electrode Coil**   **Argon / Krypton Atoms**   **Glass Tube**

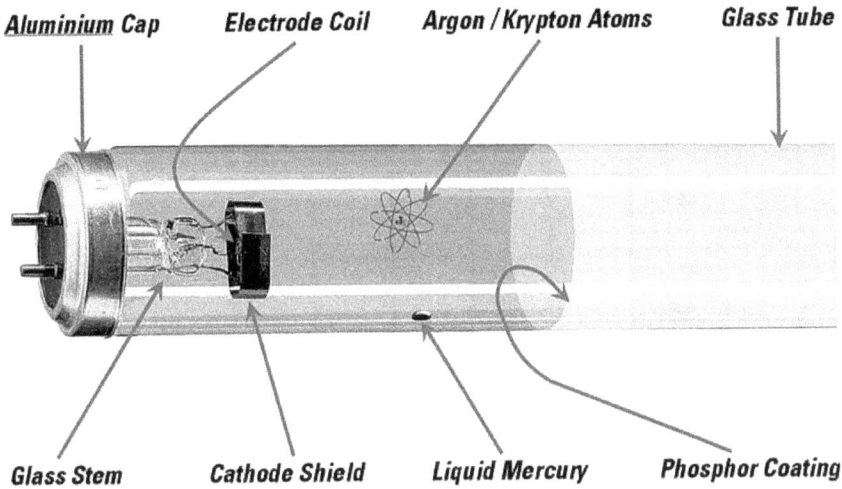

Glass Stem   **Cathode Shield**   **Liquid Mercury**   **Phosphor Coating**

Fig. 3.85.   Diagram of a tubular fluorescent lamp.

electrical energy is converted by the discharge into resonance UV radiation at 254 nm, as discharge-generated electrons hit mercury atoms. This is the main wavelength that excites the lamp phosphor and gets down-converted into visible radiation. Attainment of this high conversion efficiency requires several conditions. First and foremost, the mercury vapor pressure must be at or close to 10 microns (1/76,000 atm). This very low pressure is reached at a bulb-wall temperature of about 40°C (105 F), as seen in the plot here in Figure 3.86. This is far below the hot bulb wall temperatures of an incandescent or high-pressure mercury vapor lamp. Departures from this optimum value cause serious losses. This is easily noted in the dim light output from fluorescent lamps exposed to cold weather outdoors. Second, the lamp must be a positive-column design; i.e. one where the arc length is at least several times the bulb diameter. This minimizes the percentage of end losses at the electrodes. Third, the current-density in the discharge must be low; for example, about 1/4 ampere in a 1″ diameter bulb. All these requirements dictate the preference for a slim, high-voltage, low-current, low-wattage, cool bulb design — as the fluorescent tubelights actually are.

Fluorescent tubelights work straight off the mains voltage without any step-up transformer to boost voltage. This would, normally, preclude starting an arc in long-arc discharge tubes, were it not for a clever arrangement of some simple circuit elements connected to such a tube. This circuit is seen in Figure 3.87. The AC line passes through an inductor

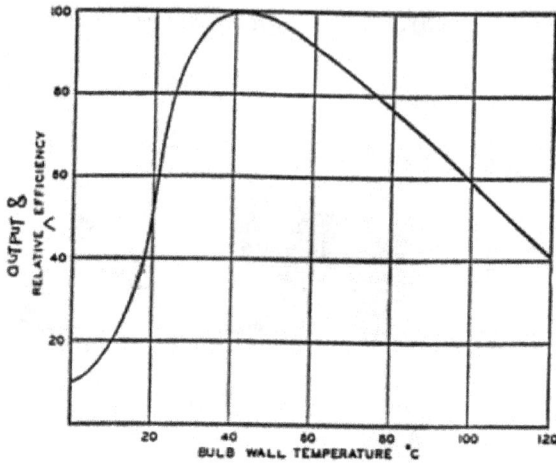

Fig. 3.86.   Relative efficiency versus tube wall temperature for a fluorescent lamp.

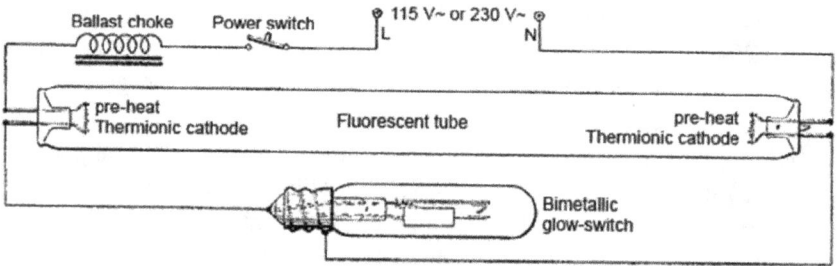

Fig. 3.87.   Electrical circuit used for starting tubular fluorescent lamps.

(choke) before running through one of the tube filaments. It then passes through a thermal bi-metallic switch (starter), heated by a neon glow discharge lamp. The circuit is then completed by going through the other tube filament and returning to the mains supply. This arrangement, dating back to the 1930s, is the oldest, simplest, and the most widely used starter circuit for conventional fluorescent tubelights. Effectively, it enables pre-heating the discharge tube to increase ion density and also applies high voltage starting pulses to initiate the discharge. The choke serves to both limit the current being fed into the discharge (hence its name), and to work with the starter to generate high voltage pulses to fire the tube. As its name implies, the starter only helps start the discharge going, and serves no purpose once the tube gets lit.

When power is first applied to the circuit by closing the switch, current flows through the choke, the two filaments, and the starter switch. The filaments heat up and begin to glow a bright red color. Thermionically-ejected electrons are injected into the tube from both ends, rapidly ionizing the argon/mercury vapor mixture inside. At the same time, a small neon glow lamp inside the starter warms up enough to trip a bi-metallic switch, opening the circuit and stopping the flow of current. A consequence of this is a rapid collapse of the magnetic field present in the choke. This, in turn, generates a high inductive kick-back voltage which appears across the tube, and attempts to initiate a long-arc discharge. If this fails at the first attempt then the cycle repeats again, as is often witnessed with older tubelights that flash a few times before turning on. More modern tubelights now employ electronic ballasts, and dispense with the starter and the large conventional choke. Once the arc is stable and the tube lit, current flows through the arc discharge, bypassing the starter entirely, which then serves no further purpose.

Fluorescent tubelights have both advantages and disadvantages, when compared to incandescent lamps. Their desirable attributes include their comparatively high efficacy and, thus, relatively low power consumption, as well as the fact that emission from extended area results in glare-free light and freedom from shadows. At the same time, tubelights also have some prominent drawbacks. Principal among these is their use of mercury, which is harmful to humans. Their extended forms also make it difficult to incorporate them in confined spaces, and to efficiently collect and direct their light. Despite these dis-advantages, tubelights and their more modern variants have been widely used for decades. Their continued survival has been driven by new developments in lamp and electronics miniaturization, together with advances in phosphor technology, so that improved fluorescent lamps continually keep coming to the market.

3.13.3 *Cold cathode fluorescent lamps*

A different type of fluorescent tube also finds specialist use for display backlighting in LCD panels. These slim tubes have lengths much longer than their diameter (see Figure 3.88), and are easily incorporated into the thin and flat backlights of LCD panels. Their operation is quite different from the hot cathode fluorescent tubes described above. LCD backlight

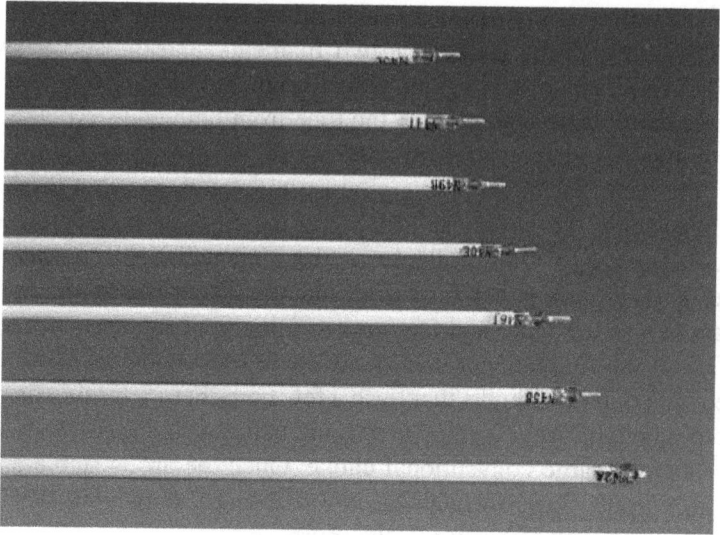

Fig. 3.88.   Cold cathode fluorescent lamps.

tubes are cold cathode devices and are, therefore, called cold cathode fluorescent lamps (CCFLs). They contain a similar fill of argon gas and mercury, together with a white fluorescent coating on the tubes' inner wall. Two metallic electrodes are fitted to each end of the tube. Using a special electronic inverter, a high DC voltage, in the range of 5–10 kV is applied to the electrodes to strike the arc. Once the arc strikes, the voltage automatically decreases to a lower value to sustain the arc. CCFL tubes emit a neutral white light, which is captured by a surrounding elliptical reflector and directed end-wise into a flat Perspex sheet that acts as a wide-area optical waveguide. Light propagating through the sheet is extracted from its surface by an embossed pattern of small dimples. This pattern is arranged in a manner such that the light extraction efficiency is lower closer to the CCFL lamp where more light is available and it progressively increases on moving away from the lamp. In this way, a uniformly lit backlight surface is created. This type of backlight is usually called an edge-lit backlight. Only one CCFL is used for small screen sizes, such as those used for computer displays, whereas larger screens utilize backlights with two CCFLs — one at each end of a flat light guide. Even larger area displays, such as larger LCD televisions usually employ several CCFL tubes, placed directly behind the backlight, whose

Fig. 3.89. LCD TV backlight, showing CCFL tubes.

light is diffused by a diffuser screen to create a uniformly-illuminated backlight, as seen in Figure 3.89. More modern LCD panels, both small and large, now use LED-based backlights which will be described later in Chapter 5.

3.13.4 *Nonlinear fluorescent lamps*

While linear fluorescent tubelights are very useful, as pointed out above, their extended shape can be a problem in certain applications. It is particularly difficult to concentrate their light into narrow beams, for example, and they also tend to take up a lot of space. To get around these issues, nonlinear versions have been developed by bending the glass tubes into various shapes. The three most common configurations are circular, U-shaped and W-shaped. The credit for the development of the first nonlinear tubelight goes to GE. The company developed the first circular tubelight in 1945. Called Circline Fluorescent, it was made available in various wattages up to 40 W. Within a few years of its development, it had also been adopted by European manufacturers, and Philips came out with its own line of circular fluorescent tubes. Later, it also started being manufactured in Japan, where Toshiba has been a leading manufacturer of circular fluorescent lamps.

Fig. 3.90.   Several styles of nonlinear fluorescent lamps. Philips N.V.

A few years after the first appearance of circular fluorescent tubes, Osram in Germany introduced a U-shaped fluorescent lamp. This went on the market in early 1950. Other manufacturers in Germany and other countries followed suit, and by the 1960s U-shaped fluorescent tubes were quite widely available. These lamps are still very popular and can be easily obtained from many retail outlets. All U-shaped luminaires have a distinct metal cross-bar connecting their two metal end caps, as can be seen in the figure here. Most such tubes have a double-layer phosphor coating to improve their light quality.

A further variant of nonlinear fluorescent tubes was brought out by Philips around the mid-1960s, with their TL-W series W-shaped tubes. The main advantage of this design was that several of these lamps could be used to illuminate large area signage panels with uniform illumination. Both U- and W-shaped tubes are made from heavy lead-alkali glass which is easier to work for making tight bends. Figure 3.90 shows the various common types of nonlinear fluorescent lamps.

3.13.5 *Compact fluorescent lamps*

While long-form fluorescent tubelights have existed since the 1930s, their more modern incarnation — the CFL only appeared first in the 1970s. Their development was an off-shoot of the energy crisis of the 1970s, in particular the oil embargo of 1973–1974, that made it expedient to come up with a significantly more efficient lamp than even the best available incandescent luminaires. By the 1970s, mercury vapor lamps had developed into several forms, including both linear and nonlinear fluorescent lamps, which were clearly more efficient, compared to even tungsten halogen lamps. Fluorescent lamp technology was, thus, obviously the way to go in order to achieve still higher energy efficiencies. However, until the fluorescent lamp could be miniaturized, it could not effectively replace incandescent bulbs. Edward Hammer of GE (see Figure 3.91), based at Nela Park in Cleveland, Ohio, is credited with this invention. Hammer was a lamp engineer in the Fluorescent Engineering Department of GE at the Lighting Business Headquarters at Nela Park in East Cleveland, Ohio. Early in his career, Hammer worked directly under Richard Thayer (1907–1992), who was a pioneer in fluorescent lamp development, as well as alongside Eugene Lemmers (1907–1992) and John Aicher (1912–1993), who also were fluorescent lamp pioneers. Hammer became interested in developing a fluorescent lamp replacement of incandescent lamps that could be retrofitted into existing incandescent lamp sockets. Such a lamp would have a very similar size and form factor as tungsten filament lamps. This was a tough problem, because until then low-pressure fluorescent lamps had only existed in rather large linear or nonlinear configurations. The task of miniaturizing the fluorescent tube while maintaining sufficient brightness required coiling a long narrow tube into a compact geometry. Hammer succeeded in creating a working prototype with help from technicians in the glass shop, but it was apparent that it was more of a concept demonstrator than a practical device. This was due to the considerable difficulty in producing the spiral tube form by hand. Thus, although he originally developed a compact version of the tubular fluorescent lamp in 1976, he did not apply for a patent on it until several years later. A big reason for this was reluctance on the part of GE management, who did not want to invest the estimated $28 million needed to build an automated production facility to manufacture the new lamp. It is possible that the lamp design, which had a special helical structure to combat reflective losses, may have been copied by other competitors as the unpatented lamp sat in full view in Hammer's office. In any case, his design rapidly spread to other companies too, but successful manufacture of helical CFLs had to await the development of appropriate machinery for shaping glass tubing into the required helical shape. This did not happen until well into the 1980s.

The modern form of the lamp comes with a miniaturized electronic power supply in the lamp base, which enables direct operation from 120 to 240 V mains AC supplies. A typical circuit for such a power supply is seen here in Figure 3.92. The resonant switch-mode power supply provides a 40 kHz AC waveform to the fluorescent lamp. To generate this drive, mains AC is first rectified to DC, filtered with filter capacitors, and

Fig. 3.91.   Edward Hammer.

Courtesy: General Electric Corporation.

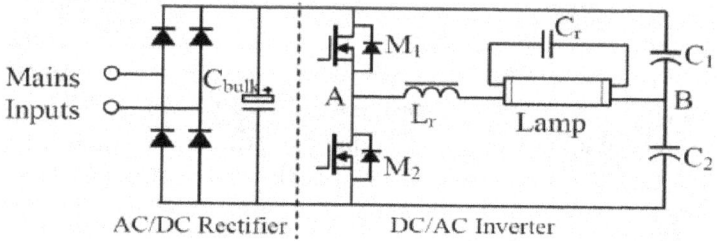

Fig. 3.92.   Embedded power supply, typically used with CFLs.

then transformed to a sinusoidal high-frequency AC voltage with a transistorized DC-to-AC inverter. This voltage then operates the discharge without any visible flicker — a big improvement over the older magnetic ballast-based supplies that were heavy, bulky, expensive, prone to audible hum, and often caused start-up problems and lamp flicker. Thanks to modern electronic power supplies, modern CFLs are instantly-on devices that come on within a fraction of a second of power being applied. They do, however, take a couple of seconds (or more at low temperatures) to reach their full brightness. A characteristic feature of the use of resonant AC-DC inverters for powering CFLs is that they tend to keep lamps operating at a constant brightness, even though the mains voltage may fluctuate over a wide range. Consequently, it is very difficult to dim CFLs using ordinary resistive or triac-based dimmers, as are used with incandescent lamps. To address this issue, special dimmable CFLs have now become available. These lamps use extra circuitry to enable dimming by conventional lamp dimmers.

Modern CFL bulbs come in several different wattages and package styles. Both screw and bayonet bases are available, and both exposed spiral as well as shrouded styles are on the market (see Figure 3.93). Furthermore, consumers have a choice of available color temperatures, for light ranging in shade from cool bluish white to warm yellowish white. With efficacies ranging from 60 to 72 lm/w, versus 8 to 17 lm/w for incandescent lamps, CFLs have a large energy efficiency advantage over tungsten filament lamps. Their longer lifespan also makes them very attractive, in comparison with incandescent light bulbs, which offsets their somewhat higher purchase cost. While ordinary tungsten filament lamps operate for anywhere between 500 and 2000 hours, typical CFLs operate for 5000–15,000 hours of continuous use. Combine this with their very respectable efficacy figures, and it is no wonder that CFLs have rapidly

Fig. 3.93.   Various forms of modern CFLs.

caught on in the developing world where they have been replacing both incandescent bulbs and tubelights. In richer, more developed countries, their use is now being gradually discouraged out of environmental concerns related to recycling mercury-containing products. In the US and the EU, the retail price of CFLs includes a small surcharge to pay for their eventual re-cycling. Importers and sellers of CFLs are under legal obligation to re-cycle used CFL luminaires, so that the small amount of mercury inside each lamp does not contaminate the environment.

3.13.6 *Black light lamps*

Another class of specialized fluorescent tube is one that is used to produce long wavelength UV radiation that falls in the UVA band. These so-called black light lamps (also called Wood's lamp after Robert Williams Wood) are of two types: those with a fluorescent UV-emitting coating as well as a visible light-blocking filter, and those with a fluorescent coating but no filter. All black light lamps are essentially low-pressure argon/mercury-filled discharge lamps, driven from a compact high voltage power supply. Their phosphor coating consists of a suitable UV-emitting phosphor such as strontium borate or strontium fluoroborate doped with europium. The

Fig. 3.94.   Fluorescent UV lamp.

peak emission of these phosphors is around 370 nm. Another common phosphor is barium silicate doped with europium, with peak emission at 352 nm. Filtered fluorescent UV lamps use a dark violet filter to filter out most visible light, and have the designation 'BLB' in their type number; standing for black light blue (see Figure 3.94). Due to the filter, these lamps have a distinct dark violet, almost black, appearance (see Figure 3.95). During operation they emit a small amount of violet light while the bulk of their emission is in the invisible near-UV region, with the exact spectral makeup determined by the particular phosphor used in the lamp. Such lamps are widely used as sources of long wavelength UV

Fig. 3.95.   Filtered fluorescent UV lamps.

Fig. 3.96.   Unfiltered fluorescent UV lamps in an insect zapping unit.

radiation for mineral prospecting, forgery detection and other similar applications. Additional uses include, lighting in discos, sun tanning booths, UV-based material polymerization (curing), and medical uses. Unfiltered UV lamps do not have a visible light-blocking filter but use similar UVA phosphors. These lamps have a clear appearance and produce both UV radiation and visible blue light. UV-generating lamps of this type are used in insect zapping equipment (see Figure 3.96), as well as in air and water purifiers.

4 Plasma Light Sources

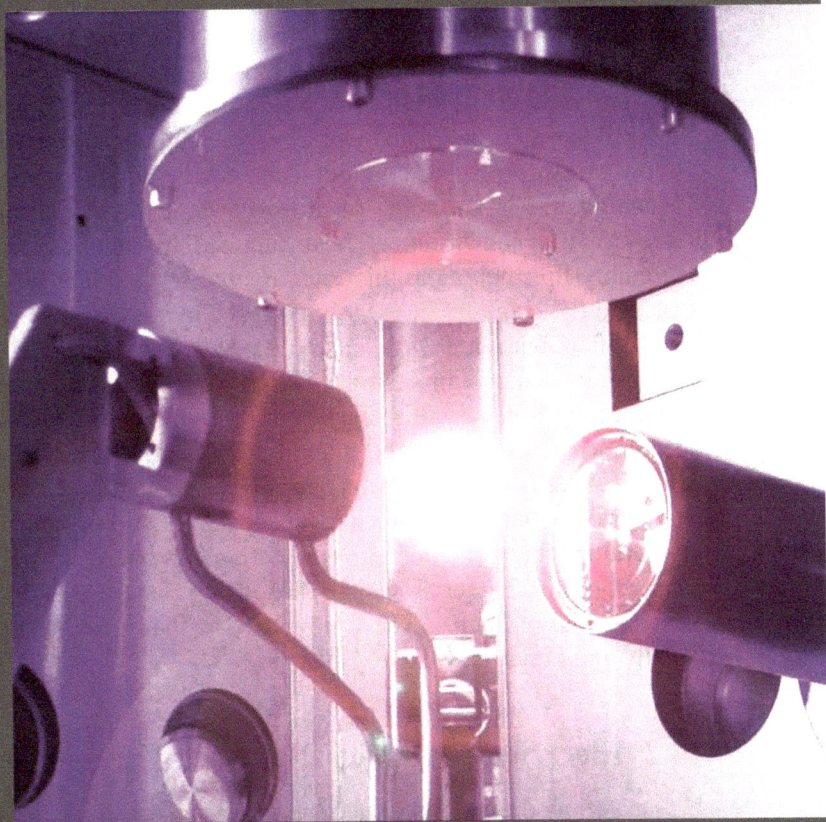

4.1 Introduction

The previous chapter dealt with electrical discharge light sources that produce radiation from discharge-based plasmas. The lamps surveyed there operate with plasmas created by passing an electric current directly through a gas or gas mixture. All such lamps, invariably, contain internal metallic electrodes to pass current directly through the gaseous environment inside the discharge tube. Plasmas in these lamps are characterized by relatively low ionization density, as well as low electron and ion energies. Due to the low kinetic energies of electrons and ions, such plasmas are termed 'cold'. This chapter examines plasma lamps where the gas is ionized in ways other than by passing an electrical current directly through a low-pressure gaseous ambient using conducting electrodes. This can be accomplished in a number of different ways, such as through radio frequency (RF) induction, or heat supplied by combustion or radiation. Ordinary fires belong to this category, but so do a number of highly specialized light sources of great importance to modern science and technology. Often, but not always, such plasmas are characterized by significant ionization density. In contrast to lamps seen in the previous chapter, sources described in this chapter do not employ phosphor-aided wavelength up-conversion, rather they rely on the direct use of emitted radiation. Indirectly excited plasma lamps have been around for many decades; first as potential light sources for space illumination, and later as radiation generators for technological applications. We'll look at both types of lamps in this chapter. The first part of this chapter examines electrically excited plasma lamps. The following part then goes on to discuss thermally and radiatively excited systems.

4.2 Electrically Driven Plasma Light Sources

A gas at low pressure is turned into a plasma when a large fraction of its atoms have become ionized. This state consists of a mixture of neutral atoms together with electrons and positive ions of various charge states. Plasma formation requires substantial energy input, which scales with the volume of the gas. This energy can be provided by various means, such as electromagnetic induction, direct thermal heating, radiative energy transfer using intense laser beams, etc. We first look at light sources where electrical energy is directly used to create and sustain a light-emitting plasma.

4.2.1 *RF-excited induction lamps*

An electrical discharge can be excited in a low-pressure gas mixture by inducing electromagnetic fields from outside of the low-pressure gas container. This is the basic operating principle behind induction lamps — also called electrodeless lamps. Field induction generally requires operation at high frequencies in the several MHz to hundreds of MHz region. Higher frequencies facilitate energy transfer through electromagnetic induction. However, this phenomenon can be easily observed even at the low frequencies, typical of AC mains lines. If an aluminum foil is stuck along the length of a fluorescent tube and connected to the AC mains live terminal, with the neutral connected to any of the pins at the end of the tube, then the entire length of the tube along the metal stripe is seen to glow with a dull white light. This process can be made more efficient by increasing the line frequency and by employing resonant induction structures to ease the transfer of energy inside the lamp envelope. Following the principle of electrical induction, energy can be transferred both capacitively and inductively. The latter is used more frequently, in practice, because it lends itself to less obtrusive and more compact excitation structures. RF inductive energy transfers are used not only for lighting applications but also in heating systems and for energizing chemical reactions in plasma-enhanced chemical vapor deposition (PECVD) and atomic layer deposition (ALD) reactors. All such systems operate at hundreds to thousands of MHz in frequency, to keep the energy transfer efficiency high. The use of high frequencies, however, necessitates very careful design of the RF parts of the system so that effects like unwanted RF leakage and power reflections are avoided.

4.2.1.1 *Tesla lamps*

Historically, the first widely seen demonstrations of induction lighting were carried out by the Serbian-American engineer and inventor Nikola Tesla (see Figure 4.1) in New York, in the 1890s. Electricity was a relatively new source of power at that time, and electrical lighting of any kind was sure to attract much interest. Tesla's demonstrations were performed more in the spirit of showmanship than with a view to promote a new and viable form of lighting. As is well known, Tesla was a great proponent of alternating current technology and, if nothing, inductive lighting must have been seen by him as a way to promote alternating over direct current

Fig. 4.1.   Nikola Tesla with his induction lamp.

systems then being endorsed by Thomas Edison. In his demonstrations, he made use of modified Tesla coils to generate large quantities of high-frequency alternating current, which was passed through a coil loosely wound around a gas-filled tube. It is said that Tesla had a wire — connected to a Tesla coil — strung around the perimeter of his New York lab. Wherever he needed light he only had to hang a gas tube in the vicinity of this high-frequency conductor. This simple arrangement was enough to excite the gas inside the tube and make it emit a bright light. But at the same time, it was clear that it was also an impractical arrangement because of the need for large Tesla coils to drive such lamps. In spite of this, it was clear that such lamps themselves had higher efficiency, better spectral output, and virtually unlimited lifetime, because of the absence of any filaments or electrodes. It is interesting to note that the perceived longevity of Tesla induction lamps also obstructed their commercialization during his lifetime. J. P. Morgan, an influential banker and financier pitted Tesla against Edison, in a famous competition, to develop the best possible electric lamp. Edison and Tesla were strong backers of their respective inventions: incandescent filament lamp and induction lamp, respectively. Of the two, Morgan chose to back Edison's lamp because he reasoned that Edison's bulbs will burn out quickly, so people would have to buy numerous replacements. Tesla's induction lamp, on the other hand, would last for half of a customer's lifetime with normal usage, greatly diminishing profits to be had from repeatedly selling bulbs to every consumer. Thus, Edison got financial backing for his lamp venture whereas Tesla's invention did not catch much attention beyond scientific curiosity. Tesla was a prolific and visionary inventor. He even had the bold and futuristic idea to illuminate the sky itself by beaming high-frequency electrical power toward the sky, where it would excite rarefied gas at high altitudes; causing it to glow and, thus, provide illumination over a very wide area. This, of course, is naturally accomplished in auroras, but with the excitation coming from energetic solar wind particles streaming outward from the sun. This proposal for creating artificial auroras illustrates his penchant for unconventional thinking. Tesla's demonstrations of inductively powered electric lighting clearly had merit but their commercial implementation had to wait many years for RF technology to develop to a sufficient extent so that large Tesla coils could be replaced by small and efficient RF sources.

During the 1940s and 1950s, some investigations were carried out by both academic and industrial laboratories to develop practical induction lamps. The main impediment was the need for high-power RF sources to drive them. In spite of the advancements made in RF technology during the Second World War, and in post-war years, hardware needed to generate decent amounts of high-frequency power was still sophisticated and expensive. Thus, the enthusiasm for induction lighting did not catch on during those years. Much later, during the 1980s interest in electrodeless lamps was re-kindled with the development of efficient solid-state RF sources. Some prototype RF induction lamps were developed in both Europe and the US, but they still suffered from high costs and somewhat suspect long-term reliability. Additionally, due to the increasing proliferation of wireless equipment in all walks of life, RF-based equipment was beginning to attract increasing scrutiny from the likes of US Federal Communications Commission (FCC). They had begun to regulate the use of both RF-emitting and RF-receiving equipment. Induction lamps inevitably radiate away some of the RF energy that they generate for lamp operation, and, thus, they fell under the purview of FCC regulations. RF generating systems had to be rigorously tested and licensed by the FCC to ensure that they did not cause undesirable interference with other RF systems in the vicinity. These additional regulatory complications also initially dissuaded lighting companies from developing induction lamps.

By the late 1970s and the early 1980s, induction lamp technology had a renaissance. There were several reasons for that, in addition to the long-standing recognition that induction lighting was much more energy efficient when compared with incandescent lighting. RF technology had, by then, matured so that small integrated RF oscillators and ballasts had become available. There was a need to diversify and energize the market with the introduction of new kinds of lighting systems, and there was also a better understanding of FCC regulations. Sylvania in the US, as well as Thorn and Philips in Europe, developed near-commercial prototypes of electrodeless lamps for space illumination. However, the technology remained expensive and this did not allow any significant market penetration until a further decade had passed.

4.2.1.2 *Fusion UV lamps*

The first commercially-successful electrodeless induction lamp was a UV-emitting source developed by Fusion UV Systems in the US in the early 1970s. This compact microwave-powered UV lamp became very popular in spite of its higher cost compared to the usual electrode-equipped UV lamps. Its stated advantages included frequent on/off operations that had no adverse effect on UV output or bulb lifetime. The bulb (which was designed to be integrated with an elliptical reflector) was also seen to maintain spectral consistency throughout its life. Several types of bulbs, differing in power output and spectral content were developed. These bulbs belonged to different series, such as F400, F600, etc. A typical Fusion UV bulb is seen here. It has a center-pinched tubular design with mounting stems at both ends (there are no electrodes). The quartz envelope (see Figure 4.2) has a characteristic filter coating to control the spectral composition of emitted radiation. Fusion UV bulbs are used within special lamp assemblies that are fed power from a separate external RF power source (see Figure 4.3). These systems are now marketed by Heraeus GmbH who acquired Fusion UV (through Spectris plc who had earlier purchased Fusion UV) in December 2012. Fusion UV-based

Fig. 4.2.   Quartz bulb of a Fusion UV lamp.
Courtesy: Heraeus Holding.

Fig. 4.3.   Power supply units for powering Fusion UV lamps.
Courtesy: Heraeus Holding.

systems are mainly intended for the UV curing and materials processing markets.

4.2.1.3 *Sulfur lamps*

While looking for a better source of ultraviolet light, scientists at Fusion Systems Inc. (parent company of Fusion UV Systems mentioned above) found that ordinary sulfur, inside a microwave-powered lamp, became the world's most efficient source of visible light. This discovery was immediately patented by the company. A spin-off company, Fusion Lighting, was entrusted with the commercial development of this technology. A commercial model called Solar 1000 was developed in 1994, followed by an improved model called Light Drive 1000 in 1997. Due to a number of technical reasons, these lamps failed to find commercial success, and, thus, their production was discontinued by the end of 1998. Later, Chinese, German and South Korean companies brought out their own sulfur lamps, based on very similar design. These have met with some commercial success, so that now there is a decent market for professional sulfur lamps.

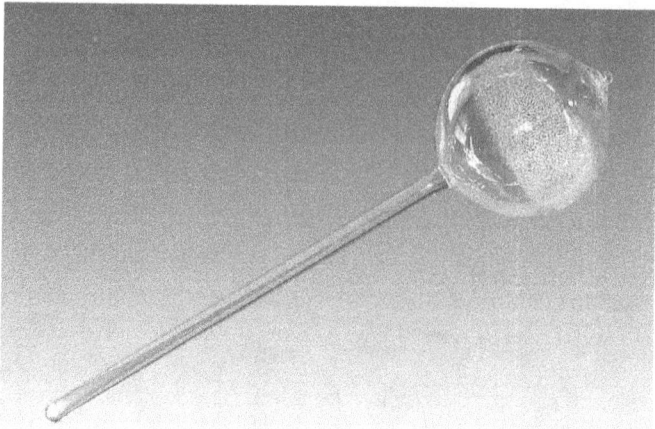

Fig. 4.4.   Stemmed quartz bulb of a sulfur lamp.

All microwave-powered sulfur lamps have followed the same design principles. This basic design was conceived of by Michael Ury, Charles Wood, James Dolan and their colleagues in the early 1990s, and later commercialized by Fusion Systems, with backing from the US Department of Energy (DOE). The lamp bulb consisted of a round golf ball-size (about 3 cm diameter) hollow fused quartz globe connected to a long quartz stem (see Figure 4.4). The bulb was filled with a few milligrams of powdered sulfur and also contained argon at about atmospheric pressure. There was no mercury present in a sulfur lamp bulb. A cylindrical metal mesh cage surrounded the bulb and was connected to a rectangular waveguide which, in turn, was fed power from a magnetron operating at 2.45 GHz. When microwave power was inductively coupled to the inert gas inside the bulb, it ionized rapidly and began to glow with a dull whitish light. Within 10 to 20 s, the temperature of the bulb rose enough to melt and vaporize all the sulfur present in the bulb. Sulfur plasma formed, which then began to emit an intense white light with a very slight greenish tinge. The one major complication with this arrangement is that because of the high heat dissipation in the bulb and the need to keep the plasma uniformly irradiated with microwaves, it had to be rapidly rotated at around 600 RPM by using the quartz stem as a spindle connected to a motor. This system can be seen in Figure 4.5 here. The motor used for spinning the bulb was also used to blow air over the magnetron to keep it cool during operation. As efficient magnetrons have only been available with power ratings of around

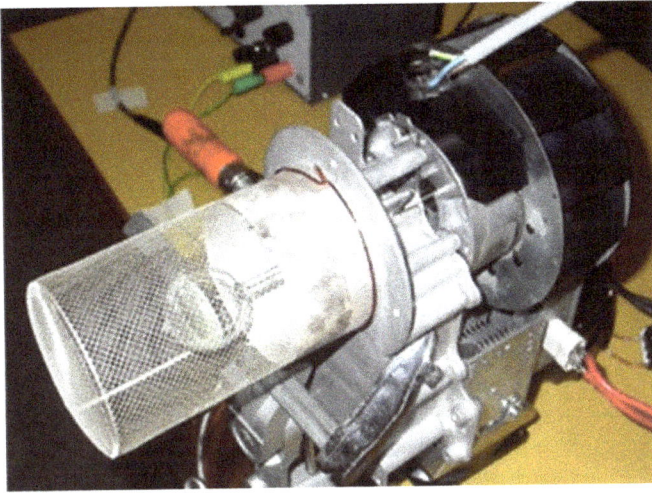

Fig. 4.5. A complete sulfur lamp system with bulb mounted in an inductive cage connected to a microwave power supply unit.

Courtesy: Solaronix SA.

1000 W, so the smallest sulfur lamps have been rated at around a kilowatt in power consumption.

During operation, the bulb runs very hot at temperatures in the range of 800°C to 1000°C with internal pressure of around 5 atm. When the microwave power is switched off, the sulfur plasma is instantaneously extinguished but the bulb keeps glowing with red hot heat for some time, as it slowly cools down. The plasma can be re-struck, if desired, after waiting for about 5 min.

The light from sulfur lamps exhibits a continuous non-Planckian spectrum with a color temperature close to 6000 K, and with very little UV or infrared radiation, as seen in Figure 4.6. This is because it is generated through a process which is neither thermal nor based on atomic transitions. Sulfur exists as S_8 cyclic octatomic molecules at room temperature. In its vapor state, however, sulfur is mainly found as the dimer, S_2. It is these dimers that absorb microwave energy in a sulfur lamp and emit radiation through S_2 molecule's rotational and vibrational transitions. This results in a continuous spectrum which peaks at around 510 nm in the green region. To the human eye, this light appears quite close to natural sunlight and, thus, sulfur lamps have also been used for horticulture applications as indoor grow lights. The spectral distribution matches human

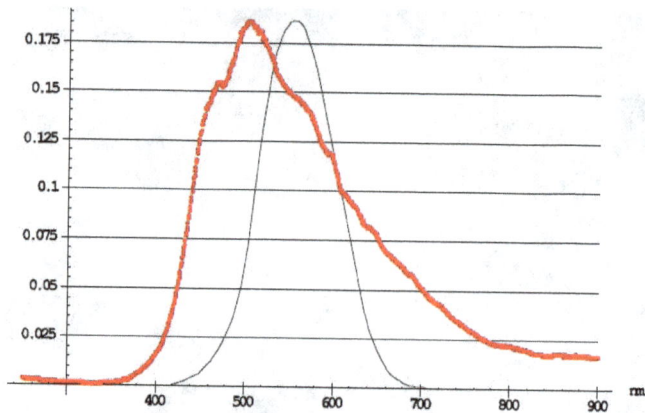

Fig. 4.6.   Spectrum of light emission from a sulfur lamp (red curve). The black curve is the spectral sensitivity of the human eye.

Courtesy: Solaronix SA.

eye's sensitivity curve quite well and this lamp, therefore, benefits from high efficacy figures in excess of 100 lumens per watt. Several additives have been tried with sulfur lamps to further improve its color rendering properties. These additives include calcium bromide ($CaBr_2$), lithium iodide (LiI) and sodium iodide (NaI).

Other advantages of sulfur lamps include freedom from mercury, extremely long bulb life of around 60,000 hours and the ability to be dimmed down to 15% of full brightness, without any significant change in spectrum. The continuous molecular spectrum of sulfur will not be accessible with a conventional electrode-equipped lamp because due to its extreme electronegativity sulfur is fiercely corrosive toward almost all metals. Thus, exposed metal electrodes will be attacked and destroyed very rapidly. Together with a desirable spectrum, sulfur lamps also provide extreme brightness, which makes them suitable for illuminating large areas. A single sulfur lamp mounted at a high location can illuminate a medium-sized warehouse. A few sulfur lamps, suitably placed, can illuminate an entire stadium. In several instances, sulfur lamps have been used with plastic light pipes to distribute light evenly over large areas. This type of installation has been used, for example, at the National Air and Space Museum in Washington DC for many years. Here, three sulfur lamps, used with 89 feet long acrylic light pipes, replaced 97 HID lamps. The horizontal 3M light pipes, in this case, are placed close to the parabolic mirrors used with sulfur lamps and with a magenta filter in between to remove the greenish tinge from the light, and give it a warmer feel. Reflective metal strips, suitably placed along the length of the light pipes, serve to extract light and direct it downward. The appearance is that of very long and bright fluorescent tube lights, placed high up in the exhibits area of the space hall. In a light pipe installation, either a mirror or another sulfur lamp is placed at the other end of the pipe.

Along with advantages, sulfur lamps also exhibit a number of major disadvantages which have limited their widespread commercial adoption. Perhaps the most prominent among these is the need to rotate the bulb at high speeds. Not only it requires additional hardware but the mechanical system required is also prone to breakdowns and, thus, poses very significant reliability issues. The bulb and magnetron require forced air cooling which generates considerable noise and requires air filtration. While the bulb can last for up to 60,000 hours, the magnetron's life expectancy only extends to around 20,000 hours after which it requires replacement. Furthermore, it is difficult to contain the microwave energy from radiating away from the installation, unless stringent shielding is put in place. This can cause serious interference issues with cell phone, WiFi and other wireless communication systems. These factors, together with the size and cost of a complete system, have meant that there have historically been few takers for sulfur lamps. Nevertheless, the fact remains that, due to their brightness and spectral characteristics, sulfur lamps are exceptionally well suited for architectural illumination applications, such as lighting stadia, warehouses, airport grounds, and dockyards.

Sulfur lamps are available from a few specialized manufacturers for lighting and scientific applications. Over the years, several improvements have been suggested to combat their prominent drawbacks. These include better, long-life magnetron designs, use of modulated microwave energy to avoid having to spin the bulb, and even the use of special coated electrodes inside the bulb to remove the need for a magnetron altogether. At present, sulfur lamps enjoy a small market share, where their advantages surpass their cost of ownership and maintenance requirements; thus keeping the technology commercially viable.

4.2.1.4 *RF-powered mercury lamps*

During the 1980s and 1990s, several electrodeless induction lamps were developed and marketed by various lamp manufacturers in Europe, US, China and Japan. These lamps had shapes and form factors similar to incandescent and compact fluorescent lamps, and were intended for both domestic and industrial uses. Their production continued well into the next century until these lamps were mostly replaced by LED lamps. Unlike the sulfur lamps, described above, these electrodeless induction lamps were excited by much lower frequency RF power (a few MHz or even hundreds of KHz), inductively coupled to the lamp gases. Inductive mercury lamps, like sulfur lamps, are available from a few manufacturers for specialized applications.

4.2.1.5 *Standing wave induction lamps*

Completely electrodeless induction lamps that use standing electromagnetic waves in a cavity resonator were developed by Luxim Corporation in the period around 2005 to 2012. These lamps consist of two distinct parts: an RF generator and a ceramic/quartz light source. The RF generator consists of an RF oscillator and an RF power amplifier that together generate several watts of high-frequency RF power. This power is then delivered over low-loss coaxial cable to an inverted U-shaped ceramic 'puck' with a sealed quartz bulb held in the top middle part. The bulb has no electrode structure of any kind inside or outside, and it is filled with a mixture of halogens and other gases at a low pressure.

The operation of the lamp is illustrated in Figure 4.7. Part 1, at the left, shows the generation of high-power RF energy, which is transported over a coaxial cable to the main lamp assembly. Part 2, in the center, shows the excitation of standing electromagnetic waves in the ceramic resonator

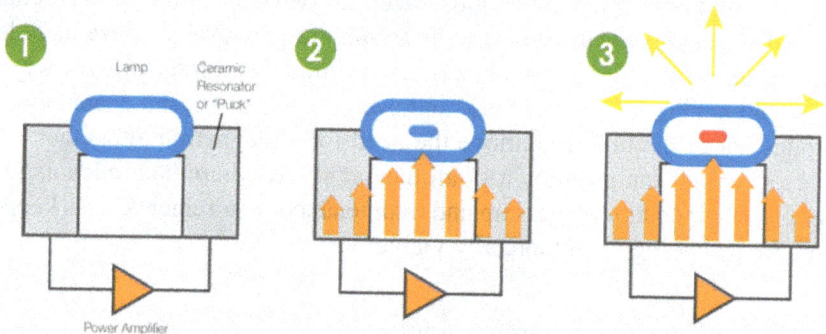

Fig. 4.7.   Schematic diagrams showing the operating principle of RF-powered Luxim inductive lamps.

Courtesy: Luma America Corporation.

with maximum amplitude located at the center. This energy excites gases in the sealed quartz bulb, causing the gas mixture to ionize and begin to glow. Part 3, to the right, shows the gases heating up, evaporating and dissociating metal halides. An intense plasma is formed which radiates light very efficiently.

A reflector placed at the bottom of the bulb ensures that all the light is directed out of the top of the assembly. The choice of the gas and halide fill in the bulb is critical in determining the spectrum of radiated light. By changing the gas composition, different spectral outputs can be tailored. Because there are no electrodes, so the bulb has a very long operating life and, thus, it can be placed in relatively inaccessible locations, with only the RF coaxial cable connecting it to the RF power generator. This lamp was designed for commercial and industrial applications that could benefit from a very bright, wide angle source of good quality white light.

4.2.2 *Light-emitting plasma light sources*

Simple electric discharge-based lamps with appropriate gas filling can be used as light sources. Visible light-emitting discharge lamps have been covered in the previous chapter. Here we look at a UV-emitting lamp technology with a cellular construction somewhat similar to that seen in plasma display panels. In contrast to other UV discharge lamps and LEDs, these lamps have a flat form factor which is useful for a number of applications.

4.2.2.1 *Micro-discharge UV lamps*

Electrically energized plasmas can emit light over a multitude of narrow wavelength regions. This has already been seen earlier in this chapter as well as in the previous chapter. In some cases, the primary emission lines lie in the UV region, and then such discharge lamps can be used as sources of UV radiation. One example is provided by micro-discharge excimer UV lamps, as seen in Figure 4.8, made by Eden Park Corporation.

The device consists of an array of quartz or fused silica cells, filled with an inert gas. Typical thickness is about 3 mm for a 5 cm × 5 cm lamp size. A high frequency (~17 kHz) discharge is passed through the gas to sufficiently energize atoms to form short-lived dimers. On breaking apart into individual atoms, these dimers emit UV radiation. The emission can be at one of several possible wavelengths (see Figure 4.9). Emissions at

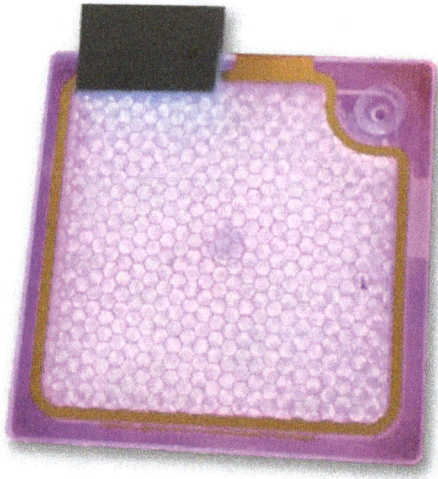

Fig. 4.8.   A flat micro-discharge excimer UV lamp from Eden Park Corporation.
Courtesy: Eden Park Corporation.

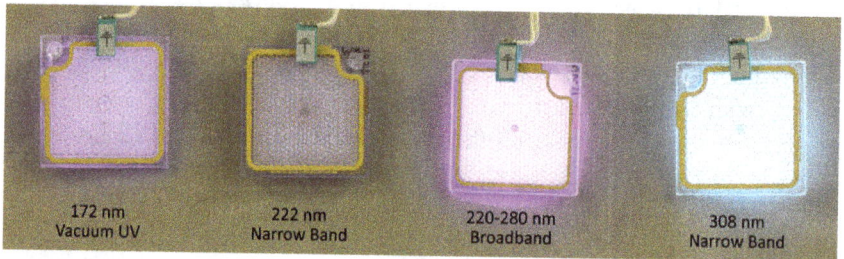

Fig. 4.9.   Micro-discharge excimer UV lamps operating at different wavelengths.
Courtesy: Eden Park Corporation.

222 nm and 172 nm are readily attained with outputs in excess of 25 mW and 200 mW, respectively.

A micro-discharge lamp consists of a sandwich structure with UV-transparent glass separated by dielectric spacers (see schematic diagram in Figure 4.10). The spacers also divide the assembly into cellular regions. On driving the two counter electrodes with a high-frequency AC voltage waveform, radiation is emitted and comes out from both sides of the lamp assembly. Peak-to-peak voltage of about 6 kV is needed for proper operation. A reflector can be placed on one side to increase the amount of radiation that comes out of the opposite side. A phosphor

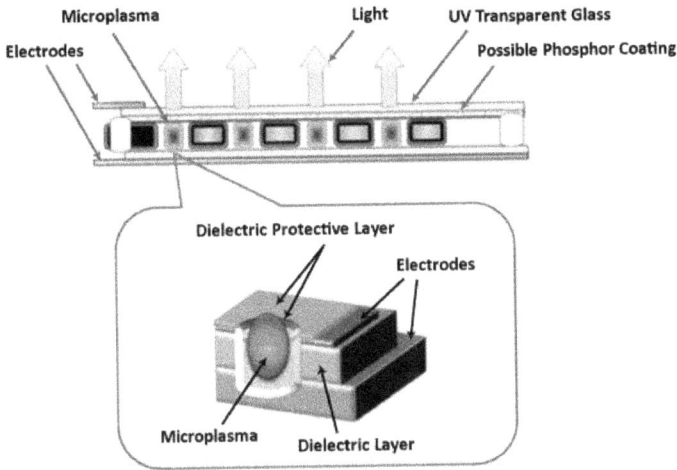

Fig. 4.10. Cutaway schematic of a micro-discharge excimer UV lamp.

coating can be placed on one of the cover plates to alter the spectral make-up of emitted radiation.

These lamps have typical efficiency of around 5%, and last for over 2000 hours of continuous operation. Their flat shape makes them useful for several applications where cylindrical or semi-spherical shape of other UV lamps causes operational difficulties. Water, food and surface disinfection are obvious applications. Other uses are in microlithography, resin curing and materials analysis.

4.3 Laser-Energized Plasma Light Sources

So far, we have looked at plasma lamps that are directly excited by electrical power inductively coupled to the lamps' fill gas. This is the simplest and the most widespread form of plasma light sources. As we have seen above, this scheme partitions the light source into two distinct components: a gas-filled lamp and its associated inductive power supply. Often such plasma lamps have their power supply integrated with the bulb itself. With this arrangement, failure of either the bulb or the power supply renders the entire system unusable. This is obviously a disadvantage but it does not make plasma sources impractical because integrated sources are usually manufactured to be cheap enough (at the expense of performance limitations) that the system could be regarded as a consumable.

Higher-end sources employ separate bulb and power supply subsystems where the failure of one does not affect the other.

Another distinct class of plasma sources employs an indirect excitation mechanism where energy is supplied to a bulb's fill gas through laser-based heating. In contrast to the sources discussed above, laser-energized plasma sources feature hot plasmas with temperatures, possibly, running into thousands of degrees Celsius. Such systems are necessarily much more costly when compared to inductive systems, met with earlier, because suitable laser systems for providing sufficient heating of the fill gas are quite expensive. The high power density of several different types of lasers makes them suitable for radiatively heating fill gases. However, to obtain stable and well-behaved plasma without unwanted instabilities, laser radiation has to be stabilized against optical power fluctuations, as well as adequately shaped and focused for efficient coupling to the fill gas medium.

In the following subsections, actual laser-excited plasma sources are described. Being expensive, such sources are used in specialized technical applications where their cost can be fully justified against the advantages they offer.

4.3.1 *Energetiq LDLS xenon plasma source*

The laser-driven light source (LDLS) from Energetiq Technologies Inc. (now a part of Hamamatsu Corporation) is a plasma light source that uses laser-heated xenon plasma to generate high-intensity broadband radiation. The raw output from this source extends from the infrared region to quite deep inside the UV region. It finds uses in applications that require a continuous broadband source of electromagnetic radiation. In this area, it competes with electrically-excited xenon lamps. An Energetiq LDLS system is seen in Figure 4.11 here. It consists of a laser housing with laser power supply, a control unit, and a xenon bulb housing. Optical power, generated by CW diode lasers, is sent to the bulb unit using a sheathed fiber optic cable. The bulb housing consists of a finned aluminum box whose outer surfaces serve as a heat sink. Radiation comes out through a round optical window set in one face of the bulb enclosure. The bulb itself is a spherical quartz globe filled with xenon gas. Laser radiation is tightly focused to a spot of around 100 μm diameter inside the xenon-filled sphere. This heats up the gas to incandescence which then emits its

Fig. 4.11. An Energetiq broadband LDLS xenon plasma system showing both power supplies and the sealed xenon plasma unit.

Courtesy: Hamamatsu Corporation.

characteristic flat broadband continuum. The bulb housing needs to be flushed continuously with dry nitrogen while the source operates. This helps keep a stable ambient for the bulb to operate in, and provides a number of benefits, such as help in heat dissipation from the outer wall of the bulb and keeping the enclosure clean, dry and dust free. Another major benefit of nitrogen purge is the reduction in ozone generation. When UV radiation at wavelengths shorter than 254 nm passes through air, oxygen molecules tend to dissociate into oxygen atoms which then combine to form tri-atomic ozone molecules. This is a very undesirable process because not only does it result in considerable absorption of short wavelength UV radiation, ozone is also biologically harmful, and is a strong chemical pollutant. Nitrogen purge maintains a nitrogen atmosphere in the bulb enclosure, avoiding the problem of ozone generation.

LDLS bulb housings are available in both free space coupling (windowed) and fiber-coupled varieties. The latter is shown in Figure 4.12.

Larger systems with much higher brightness, intended for industrial lithography and other applications requiring large optical output powers, are also available from Energetiq Inc. Figure 4.13 shows such a system for industrial applications. This EQ-10 electrodeless Z-pinch source generates up to 10 W of radiative power in the extreme UV/soft X-ray region. This is a pulsed system which operates at pulse repetition rates as high as 2 kHz, while delivering up to 10 W of optical power into a

Fig. 4.12.   Close-up of an Energetiq xenon plasma unit.
Courtesy: Hamamatsu Corporation.

2π steradian cone. During operation, high current pulses are injected into a gas manifold where plasma is generated and pinched by large magnetic fields to a small region around 1 mm in size. Other similar systems with power outputs up to 20 W are also available.

4.3.2 *Laser-produced plasma (LPP) extreme ultraviolet (EUV) light sources*

Over the years since its inception in the 1960s, optical lithography in the semiconductor industry has shifted to using shorter and shorter wavelengths. Until recently, krypton fluoride (KrF) and argon fluoride (ArF) excimer lasers, emitting at 248 and 193 nm, respectively, had been the primary light sources for semiconductor lithography. Now, however, even such deep UV (DUV) wavelengths are inadequate for patterning features that are often less than 10 nm in minimum dimension. With the transition to the 7 nm technology node, sub-100 nm wavelengths are needed for

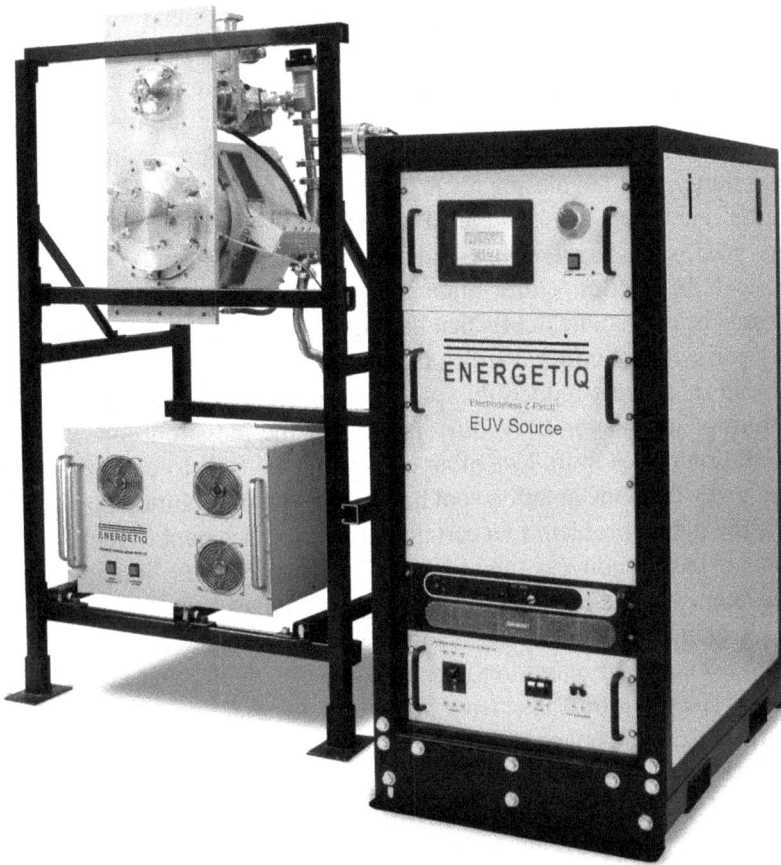

Fig. 4.13.   An Energetiq plasma-based high flux EUV source.
Courtesy: Hamamatsu Corporation.

effective wafer patterning. It is not possible to obtain such very short wavelengths using ordinary plasma sources because electron transitions within very high charge-state ions will be required, and such ions require extremely high energy inputs to produce in adequate numbers. Furthermore, quite high output powers are needed at very short wavelengths to guarantee sufficient wafer throughput in a mass-manufacturing setup. Ideally, a source with output power nearly two orders of magnitude higher than the capability of synchrotron accelerators is needed. Starting in the late 1970s, high-power plasma point sources were investigated for generating large power outputs at very short wavelengths in the UV region. Intense laser beams, usually in the infrared region, focused on various metallic targets, such as gold plates and wires, were first developed as experimental sources. Heat deposited by laser radiation causes the metal to melt, vaporize and then form a plasma. Further interaction of laser energy with the plasma results in radiation emission at various wavelengths, including some in the EUV region. While effective at their primary goal, such laser-produced plasma (LPP) sources pose a number of practical difficulties. Chief among these is the generation of copious amounts of particulate debris that damages radiation collection optics and other parts of the emitter. Various ingenious schemes have been developed to avoid or at least reduce this problem, in order to lengthen the useful lifetime of multilayer EUV mirrors used in the radiation collection and collimation optics train. Use of an entraining gas flow to carry metal particles away from the ablation spot has been a notable solution that circumvents the debris problem to a certain extent. Replacement of metal targets with a frozen xenon gas target has also been tried with limited success. This, later evolved into the use of a Xe gas jet target, first developed at the Sandia National Laboratories. However, it was found that high energy ions in a Xe plasma caused erosion of collection optics, and even this was not a perfect solution. Decisive change in the state-of-the-art for EUV generation came from an academic laboratory which led to the first commercially-viable sources of EUV radiation.

G. O'Sullivan and P. K. Carroll from the Department of Physics at University College Dublin reported in 1981 that strong resonance emission was generated when various rare-earth and transition metal targets were irradiated with intense laser beams. These emissions were, principally, found to lie in the wavelength range of 4 to 20 nm and were attributed to 4d–4f transitions taking place in metal ions present in the plasma. Through a systematic study, the researchers found that with increase in atomic number of the target, the peak wavelength of these narrowband emissions moved monotonically toward shorter wavelengths. They also found that a tin (Sn) plasma, in particular, emits a very strong band with the spectral peak at 13.5 nm. Initially, this report was not taken seriously because a solid tin target was known for generating a tremendous amount of debris. Additionally, the conversion efficiency of radiation generated from a tin target was considerably lower than that generated from a gold target. Many years after O'Sullivan and Carroll's publication, interest in tin plasma-based EUV sources was rekindled with the realization that the radiation produced was intense, narrowband, and at a wavelength that made it particularly suitable for EUV lithography. Several laboratories and companies then started working on developing practical, reliable, and long-lived tin-plasma EUV sources. Such efforts culminated in the development of systems based on tin droplets that fire a steady stream of small (~30 μm diameter) highly purified liquid tin droplets at rates of up to several tens of thousands per second, at speeds around 70 m/s. These droplets are irradiated with the beam from a CO_2 laser (wavelength: 10.6 μm) at power levels of 10 to 30 kW. Such a system, demonstrated by Cymer (now a division of ASML B.V. of Netherlands) several years ago, succeeded in generating 11 W of power at 13.5 nm wavelength. More recent systems have demonstrated EUV output powers well in excess of 100 W while modules producing 250 W have also been developed, for use in production-level silicon wafer manufacturing. Figure 4.14 is a cutaway drawing of a Cymer EUV source, showing all the principal parts, including the tin droplet generator, waste droplet catcher, collection mirror, various diagnostic elements and turbo pumps needed to maintain the required ultra-high vacuum conditions in the source chamber. Note that

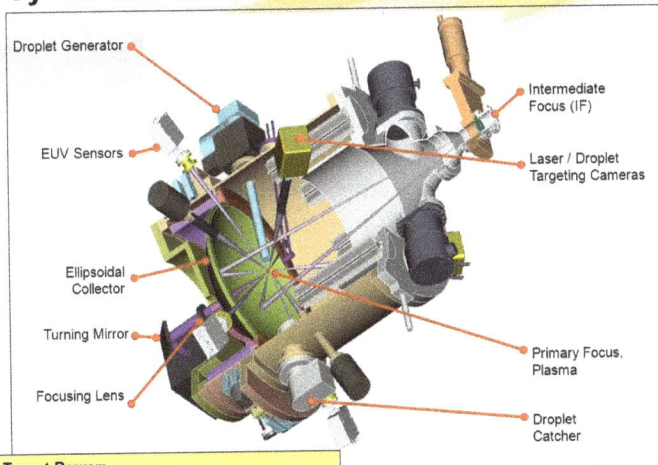

Cymer LPP EUV Source Vessel Architecture

Droplet Generator

Intermediate Focus (IF)

EUV Sensors

Laser / Droplet Targeting Cameras

Ellipsoidal Collector

Turning Mirror

Primary Focus, Plasma

Focusing Lens

Droplet Catcher

Target Power:
100W for process development tools
250W for high-volume manufacturing

CYMER

Fig. 4.14.   Cutaway diagram of a Cymer LPP EUV source.

Courtesy: Cymer.

the CO_2 laser itself is not shown as it is a separate part of the overall system, delivering high-power infrared radiation to the source vessel through a special beam delivery tube.

At the excessively short wavelength of EUV radiation, all materials, including ambient air, become strong radiation absorbers. Thus, all-optical manipulation from radiation collection and beam shaping to beam delivery and final focusing is performed in ultra-high vacuum; making exclusive use of dielectric mirrors. These reflective elements are made by depositing hundreds of precisely-controlled alternating layers of silicon and molybdenum, each about 3 nm thick, on appropriately shaped substrates. Due to the very short wavelength of EUV radiation, mirror surfaces need to be smooth down to $\lambda/30$ grade, which translates to around 0.45 nm rms surface smoothness at $\lambda = 13.5$ nm. Even with such exceptionally smooth mirrors, reflectivities of individual mirrors typically do not exceed 70%. Because a large number of such mirrors are used in a complete system, so the optical power throughput from the source to the lithography tool becomes quite small. Current EUV lithography systems typically contain at least two condenser multilayer mirrors, six projection multilayer mirrors, and a multilayer mask object. Together, these surfaces absorb more than 90% of the radiation that is delivered by the EUV source. Thus, to enable commercially-feasible high volume manufacturing (HVM), much larger amounts of EUV power needs to be generated than is actually needed at the point of use. If the EUV source brightness is not increased, then the resist exposure times will become too long; leading to unacceptably low wafer throughput. Consequently, much effort has been devoted in recent years into increasing the brilliance of EUV sources by incorporating a number of ingenious concepts. As one example, Cymer has demonstrated that mismatch of tin droplet size with the focused spot size of CO_2 laser — a prominent cause of low conversion efficiency in LPP systems — can be obviated by using a pre-pulse approach. This technique makes use of a separate laser to irradiate tin droplets prior to their exposure to the main laser beam. Pre-pulsing causes the droplets to distort into the shape of flattened disks which present a much larger cross-section to the main plasma-generating laser. Thus, almost all of the power in the focused spot is utilized for plasma generation — greatly increasing the infrared to EUV conversion efficiency. Without pre-pulse, only an estimated 0.33% of a tin droplet is converted to the plasma phase whereas with pre-pulse irradiation up to 10% ablates to form tin plasma. Figure 4.15 shows an EUV source module vessel in development at Cymer's facilities in Veldhoven, The Netherlands.

Fig. 4.15.   A Cymer EUV source module.
Courtesy: Cymer.

Other refinements added to tin droplet-based EUV sources include magnetic plasma confinement for protecting collection optics from energetic ion bombardment and beam shaping of CO_2 laser beam for improving radiative conversion efficiency.

Besides Cymer, Gigaphoton (a subsidiary of Komatsu Corporation) in Japan is also active in commercial R&D on similar EUV sources. In July 2016, they announced the completion of a 250-W EUV source with conversion efficiency of 5%. Gigaphoton's experimental GL200E EUV system uses a solid-state diode-pumped Nd:YAG pre-pulse laser operating at 1064 nm wavelength with a CO_2 main pulse laser. Figure 4.16 shows the layout of a typical Gigaphoton LPP EUV light source. In addition to the collector mirror, tin droplet generator and various diagnostic systems, the source vessel also incorporates a superconducting magnet-based debris mitigation system which deflects away tin ions and, thus, greatly extends the life of the collection mirror. Gigaphoton favors a smaller tin droplet diameter of 20 μm, as it significantly reduces debris from that portion of tin which does not ionize and is, thus, unaffected by the debris mitigation magnetic field. This is in spite of the fact that a smaller droplet size leads to less plasma being generated and, thus a lower

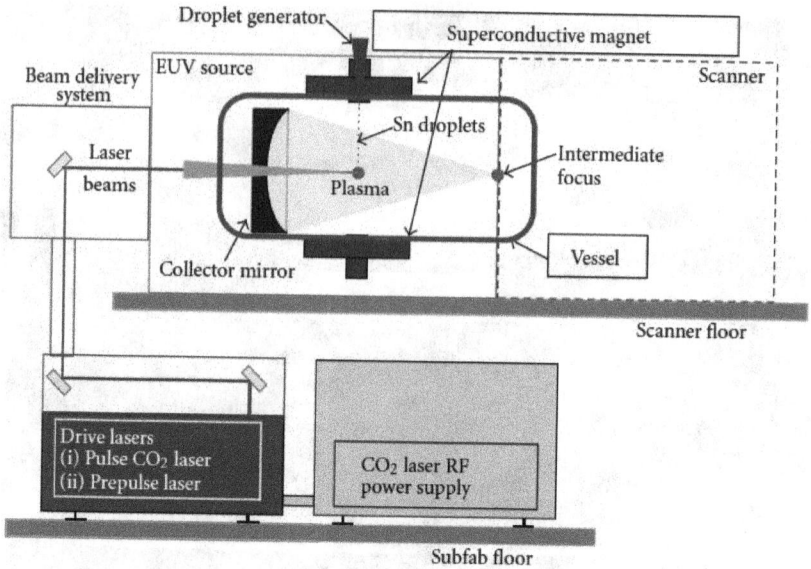

Fig. 4.16.   Schematic of a Gigaphoton LPP EUV source.
Courtesy: Gigaphoton Inc.

EUV output. Gigaphoton has been optimizing the size of the tin droplet to achieve both high EUV power as well as good debris mitigation. Additionally, the Gigaphoton EUV system is also equipped with a gas-phase chemical etching mechanism designed to remove any remaining tin atoms that are not trapped in the magnetic field and, thus, can potentially be deposited on the collector mirror surface, or other optically sensitive surfaces. The company has been making steady progress in improving its EUV source and in September 2017 it announced the development of Spectral Width Control Technology called hMPL, which enables the spectral width of 13.5 nm EUV radiation to be reduced from 300 fm to 200 fm. Reduced spectral width significantly improves lithographic performance.

One other company that has developed LPP EUV sources of broadly similar design is Adlyte Inc. of Zug, Switzerland. A spin-off from ETH, Zurich, Adlyte has demonstrated relatively small but very reliable EUV systems aimed at EUV mask inspection applications.

While LPP is, by now, the established approach for generating substantial amounts of EUV radiation, direct current electrical discharge can also be used to produce EUV energy. This so-called discharge-produced plasma (DPP) technique is very different from LPP. DPP EUV sources generally cost less to build but are not easily scaled to high power levels. Consequently, manufacturers of EUV sources for lithography have not considered DPP as a serious alternative to tin-based LPP sources. However, some companies have been working on lower power, low cost-of-ownership DPP EUV sources for niche applications that do not require large amounts of EUV power. Mask inspection and reticle metrology in EUV lithography, for instance, require low-power EUV sources of a few tens of watts. DPP sources of EUV radiation generally employ a pure xenon or helium/nitrogen/xenon mixture jet confined inside a ceramic capillary inside which a high-temperature plasma is struck by high current pulses. The spectrum of the resulting discharge is shown in Figure 4.17 which shows EUV emission lines from different multiply ionized states of xenon. Emission from one excited state of Xe^{+10} ions is centered at 13.5 nm and this line can be used for 13.5 nm EUV lithography.

A conceptual diagram of a DPP source is shown in Figure 4.18. This configuration was first investigated at the Sandia National Labs. Several such sources, each generating a small amount of EUV power, can be combined to produce a reasonably large amount of EUV flux. Such an approach is used in a DPP EUV source developed by Nano-UV, called Hydra, which is made up of an assembly of 10 individual DPP cells

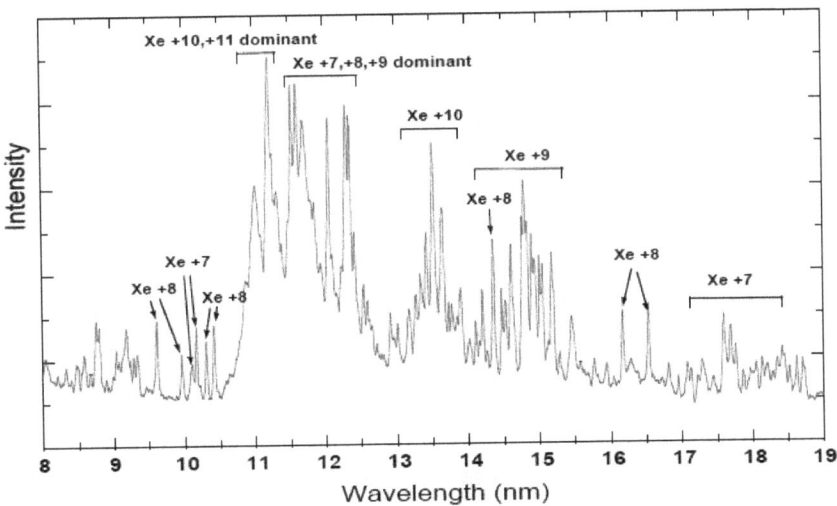

Fig. 4.17. Emission spectrum from xenon LPP EUV source showing emission lines from different charge states of xenon ions.

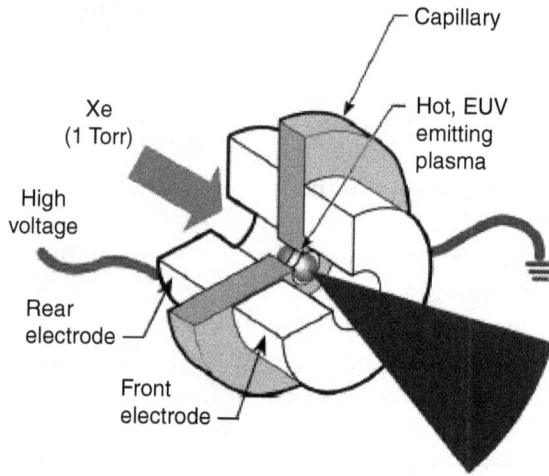

Fig. 4.18.   Diagram of a DPP EUV source.
Courtesy: Sandia National Labs.

called Cyclops. This system, targeted for EUV mask inspection and metrology systems, features an irradiance of 10^{18} photons/cm^2/s, when all Cyclops cells are operating optimally. Nano-UV's EUV source does not use a collection mirror, instead it relies on spatial multiplexing (using multiple small sources) to build up radiation intensity. However, like all electrical discharge-based radiation sources, the discharge cells are prone to electrode erosion and, thus, require periodic maintenance. A very different approach to EUV DPP implementation has been demonstrated by Philips in the Netherlands, which does not suffer from electrode erosion problem. As shown in Figure 4.19, this technique employs rotating liquid tin electrodes that continuously present fresh tin surfaces, thus avoiding electrode erosion. Two tin-coated drums are connected to a high voltage capacitor bank which serves as the discharge energy source. The closest separation between the drums is the 1–2 mm spark gap. This equipment is contained inside a vacuum vessel. The discharge is initiated by firing a trigger laser at one of the tin drums. Cooling channels embedded inside each drum remove the heat produced in the discharge process. EUV radiation is emitted from a plasma volume about 1.3 mm in diameter. With this arrangement, Philips demonstrated EUV power generation in excess of 100 W, with 2% conversion efficiency, at 40 kHz pulse repetition rate.

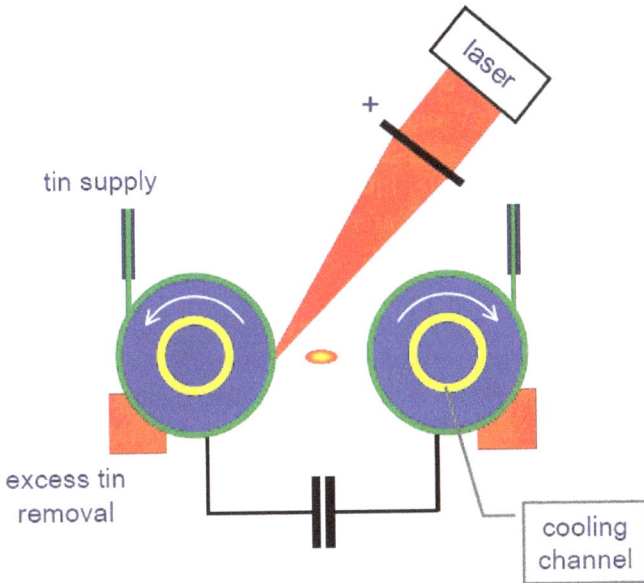

Fig. 4.19. Diagram of a liquid tin electrode EUV source.

Courtesy: Ushio Corporation.

Fig. 4.20. ASML's NXE3300B lithography stepper in IBM's Albany fab.

Courtesy: ASML Inc.

There have even been attempts to combine both DPP and LPP approaches for EUV generation. Xtreme Technologies GmbH of Aachen, Germany, and its parent company — Tokyo-based Ushio — reported the development of an EUV light source that combines both LPP and DPP into a hybrid called laser-assisted discharge-produced plasma (LDP) technology. This source was later integrated with an ASML stepper at IMEC, Belgium and used in trial wafer runs. Subsequently, it has also been installed at TNO laboratories in The Netherlands for performance evaluation studies.

EUV-based lithography systems have been introduced into production beginning from the year 2019 and are being used for the most critical layers of pattern imaging on semiconductor wafers. Technology nodes below 7 nm rely almost exclusively on EUV radiation for wafer patterning. For this reason, continuing advances in this technology are being carefully monitored by semiconductor fabs, with some chip manufacturers already relying on EUV-based lithography systems for most critical patterning layers. Figure 4.20 shows an experimental lithography system based on ASML's NXE3300B lithography stepper, employing Cymer's EUV source described above, installed in an IBM fab in Albany, New York.

5

Light-Emitting Diodes

5.1 Introduction

In the previous chapters, we have looked at the generation of light from hot, solid objects, and from electrical discharges in low pressure gases. For most of the 20th century, light sources based on those technologies were the main devices used for most lighting applications. Incandescent and discharge-based lamps had been developed to their utmost perfection by the last decade of the last century. As a result, such lamps became very inexpensive and widely available. While they represent a triumph of human ingenuity, these traditional light sources suffer from gross inefficiencies in converting electrical energy input into visible light. Incandescents are widely known to be highly inefficient light sources; generating mostly heat, with visible light being just a by-product of heat production. While the modern tungsten-halogen lamps are definitely much more efficient when compared to the old tungsten filament lamps, they still produce more heat than light. Discharge lamps do much better in this regard but their conversion efficiencies too fall far short of what would be desirable. In recent decades, concerns related to environmental protection, use of fossil fuels, our carbon footprint and waste reduction have assumed much importance. The resulting socio-political debates are increasingly influencing the choice of energy conversion devices. Such discourse has resulted in renewed interest in further lowering the environmental impact from lighting our surroundings — both indoors and outdoors. Highly efficient light sources utilizing fewer, if any, hazardous materials in their construction (and, thus, easily recyclable) are much more desirable in the world we live in today than was the case just a few decades ago. Consequently, the focus of electric lighting innovation has shifted to radically different technologies. Foremost, and the most successful, among these are sources based on light-emitting semiconductors. Lamps made from semiconductors are cold, non-discharge-based light sources that offer many benefits over the more traditional incandescent and discharge lamps. Their principal strength is their very high energy conversion efficiency, when compared with traditional luminaires (Figure 5.1). Add to that their use of environmentally friendly recyclable materials and long service life, and they begin to appear as the long sought-after ideal light sources. Their history and technology are both as interesting as that of the other lamps that we have encountered in earlier chapters. Because of their relatively recent appearance, semiconductor lamps have still a long way to go, and, thus, it is very worthwhile to spend some time in studying the science and technology

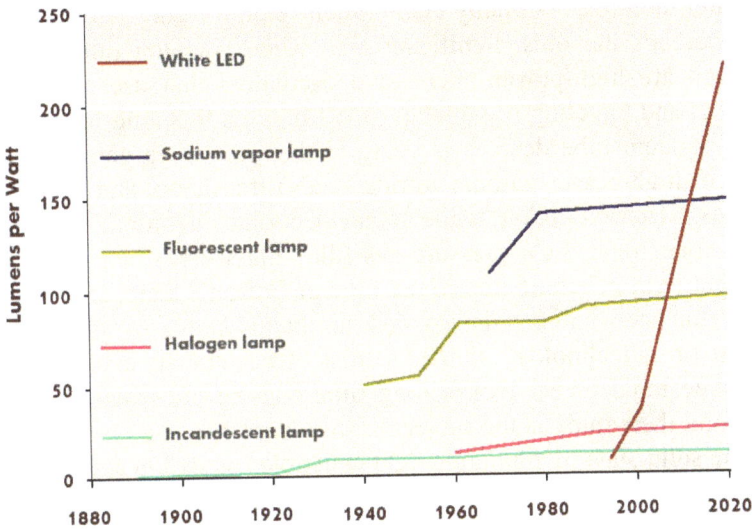

Fig. 5.1. Increase in luminous efficacy of various light sources over time.

that underpins their operation. The rest of this chapter looks at solid-state lighting — as this new technology is broadly called, based as it is on the solid-state semiconductor diode.

5.2 Light from Solid-State Devices — Some Early History

The 1940s saw the emergence of solid-state electronics. The germanium bipolar transistor invented at the Bell Labs in 1947 started a trend that saw the gradual replacement of vacuum tubes with semiconductor crystal-based replacements. By the late 1960s, this transformation was nearly complete so that vacuum tube 'valves' were only left in a few highly specialized applications. The most visible sign of the dawning of a new era was seen in the proliferation of 'transistor' radios. Some of the remaining strongholds of vacuum tubes, however, proved surprisingly resistant to the semiconductor revolution. Notable among these was the cathode ray tube (CRT), as the main display device for visual output. Only by the 1990s, liquid crystal displays (LCDs) had developed sufficiently to challenge the entrenchment of CRTs in televisions and computer monitors — finally ending their reign for good. At the present time, perhaps the only significant area where vacuum tubes are still prevalent are high-power microwave oscillators that use magnetrons, klystrons and traveling wave tubes (TWTs). Even this long-held monopoly of vacuum tube devices is being challenged by on-going developments in high-power gallium nitride (GaN) transistors and microwave modules. Electric lighting, while not an electronic operation, also mirrors this historic trend. Low pressure gas-filled glass tubes have dominated electrical light generation since the time of Edison's lamp. Incandescent lamps, tungsten halogen lamps and discharge lamps — all could be thought of as technology similar to that of vacuum tube valves. Their solid-state replacement took a long time coming but eventually it did arrive and their study is the subject of this chapter.

The solid-state revolution in light generation started in a small way in the 1960s, with the invention of the first light-emitting semiconductor diode. This was a device made from the compound semiconductor: gallium arsenide (GaAs). It emitted infrared radiation when the diode was forward biased so that an electric current could easily flow through it. However, much earlier, in 1907, Henry Joseph Round (Figure 5.2) — an English radio engineer and assistant to the Italian radio pioneer

Fig. 5.2.   Henry Joseph Round.

Fig. 5.3.   Guglielmo Marconi.

Guglielmo Marconi (Figure 5.3) — had already reported the emission of a bright yellow-green light from 'cat's whisker' diodes made by pressing a sharp metal wire tip against a piece of carborundum (silicon carbide (SiC)). Whenever around 10 V was applied between the wire (acting as the cathode) and the carborundum crystal, a spot of yellow-green light was seen at the point of contact (see Figure 5.4). This was the first ever report of the emission of light from a semiconductor material (the metal-SiC arrangement forms what is technically called a point contact Schottky diode). Later, a very similar observation was also reported by Oleg Vladimirovich Losev — a Russian radio technician. He carried out detailed investigation of this phenomenon in 'crystal' rectifiers used in radio sets (see Figure 5.5) around 1927 and published his findings in several Russian journals. Among the topics he studied was the least current (threshold current) for light emission to be observed and the dependence of the light's intensity on current through the diode. It is remarkable that he was able to publish such detailed investigations in spite of having little formal training in physics or engineering. While Losev's observations were the first reports of this new phenomenon, proper understanding of the mechanism of light emission from semiconductor devices had to await the development of the quantum theory of solids. The quantum mechanical energy band description of semiconductors is central to understanding

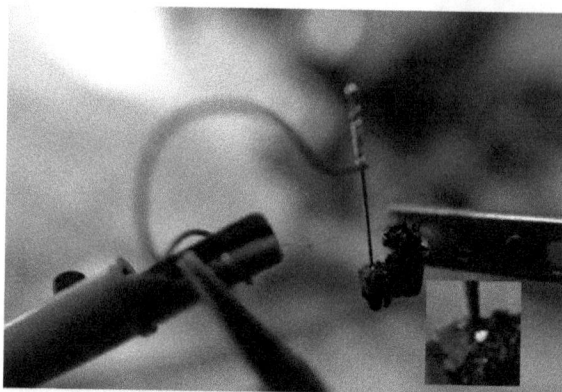

Fig. 5.4.   Light emission from a metal wire tip in contact with a carborundum crystal.

Fig. 5.5.   Early 'cat's whisker' rectifier.

light emission from semiconductors and for building practical solid-state light emitters.

From a historical perspective, the first proper light-emitting diodes (LEDs) were infrared-emitting devices. In 1955, Rubin Braunstein — a physicist, employed by the Radio Corporation of America (RCA), published the first accounts of infrared radiation emission from semiconductor diodes made from such compound semiconductors as GaAs, gallium antimonide (GaSb) and indium phosphide (InP). These are all so-called III–V compound semiconductors, as their constituent

Fig. 5.6. Nick Holonyak Jr.

elements derive from groups III and V of the Periodic Table. Even today, almost all commercial LEDs are manufactured from these and other similar III–V compound semiconductors. A few years later, in 1961, Gary Pittman and Bob Biard from Texas Instruments (TI) also reported the emission of infrared radiation from GaAs diodes conducting an electric current. They also obtained a patent for infrared LEDs made of GaAs, with the title: Semiconductor Radiant Diode. A year later, in 1962, Nick Holonyak Jr. (Figure 5.6) working at General Electric research labs in Syracuse, New York, developed the first visible LED, emitting red light. This device was made using gallium arsenide phosphide (GaAsP) — a 'ternary' alloy semiconductor of gallium, arsenic and phosphorus that has a suitable band gap for the emission of red light. Ten years after the demonstration of the first red LED, a yellow-emitting device was developed by M. George Craford (Figure 5.7), a former graduate student of Holonyak Jr. at the University of Illinois. At the time of this development, he was working at the Monsanto Chemical Company. Craford went on to develop a range of further LEDs emitting visible light of different colors, made from such semiconductors as gallium arsenide phosphide, aluminum gallium arsenide (AlGaAs) and aluminum indium gallium phosphide (AlInGaP). These materials are still used for making visible LEDs that emit red, orange, yellow and green light. Around the same time as the development of the first yellow LED in the early 1970s, Jacques Isaac Pankove (Figure 5.8) at the RCA labs in Princeton, New Jersey, demonstrated the first working blue LED made from GaN. This early device was not particularly bright and further

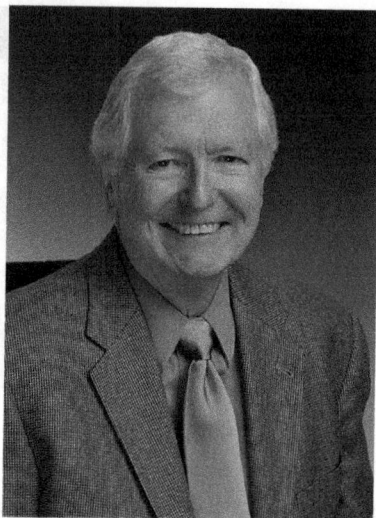

Fig. 5.7.   M. George Craford.

Fig. 5.8.   Jacques Isaac Pankove.

progress in blue-emitting LEDs had to await new developments in material growth and processing that came many years later.

The first commercial LEDs were commonly used as replacements for
incandescent indicators, and in seven-segment displays — first in

expensive equipment such as laboratory and electronics test equipment, then later in such appliances as TVs, radios, telephones, calculators, and even watches. However, these LEDs were not bright enough for use in space illumination. That application had to wait for further advances in high brightness (HB) LED technology. At this point, we must first familiarize ourselves with the basics of some semiconductor physics in order to properly understand the manufacture and operation of semiconductor-based light sources.

5.3 A Look at Semiconductors and Semiconductor Diodes

Before we start discussing LED technology, it is essential that we spend some time in going through the basics of semiconductors, these devices are made of. Semiconductors could be elements, such as silicon, germanium, etc. or compounds, such as gallium aresenide, indium phosphide, etc. Compound semiconductors are also known as semiconductor alloys, especially if these are composed of three or more different elements. Almost all inorganic semiconductors are crystalline solids that show electrical conductivity that falls roughly half way between that of good and bad conductors. However, while this description is the basis of the term 'semiconductor', it is not a good scientific definition for this distinct class of materials. Unless prepared in exceptionally pure form, the 'semiconducting' property of semiconductors is not apparent at all. This is because the electrical conductivity of semiconductors is extremely sensitive to the presence of impurities. Even minute amounts of impurities could greatly enhance the conductivity of a semiconductor sample (or in rare cases, suppress it). This is the reason; semiconductors were not identified as a distinct class of materials until well into the 20th century. Only by the 1920s, material refining techniques had been perfected to an extent that very pure crystals could be grown and their intrinsic properties properly studied.

5.3.1 *The band theory of solids*

While the fact that semiconductors are a distinct class of materials, separate from metals (such as iron, copper, aluminum, etc.) and non-metals (such as boron, sulfur, iodine, etc.), was established by the 1930s, not much else was known about them. True understanding of semiconductors only developed when quantum mechanics was applied to understand the

Fig. 5.9.   William Shockley.

physical properties of solids. This took place during the 1930s and 1940s and, in the case of semiconductors, was pioneered by William Shockley (Figure 5.9), while employed at Bell Labs. He developed the modern quantum theory of semiconductors which was instrumental in the development of all semiconductor devices. His 1950 book — *Electrons and Holes in Semiconductors* — was a classic of semiconductor physics and remains in use today. All textbooks and research papers written on semiconductor materials and device physics to date continue to make use of the formulism that was originally described in that book. In 1947, theoretical understanding of semiconductors developed by Shockley enabled him and his co-workers to develop the first solid-state amplifying device — the transistor — which ushered in the era of all-solid-state electronics. William Shockley, together with his Bell Labs colleagues John Bardeen and Walter Brattain, received the Nobel Prize in Physics, in 1956, for the development of the bipolar transistor. In later years, after leaving Bell Labs, Shockley started a semiconductor device company in Palo Alto, California, which in turn gave rise to other electronics companies in that region. Figure 5.10 shows the location (391, San Antonio Road) where his company once stood. For this reason, William Shockley is said to be the man who 'brought silicon to the Silicon Valley'.

It turns out that the key to understanding the properties and behavior of all solids — including semiconductors — lies in their energy level

Fig. 5.10.   Location of Shockley's semiconductor company in Palo alto, California.

structure. This is separate, but inextricably linked, to the physical structure of solids, i.e. the physical three-dimensional placement of atoms or ions on the crystal lattice. Together, the structural and energy level descriptions provide complete details about the nature and behavior of solids.

Atoms of any given element have a distinct energy level structure where electrons occupy states of different electrical potential energies. The atomic energy levels of isolated atoms are very sharply defined. When atoms come together to form molecules then each atom's energy level is perturbed due to the presence of other nearby atoms. It can be shown that this results in one energy level splitting into the same number of levels as the number of atoms that are close together. Thus, larger the number of atoms close to each other, the larger the number of split energy levels and the closer their spacing to each other. With a very large number of atoms coming together to form any macroscopic piece of material, the resulting manifolds of extremely closely spaced energy levels — each deriving from a single energy level of an isolated atom — are said to form energy bands. Thus, the coming together of a vast number of atoms to form a piece of bulk matter results in each atomic energy level changing into a somewhat wide band of very closely-spaced energy levels. Energy bands are a characteristic feature of bulk matter. Figure 5.11 illustrates the scheme of energy band formation as a large number of atoms come together to form a macroscopic piece of matter.

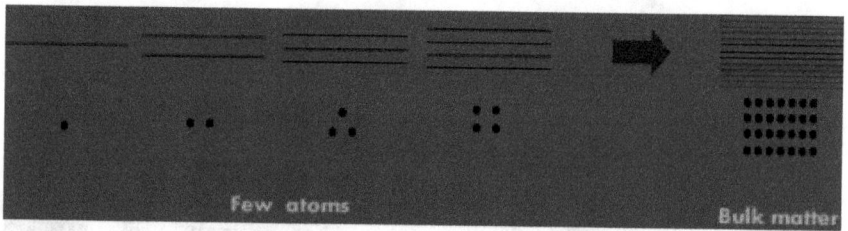

Fig. 5.11.   Formation of energy bands in solids from individual atomic energy levels.

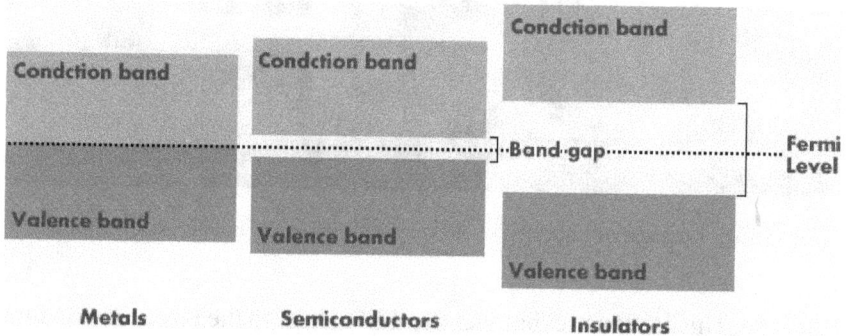

Fig. 5.12.   Energy band structure of metals, semiconductors and insulators.

Just as atomic energy levels are separated by gaps, the energy bands formed from the broadening of atomic energy levels are also interrupted by energy intervals that are devoid of energy levels. These interruptions are called energy gaps or band gaps. The system of energy bands separated from each other by energy gaps, where no energy levels exist, is central to understanding the electronic properties of solids. We will rely on this picture to understand how semiconductor diodes work to emit light.

The quantum mechanical description of the energy level structure of bulk solids has been a crowning achievement of modern physics. It has allowed solids to be properly classified as electrical conductors, non-conductors and semiconductors, on the basis of their energy band structure, as is seen in Figure 5.12. Good electrical conductors (metals) have a partially-filled energy band with empty energy levels just above the highest filled energy level (called the Fermi energy level). This allows electrons to easily gain energy from an applied electric field, move to a higher energy level and travel through the conductor carrying an electric current. Non-conductors, in contrast, have a completely filled energy band lying below a completely empty energy band with a large energy gap (several electron volts), separating the band edges from each other. The filled band is called the valence band whereas the empty band right above is called the conduction band. The valence band is formed from energy levels belonging to valence electrons of the solid whereas the conduction band represents energy levels for electrons that have been released from atoms to become free electrons. In the case of semiconductors and insulators, the Fermi level is taken as passing exactly half way through the band gap. The large energy separation between the edges of the conduction and valence bands does not let electrons jump easily from the former to the latter, as a result of gaining energy from an applied electric field. This, in effect, stops them from moving under the influence of an applied electric potential difference and, thus, no electrical conduction takes place.

The energy band picture, finally, makes it possible to understand semiconductors. These materials are characterized by the existence of a relatively narrow gap in their energy spectrum, i.e. their energy band diagrams show a narrow discontinuity between the energy levels belonging to the conduction and valence bands. This forbidden gap is around 1–2 eV wide in typical semiconductors. Note that the exact width of a band gap is slightly temperature-dependent, but we use its value at room temperature for most discussions. The narrow energy gap separating the conduction and valence band states in semiconductors endows these materials with some very interesting properties. Foremost among these is the small but highly temperature-dependent electrical conductivity that is seen in semiconductors. This is a result of thermally-activated charge carrier excitation across the energy gap from the valence to the conduction band. At room temperature, sufficient thermal energy is available to send some electrons from the valence band to the conduction band. Being a thermal across-the-gap process, the carrier density, thus, produced is an exponential function of temperature. This property makes devices made from semiconductor materials susceptible to failure at high temperatures. This happens because increase in temperature produces more charge carriers, which carry more current, further increasing the temperature due to increased power dissipation, which then creates even more carriers, creating a positive feedback effect called thermal runaway. This places an upper limit on the operating current of all semiconductor devices.

5.3.2 *Doping of semiconductors*

The other interesting property displayed by semiconductors is their remarkable sensitivity to the addition of even minute amounts of impurities. This has already been alluded to above where it was cited as the reason for the late discovery of semiconductors. Most elemental impurities cause a semiconductor's conductivity to rise precipitously. Impurities (more commonly called dopants) create additional charge carriers (with energies lying in either the valence or the conduction band), which leads to increased electrical conductivity. Remarkably, one can select dopants to either seed the valence band or the conduction band with charge carriers (called holes and electrons, respectively). At this point, it is instructive to note that the only real difference between non-conductors (also called insulators or dielectrics) and semiconductors is the difference in their band gaps — materials with band gaps below about 5 eV are generally considered semiconductors whereas materials with wider band gaps are categorized as insulators.

To understand the effects of doping further, let us consider silicon which is a typical semiconductor with a room temperature band gap of 1.12 eV. It belongs to group IV of the Periodic Table and can be prepared in a monocrystalline form, with the diamond crystal structure. Addition of miniscule amounts of boron (an element from group III of the Periodic Table) causes electron vacancies to appear in the crystal. This happens because each atom of boron has three valence electrons — one less than the four valence electrons possessed by a silicon atom. When a boron atoms substitutes for a silicon atom then it can provide only three electrons to bond with the neighboring silicon atoms. This leads to a shortage of one electron, as is seen on the right in Figure 5.13. These electron

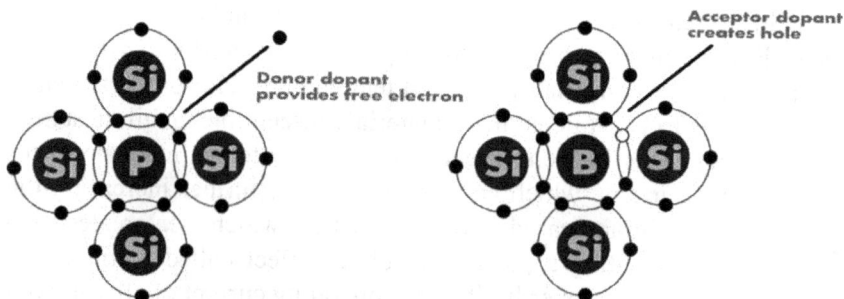

Fig. 5.13. Donor (left) and acceptor (right) doping in silicon.

vacancies behave as positively charged mobile carriers called holes. An electron from a neighboring atom can fill an electron vacancy and in this process the vacancy moves to the location of that electron. In this manner, through successive electron occupations, a hole can move through the crystal. Such movement of holes is equivalent to the movement of positive charge carriers.

Conversely, addition of tiny amounts of phosphorus (an element from group V of the Periodic Table) causes the appearance of additional electrons in the conduction band, as phosphorus atoms have one more valence electron compared to silicon atoms. This electron is released into the crystal and becomes a free electron — capable of carrying an electric current, as can be seen at the left in Figure 5.13. Thus, addition of either trivalent atoms (having three valence electrons) or pentavalent atoms (having five valence electrons) to silicon generates extra charge carriers, increasing the electrical conductivity of silicon. In the parlance of semiconductor physics, such electrically-active atoms, added to a semiconductor, are called dopants, as mentioned above. Free electron-producing dopants are called donors whereas hole-producing dopants are called acceptors. Phosphorus, arsenic and antimony — elements belonging to group V of the Periodic Table — are commonly used donor dopants for producing electrons in the conduction band whereas boron, aluminum and gallium — elements belonging to group III of the Periodic Table — are commonly used acceptor dopants for producing holes in the valence band. Furthermore, when a donor dopant atom releases a free electron it acquires a positive charge, i.e. it becomes a positive ion which is fixed (immobile) in the semiconductor crystal. Thus, each donor doping event produces both a mobile electron and an immobile positive ion. Conversely, each acceptor doping event produces a mobile hole and an immobile negative ion. This is how charge neutrality is maintained in a doped semiconductor.

A semiconductor without any doping is referred to as 'intrinsic' while once doped it is termed 'extrinsic'. Any semiconductor doped with donors to produce an excess of electrons is called '*n*-type' but if doped with acceptors to produce an excess of holes, it is called '*p*-type'. In addition to doping-generated charge carriers, every semiconductor sample also contains a temperature-dependent quantity of equal numbers of electrons and holes. Physically, this comes about because thermal energy breaks a few bonds in the crystal, with the electrons, thus, released residing in the conduction band. At the same time, the broken bonds become sites of a missing electron, i.e. electron vacancies or holes, residing in the valence band. Thus, thermal energy always creates a certain population of equal numbers of electrons and holes in any semiconductor crystal. This 'natural' population of electron–hole pairs is always present, with their numbers increasing rapidly (exponentially) with rise in temperature. Note that intentional doping, with either donors or acceptors, only boosts the number of either electrons or holes — not both. Doped (extrinsic) semiconductors contain either a far greater number of electrons (*n*-type material) or a much larger number of holes (*p*-type material), when compared with an intrinsic semiconductor sample at room temperature. Notice that, while *n*-type semiconductor materials are characterized by the presence of free electrons as charge carriers, there are also a much smaller number of holes present there due to thermal electron–hole pair generation. In this case, the electrons are called 'majority carriers' whereas the holes are called 'minority carriers'. With increase in temperature, the number of minority carriers increases so that the difference between the number of electrons and holes becomes smaller at elevated temperatures. For *p*-type material, the situation is exactly reversed, with holes being majority carriers and a much smaller number of electrons being present as minority carriers.

5.3.3 *Electronic transitions between conduction and valence bands*

Electronic transitions can take place between the conduction and valence bands of a semiconductor, just like transitions between the energy levels of an isolated atom. For example, an electron in the conduction band can fall into the valence band. The decrease in energy then appears as either heat or the emission of a photon. From a structural point of view, the free electron simply fills up a hole, i.e. heals a broken bond. This results in the disappearance of both the free electron and the hole in a process called radiative electron-hole recombination. This is the basic mechanism of light generation in all semiconductor light-emitting devices. Conversely, it is also possible for an incoming photon to be absorbed by a semiconductor with the energy, thus, gained causing an electron to jump from the valence band to the conduction band. Again, from a structural point of view, this amounts to the photon's energy causing a structural bond to break, creating a free electron and an electron vacancy, i.e. a hole. This electron–hole pair generation process is the basis of all semiconductor radiation detectors and solar cells. These processes can be seen in the diagrams in Figure 5.14.

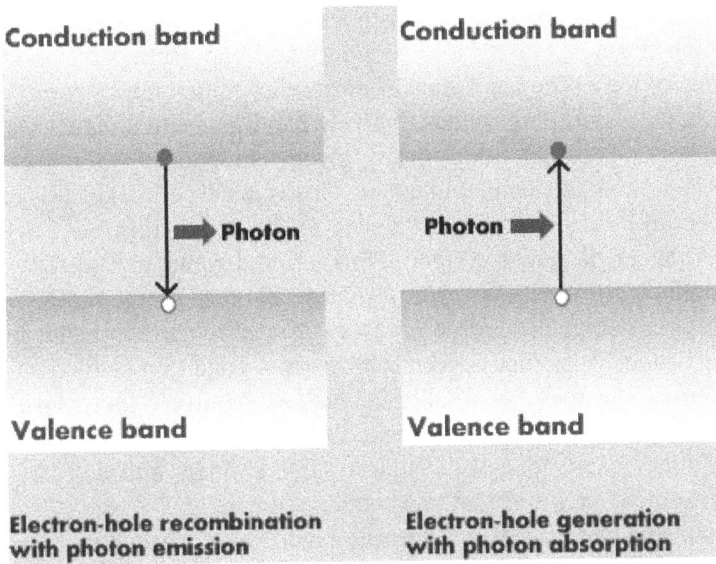

Fig. 5.14.   Electron transition between conduction and valence bands due to the emission and absorption of photons.

At any temperature above absolute zero, thermal energy keeps breaking random covalent structural bonds; producing electron hole pairs. This accounts for the small electrical conductivity that all semiconductors possess, and which increases with increase in temperature. Note that an electrical current can be carried by both electrons and holes (which behave as positively charged carriers) in a semiconductor. This 'bipolar' conduction contrasts with the 'unipolar' electric conduction in metals where only free electrons exist to carry electric current. Thermally generated electron–hole pair creation in semiconductors is balanced by a continuous process of electron–hole recombination. The two processes operate at the same rate (which increases with rise in temperature) so as to keep the population of free electrons and holes constant at a given temperature, for a given semiconductor.

The energy gap separating the lower edge of the conduction band from the upper edge of the valence band is one of the most important distinguishing characteristics of any semiconductor. This gap is determined by the identities of the atom or atoms that make up the semiconductor and their structural bonding arrangement. To get a feel for the sort of values one finds for semiconductor band gaps, let us consider two semiconductors with extreme band gaps. Indium antimonide (InSb) is a very narrow band gap semiconductor with a band gap of only 0.17 eV whereas aluminum nitride (AlN) is a very wide band gap semiconductor with a band gap of 6 eV. The band gaps of most other semiconductors fall somewhere in between these values. Narrow band gap semiconductors show higher intrinsic conductivity at any given temperature when compared with wider band gap semiconductors. This is because it takes less energy to excite an electron from the valence to the conduction band across a narrow gap in the band system. Thus, around room temperature, such semiconductors possess a relatively large number of thermally-created electrons and holes. The large numbers of charge carriers result in substantial conductivity that is seen with narrow band gap semiconductors. Such semiconductors are suitable for making relatively long wavelength emitters and detectors, such as those that operate in the infrared region. On the other hand, to make visible or UV emitters and detectors, one needs a somewhat wide band gap semiconductor.

A suitable band gap is not the only qualification needed in a semiconductor to make radiation-emitting devices from. Silicon — the most common semiconductor around, which is the basis of almost all integrated circuit (IC) devices — provides a case in point. Light-emitting semiconductor devices cannot be made from silicon because of a feature called the indirect band gap. Essentially, what this means is that for electronic transitions between conduction and valence bands in silicon, resulting in the absorption or emission of a photon, both energy and momentum cannot be simultaneously conserved. But, the conservations of both are essential for any allowed physical process. This makes silicon, and other indirect band gap semiconductors, incapable of efficiently detecting and generating photons. In devices made from silicon, electrical energy can only get dissipated as heat; unless some extraordinary arrangements are made. We have to go to other semiconductors, such as GaAs, GaN, InP, etc., with so-called direct band gaps, to make efficient photon detectors and emitters. We will look more closely at such semiconductors and ways to make them in the next section. For now, we turn our attention to device structures needed for making light-emitting devices. The most basic device is what is called a *pn-junction diode* so we will examine it in some detail here.

5.3.4 *The pn-junction diode*

It is possible to prepare a semiconductor crystal in such a way that part of it is *n*-type while the rest is *p*-type. Interesting things happen where these two distinct types of regions meet. This junction between *n*- and *p*-type materials is called a *pn*-junction. Such a junction has several very interesting properties that arise from the confluence of semiconducting regions with dissimilar charge carrier populations. Because the *n*-type region contains many free electrons in comparison with the *p*-type region so there is a natural tendency for mobile electrons to cross the junction and enter the *p*-type region. Similarly, holes from the *p*-type region tend to cross over to the *n*-type region. Natural diffusion of electrons from the *n*-type to the *p*-type region and of holes in the opposite direction leave un-neutralized dopant ions behind. These ions, on either side of the junction, are then no longer neutralized by mobile carriers of opposite polarity and, thus, give rise to an electric field in the vicinity of the junction region. The direction of the electric field is such that it opposes the natural diffusion of electrons and holes. The field keeps building in strength as electrons and holes diffuse across the junction, until the field strength is sufficient to stop any more carriers diffusing. A distinctive region then develops at the junction which lies partly in the *n*-type material and partly in the *p*-type material. This region contains fixed ions but no mobile carriers. Due to the absence of current carriers, it is called the 'depletion region'. It is also, less commonly, called the space charge region (see Figure 5.15). The depletion region is the most important part of any semiconductor *pn*-junction diode. Electrons and holes recombine with each other in this region and most of the heat and light generation in any diode takes place in the depletion region.

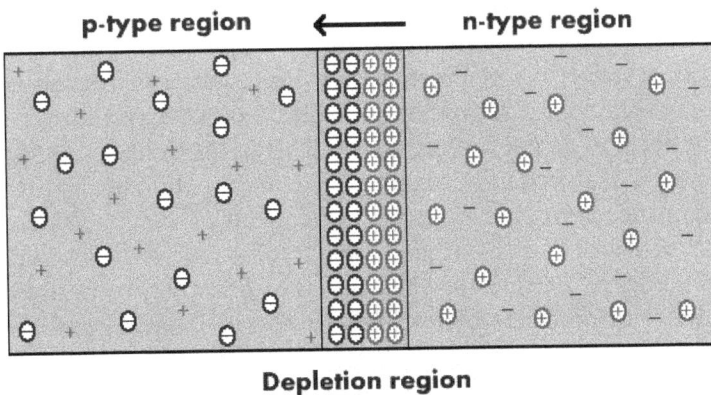

Depletion region

Fig. 5.15.   Structure of a *pn*-junction diode.

The existence of an electric field in the depletion region is equivalent to having a built-in potential difference (voltage) across this region. The value of this potential depends on the band gap of the semiconductor and on the amount of doping of the *p* and *n* regions. For *pn*-junction structures made from silicon, with typical doping concentrations, the built-in voltage is usually between 0.6 and 0.7 volt. If germanium is used as the semiconductor, then, due to the smaller band gap of germanium, the built-in potential is usually around 0.3 volt. The depletion region behaves as a natural built-in battery with a polarity that opposes both electrons and holes from entering this region. This means that it acts as a barrier and charge carriers cannot flow through such a series arrangement of *p*- and *n*-type semiconductor regions. However, if the *p* side of a *pn*-junction is connected to a higher electrical potential compared to the *n* side (using a battery or power supply), then the built-in voltage can be effectively canceled. If the externally-applied potential difference across the *pn*-junction is higher than the value of the built-in potential then electrons and holes can cross the junction and flow to the opposite side. In this way a current can flow through the structure. This type of electrical connection or biasing, where the *p* side is held at a higher potential compared to the *n* side is called forward bias. In the opposite case of reverse bias, where the *p* side is held at a lower potential with respect to the *n* side, the electric field due to the external bias reinforces the built-in depletion layer field and carriers cannot flow across the junction. In this case, only a small leakage current flows on the application of a reverse bias. We see that the amount of current conduction through a *pn*-junction depends on the direction of externally-applied bias: forward bias causes a large current flow whereas reverse bias results in only a very small leakage current. Thus, a *pn*-junction behaves very much like a one-way valve — allowing easy current conduction in one direction only. For this reason, it is also called a diode, which means a two-terminal electronic component that allows easy current flow in one direction only. This is also reflected in its schematic symbol that looks like an arrow — current only flows easily in the direction of the arrow.

The so-called current–voltage characteristic of a typical diode is shown in Figure 5.16. Forward bias is at the right whereas reverse bias is at the left. Forward bias corresponds to positive currents and voltages while reverse bias represents negative voltages and currents. When a diode is forward biased, very little current flows through it, unless the bias voltage is increased to a value V_d called the forward turn-on or cut-in

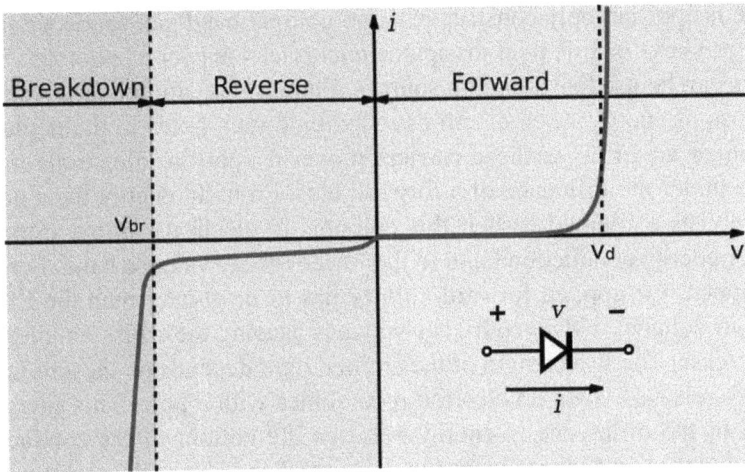

Fig. 5.16. Electrical (current–voltage) characteristics of a *pn*-junction diode.

voltage. This value is almost the same as the built-in voltage of the diode's *pn*-junction. For voltages greater than V_d, the forward current through the diode increases very rapidly. In the case of LEDs, this feature can easily destroy the diode due to excessive heat generation, unless some external component, such as a series current limiting resistor, is employed to limit the current to safe levels. In reverse bias, a small reverse leakage current flows through the diode. This current increases slowly with increase in reverse bias voltage. Beyond a certain reverse voltage, called the breakdown voltage V_{br}, a large reverse current can suddenly start to flow through the diode. Again, this current can cause excessive heat dissipation and can potentially destroy the diode. In practice, reverse biasing is avoided with LEDs so that this situation does not arise. An LED can be driven by a simple arrangement of a power supply and a current limiting resistor — all in series. In practice, often more involved circuit designs using special LED driver ICs are used for better power efficiency, controllability and component protection.

5.3.5 *Light emission from diodes*

In ordinary semiconductor *pn*-junction diodes, also called rectifier diodes (usually made from silicon, although germanium diodes are also available), current flow simply causes the diode to heat up. However, when a diode is appropriately constructed from a direct band gap semiconductor, such as GaAs or InP, then dissipated energy can appear as photons. Such LEDs can be used as radiation sources. Photons are emitted as a result of the loss of energy when electrons recombine with holes in the depletion region of an LED, as these carriers move in opposite directions in the diode under the influence of a forward bias. From an energy band point-of-view, as explained earlier, this amounts to electrons falling from the higher energy conduction band to the lower energy valence band. For this to happen, the applied forward voltage has to be greater than the LED's built-in voltage, with even higher voltages causing the emission intensity to increase. The wavelength of the emitted light depends on the amount of energy released when an electron recombines with a hole. This energy is equal to the difference in energy between the bottom of the conduction band and the top of the valence band, i.e. the band gap of the semiconductor. As an example, for GaAs *pn*-junction diodes (band gap = 1.4 eV), the emission wavelength can be readily calculated from the expression, $hc/\lambda = E_g$, where E_g is the band gap. This gives $\lambda = 886$ nm, which is a near-infrared wavelength. Replacing some gallium with aluminum, one can obtain aluminum gallium arsenide (AlGaAs), a ternary semiconductor with a wider band gap than GaAs. The exact value of the band gap depends on the relative proportions of gallium and aluminum, but as expected this will lie between the band gaps of GaAs (1.4 eV) and AlAs (2.1 eV). The wider band gap of AlGaAs makes LEDs made from this material emit light at shorter wavelength — even visible red — if the alloy is made with a wide enough band gap.

Electron–hole pair recombination across band gaps produce photons in direct band gap semiconductors, but as a means of producing useful light this process suffers from a number of inefficiencies. An understanding of these has enabled the construction of modern high efficiency LEDs. The electrical-to-optical power conversion efficiency, also called the wall-plug efficiency, of an LED can be written as a product of the so-called external efficiency of the LED and the input efficiency, as

$$\eta = \eta_{\text{ext}} \times \eta_{\text{input}}$$

Here, η_{input} is the ratio of the energy of the photon emitted $h\nu$ and the total energy of an electron–hole-pair:

$$\eta_{\text{input}} = h\nu/qV$$

q here is the value of the electron's charge, 1.6021×10^{-19} coulomb and V is the power-supply voltage.

The external efficiency of the LED is given by the product of three factors as

$$\eta_{ext} = \eta_{inj} \times \eta_{int} \times \eta_{extraction}$$

Here, η_{inj}, the injection efficiency, is the ratio of electron–hole pairs injected into the central luminescence-producing region of the LED to the number of electrons actually injected into the device from its external cathode contact. Loss of carriers, primarily through various leakage mechanisms, causes this factor to be less than unity. η_{int} is the internal efficiency which measures the number of photons created with a certain number of electron–hole pairs present in the active region of the LED. Non-radiative electron–hole recombinations at defects and dislocations are the primary causes that results in the loss of internal efficiency. Finally, $\eta_{extraction}$ is the extraction efficiency, which accounts for the actual brightness produced by an LED. It is simply the ratio of the number of photons that are able to escape the LED versus the total number of photons internally generated through radiative electron–hole recombinations.

Use of good device fabrication processes, together with high purity materials, optimized device structures and proper packaging, all contribute toward raising the various efficiency figures given above. Continued advances on all these fronts have resulted in the commercial availability of high efficiency LEDs in today's market.

The light emitted by an LED is usually narrow in spectral spread and can be practically considered as light of a single color. Figure 5.17 illustrates the concepts of peak wavelength and wavelength spread for an actual LED emission spectrum for a typical blue LED. The emission width is generally measured as full width at half maximum (FWHM), i.e. the width of the emission spectrum half way between the top and the bottom of the emission envelope. The broadening of LED emission is caused by a number of factors, such as band gap variation due to spatial variation in alloy stoichiometry, thermally-induced component of the kinetic energy of electrons and holes, coulomb interaction energy between electrons and holes and zero-point fluctuation of electrons and holes at their respective band edges.

Most common LEDs emit radiation in the so-called Lambertian pattern, seen in Figure 5.18. This means that the light intensity is at its

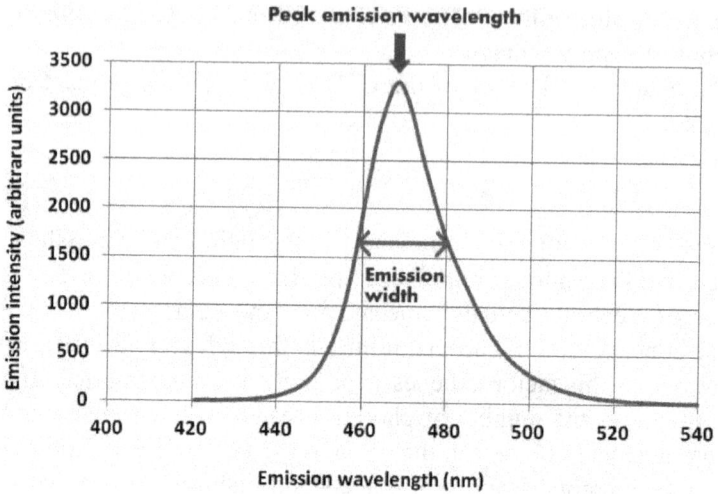

Fig. 5.17.   Emission spectrum of a typical blue LED.

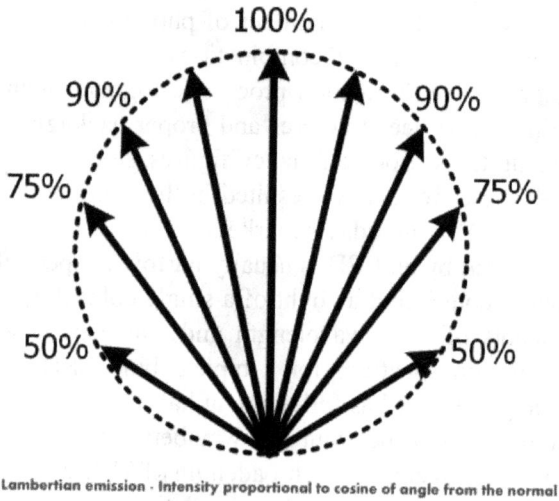

Fig. 5.18.   Angular emission pattern of light from a typical LED.

highest when viewed directly from above the LED. As one moves increasingly sideways, the intensity falls in proportion to the cosine of the angle from the normal to the LED. This essentially means that the light is

Fig. 5.19. LED package with integral reflector.

emitted without any directional bias. However, it is possible to alter the angular distribution of light intensity with the use of appropriate materials or structures. These are sometimes a part of the LED package itself (see Figure 5.19) and at other times may be supplied as external optical components, such as shaped enclosures, reflectors and lenses. With suitable optical arrangements, the light cone can be made to have any angular width from 180° to less than 10°.

5.4 LED Technology

It has been more than 60 years since the first semiconductor diodes with the ability to emit light were first demonstrated. During this time, LEDs have evolved from tiny devices emitting only a glimmer of light to veritable light engines capable of putting out several watts of visible light. The first LEDs were made from such compound semiconductors as GaP and aluminum gallium arsenide phosphide (AlGaAsP). Those devices emitted red, yellow and light green light, and were only used as small visual indicators in electronic instruments. LEDs began to be seriously considered for space lighting applications only when bright blue-emitting LEDs based on the semiconductor, GaN, were developed in the 1990s. That key enabling development won three Japanese scientists: Shuji Nakamura, Hiroshi Amano and Isamu Akasaki, the 2014 Nobel Prize in physics. The development of the blue-emitting LED completed the trio of red/green/blue emitters that was needed for full-color lighting. Without exception, all commercial LEDs emitting visible light are made from the so-called III–V compound semiconductors. These are binary, ternary or quaternary compounds made by combining one or more of aluminum, gallium and

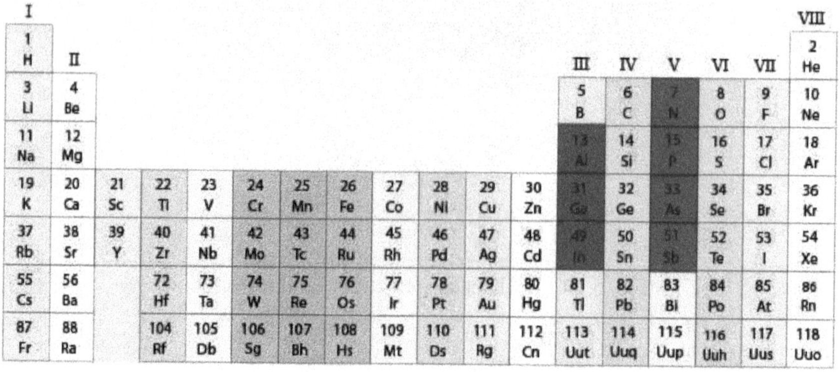

I																III	IV	V	VI	VII	VIII
1 H	II																				2 He
3 Li	4 Be															5 B	6 C	7 N	8 O	9 F	10 Ne
11 Na	12 Mg															13 Al	14 Si	15 P	16 S	17 Cl	18 Ar
19 K	20 Ca	21 Sc	22 Ti	23 V	24 Cr	25 Mn	26 Fe	27 Co	28 Ni	29 Cu	30 Zn					31 Ga	32 Ge	33 As	34 Se	35 Br	36 Kr
37 Rb	38 Sr	39 Y	40 Zr	41 Nb	42 Mo	43 Tc	44 Ru	45 Rh	46 Pd	47 Ag	48 Cd					49 In	50 Sn	51 Sb	52 Te	53 I	54 Xe
55 Cs	56 Ba		72 Hf	73 Ta	74 W	75 Re	76 Os	77 Ir	78 Pt	79 Au	80 Hg					81 Tl	82 Pb	83 Bi	84 Po	85 At	86 Rn
87 Fr	88 Ra		104 Rf	105 Db	106 Sg	107 Bh	108 Hs	109 Mt	110 Ds	111 Rg	112 Cn					113 Uut	114 Uuq	115 Uup	116 Uuh	117 Uus	118 Uuo

Fig. 5.20.   Periodic Table of elements.

indium from group III with one or more of nitrogen, phosphorus, arsenic and indium from group V of the Periodic Table (Figure 5.20). With proper compositional control, it is possible to obtain a range of band gaps and, thus, emission wavelengths. Not all chemical compositions, however, are suitable for this purpose because some compositions may lead to material with indirect band gap, which is unsuitable for making LEDs. Still other compositions may not be suitable for crystal growth and, thus, could not be used for making practical devices. By choosing suitable materials and compositions, it is now possible to fabricate LEDs that emit radiation from the mid-infrared region all the way to the deep ultraviolet. Research continues to improve various performance metrics of these LEDs, so that their state-of-the-art continues to advance with time.

At first, red, green and blue (RGB) LEDs were used for producing white light through color mixing, but later phosphor-based white LEDs were developed, and that really got the solid-state lighting revolution going. Generating white light by combining the light from LEDs of three (or more) colors is difficult because of the need to properly mix together light from different emitters. Those who have tried this, know the difficulty involved in proper color mixing. A single LED solution for this purpose is, therefore, much preferable. Today's 'white' LEDs consist of a blue LED chip coated with a suitable luminescent material. This material, called a phosphor, converts some of the blue light from the LED into red, orange and yellow light. These colors, in combination with the residual (unconverted) blue light from the LED, give the impression of white light to human eyes. Phosphors are, thus, central to the working of all white LEDs. In fact, the quality of white light produced is almost exclusively dependent on the properties of the phosphor used for making a white LED. Making better illumination-quality LEDs, thus, comes down to developing better phosphors, more than anything else.

The construction of a typical phosphor-converted LED can be seen in Figure 5.21. First, a small GaN/AlGaN blue-emitting LED chip measuring from 1 mm × 1 mm to 3 mm × 3 mm is attached to a suitable chip carrier made of plastic or ceramic. Next, the anode and cathode of the LED chip are gold wire-bonded to wire bond pads on the chip carrier package. After that, a precise amount of phosphor slurry is dispensed on the chip. This may be followed by the addition of a polycarbonate or silicone dome lens in liquid form. The phosphor + lens structure is then cured to a semi-rigid state with UV radiation while the package sits under vacuum to remove any bubbles that might form during the resin curing process. At the end, the packaged LED is tested for proper functioning and then packaged into trays or tubes for shipment to its destination. The photograph of an actual commercial white LED appears below in Figure 5.22. The yellow phosphor coating on the LED chip and the silicone dome lens

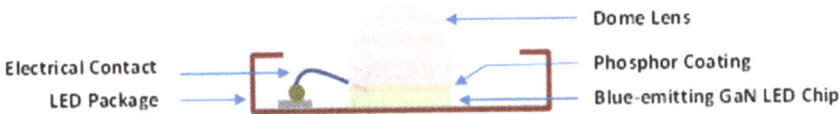

Fig. 5.21.   Diagram showing the construction of a typical phosphor-coated LED.

Fig. 5.22.   Typical commercial white LED.

on the top can be seen in this figure. This is a 1 W power LED in a plastic surface mount package with a metal underside (not visible in this picture).

5.5 The Blue LED

Once blue and phosphor-based white LEDs were developed, attention turned toward making them brighter so that these devices could be used for lighting applications. This required many innovations, ranging from new materials and phosphors to novel device packaging schemes. In order to appreciate these developments, it is worthwhile to start by taking a look at the structure of typical modern blue-emitting LED chips.

Blue LEDs are now almost universally made from GaN and indium gallium nitride (InGaN) deposited on either sapphire or SiC wafers (used as substrates). These two different, but related, semiconductors are used in order to take advantage of some clever engineering that is possible when two separate semiconductors are layered on top of each other. This approach is known as 'heterostructure' engineering because of the use of dissimilar materials. To understand the idea, take a look at Figure 5.23 which shows the band profiles of two different semiconductors in contact. The band gaps of the two types of semiconductors are shown shaded in different colors. If a semiconductor with a relatively narrow band gap, δE, is sandwiched between layers of a different semiconductor with a wider band gap, ΔE, then the situation shown in this figure arises. Here, a well-like profile is created from the line-up of band edges (shown at the top). This structure is known as a quantum well, and it plays a central role in all modern LEDs. While an LED is essentially just a *pn*-junction diode, by placing one or more quantum wells at the center of the diode one can greatly enhance the light generation efficiency of the diode. This happens because quantum wells can accumulate charge

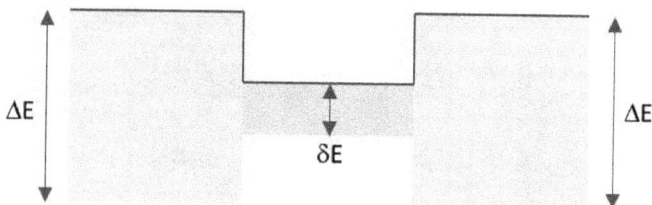

Fig. 5.23.   Schematic illustration of the band profile of a semiconductor quantum well.

carriers inside the well regions, with electrons pooled at the edge of the conduction band and holes pooled at the edge of the valence band. The carriers are stopped from dispersing easily because of the presence of energy barriers on either side of the well. The large population of electrons and holes makes their radiative recombination much more likely — boosting the net rate of photon generation and making the device appear much brighter than would be the case if a quantum well structure was not used. Most modern commercial LEDs contain five quantum wells in series, placed right at the center of the diode structure, sandwiched between thicker *p*- and *n*-type layers. For blue LEDs the standard practice is to use GaN for the barrier layers and the narrower band gap InGaN for the well layers. The exact band gap of InGaN depends on the alloy stoichiometry, with larger amounts of indium leading to narrower band gaps (and thus longer emission wavelengths).

5.5.1 *Blue LED wafer fabrication*

A typical blue LED wafer consists of a number of very thin layers of GaN and related compounds that are deposited on a typically 3-inch diameter sapphire (Al_2O_3) substrate through a process called metal organic chemical vapor deposition (MOCVD). This is also alternatively known as metal organic vapor phase epitaxy (MOVPE). Sapphire wafers are used for GaN layer growth because of their low cost and compatibility with the growth process. Figure 5.24 shows a set of sapphire wafers meant for GaN LED

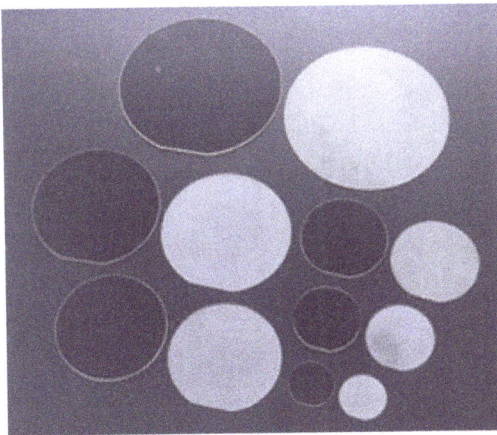

Fig. 5.24.   Sapphire wafers for MOCVD GaN layer growth.

layer growth. One side of the wafers is finely polished to a mirror-like finish. This is the surface that is used for MOCVD growth of LED layers. Ideally, one would want to use GaN wafers for the growth of GaN-based blue-emitting LED structures as this offers true 'epitaxial' (same crystal structure) growth. Epitaxial growth minimizes defects at the interface of the base wafer (substrate) and the over-lying structural layers. However, one component of GaN (nitrogen) being a gas, there is no straightforward way to produce bulk crystals of the compound GaN by starting from the constituent elements, i.e. gallium and nitrogen. Very high-pressure synthesis routes are possible but those techniques yield only small crystals of GaN. Given this situation, sapphire wafers — sawed off from large synthetic boules of factory-grown sapphire — provide the best alternative. New techniques have been developed to produce GaN substrate wafers using hydrothermal processes that make use of ammonia at high temperature and pressure. These methods yield high quality 'native' GaN pieces from which wafers can be produced. Figure 5.25 shows such a 'free standing' GaN wafer. While GaN wafers are now commercially available in several sizes, their cost is far too high when compared with sapphire wafers. Although native GaN wafers offer all the advantages that are associated with epitaxial growth on the same material substrate, their very high cost has not allowed them to be used for commercial blue LED production so far. These wafers, instead, are being used for the growth of GaN laser diodes which demand a better match between the substrate and epitaxial layers grown on it. In that case, the higher price of laser diodes allows the higher cost of free-standing GaN wafers to be absorbed in the

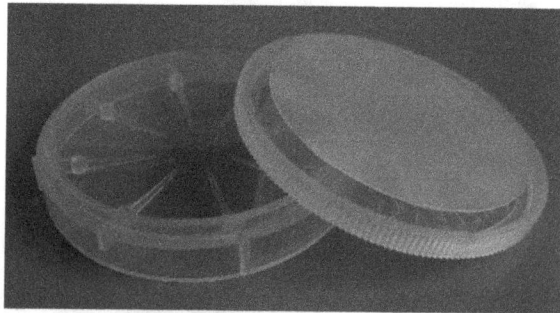

Fig. 5.25.   Free-standing (bulk) GaN wafer.

production process. If GaN wafers are used for LED growth in the future, then large HB LEDs will probably be the first devices that will see their use.

The actual epitaxial growth of structural layers that will form blue-emitting LEDs is carried out through an MOCVD process, as mentioned earlier. This consists of passing reactant gases, called precursors, over heated wafers, in an otherwise evacuated chamber. A schematic diagram of an MOCVD reactor, for growing layers of GaN and related semiconductors, is shown here in Figure 5.26.

MOCVD reactors for commercial GaN LED wafer production are manufactured by several companies. Prominent among these, at the time of this writing, are Aixtron and Veeco. The core of any MOCVD system consists of a cylindrical stainless-steel chamber with a hermetically-sealable top cover. Inside the chamber there is a round platter with circular depressions where growth substrates can be placed. This platter is called a susceptor and is usually made from an inert high melting point material,

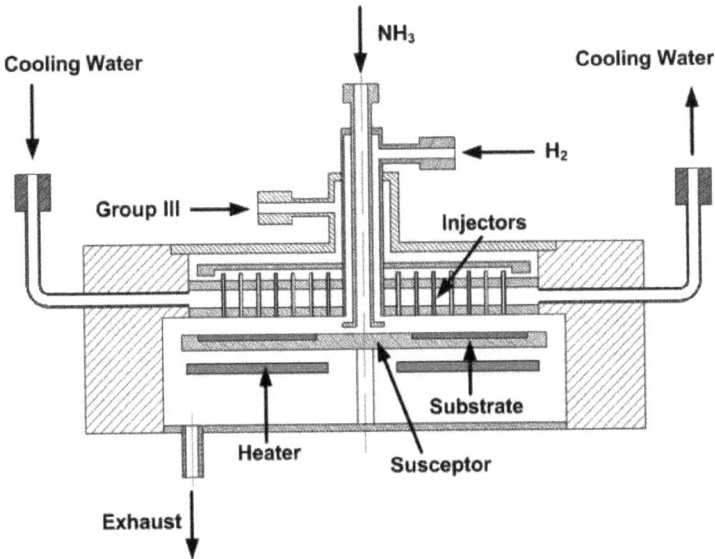

Fig. 5.26. Diagram of an MOCVD reactor used for the growth of III-nitride epitaxial material.

Fig. 5.27.   Susceptor (wafer carrying platen) seen inside an MOCVD reactor chamber.

such as SiC or graphite. A susceptor can be seen in Figure 5.27, where a system that can accept five wafers at a time is shown. Larger systems can take twelve or more wafers at one time, and are generally preferred for their high throughput and use of less reactant per processed wafer. The susceptor is capable of heating up to several hundred degrees Celsius by precision-controlled electric heaters situated at the bottom. The susceptor temperature can be varied during the growth process, as required by the recipe used for a particular epitaxial layer sequence. Reactants, in the form of gases, are introduced from the top through a gas distribution manifold. Group III elements are provided using a suitable organometallic compound, such as trimethyl gallium (TMG). Nitrogen is provided by using ammonia. Specially-designed baffles and injectors ensure uniform distribution of all reactants at wafer surfaces. The temperature is very carefully monitored at various locations inside the growth chamber, including the wafer surfaces, using strategically placed thermocouples and pyroelectric temperature sensors. Cooling is provided by circulating chilled water through components that may get very hot. Finally, the background pressure inside the chamber is maintained with suitable pumping systems and a dry nitrogen-based re-pressurization system is used whenever the chamber is opened to atmospheric pressure after a growth run.

Gallium nitride LED wafer growth is carried out in a careful step-by-step manner after cleaned substrate wafers that have undergone appropriate surface preparation have been loaded into the MOCVD growth chamber. Once the correct temperature, pressure and gaseous ambient has

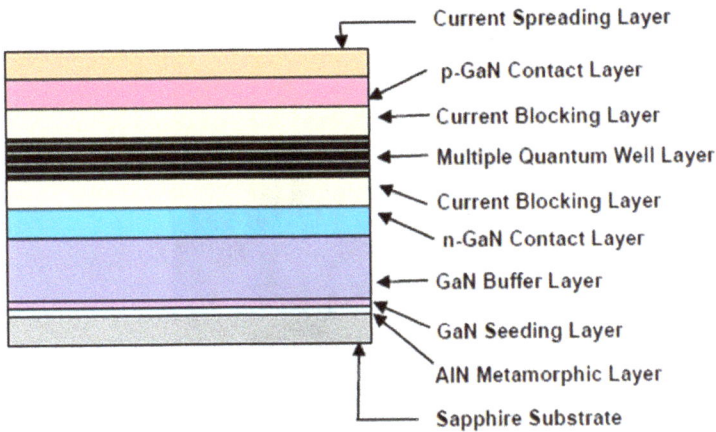

Fig. 5.28. Typical epitaxial layer structure for GaN LEDs.

been attained in the chamber, a previously-validated layer scheme is deposited on the substrates in a sequential fashion. Each layer is deposited at its own growth temperature and reactant flow rates, with the entire sequence controlled by a pre-programmed process control computer. A typical layer structure for blue-emitting gallium nitride LEDs is shown in Figure 5.28.

The very first layer consists of AlN which makes it possible to deposit subsequent layers. AlN is a suitable choice for this purpose because it has one element (aluminum) which is common with the substrate (sapphire is aluminum oxide) while the other element nitrogen is common with the GaN layer which comes next. The purpose of this so-called metamorphic layer is to make a gentle transition from the crystal structure of aluminum oxide to that of GaN. Isamu Akasaki and his colleagues came up with this idea at Nagoya University in Japan in 1986. It allowed successful epitaxy of the so-called III-nitride material system on sapphire substrates even though there is a 14% lattice parameter mismatch between sapphire and GaN. Next comes a very thin GaN layer called a seeding layer which is grown at a relatively high temperature. A thick buffer layer of un-doped GaN is deposited next. Its purpose is to terminate any threading disloca- tions (Figure 5.29) that may have started growing through the seeding layer. Threading dislocations are line-like vertical defects in crystal struc- ture that form at layer interfaces due to the mismatch of lattice constants, i.e. the sizes of crystals' unit cells. Use of an AlN inter-layer between the

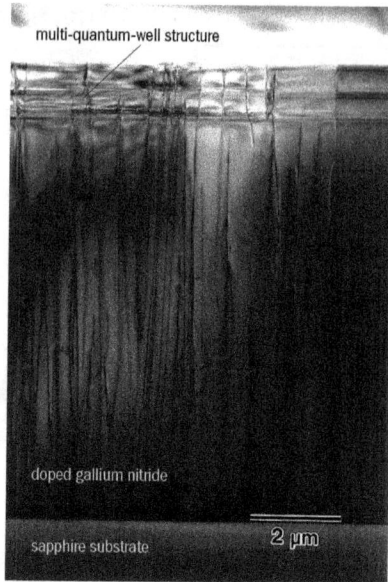

Fig. 5.29.   Transmission electron micrograph showing threading dislocations in a GaN LED wafer.

sapphire substrate and the over lying III-nitride layers very significantly reduces the density of threading dislocations, but does not completely eliminate them. If allowed to propagate to the active LED layers (quantum wells), the dislocations can result in excessive non-radiative recombination of electrons and holes, which generate heat rather than photons. Next comes an *n*-type GaN layer (containing silicon as dopant, introduced through the use of silane gas: SiH_4). This is followed by a series of (usually five) quantum wells, each consisting of a thin layer of InGaN sandwiched between GaN layers. Finally, a layer of *p*-type GaN (doped with magnesium, provided by the use of *bis*(cyclopentadienyl)magnesium: Cp_2Mg) is deposited at the top. Note that the *p*-GaN layer is deposited at the very last in the layer growth process because magnesium atoms have a tendency to diffuse through layers that may be grown on top of magnesium-containing layers.

After the layer growth is complete, the wafers are removed from the MOCVD chamber and subjected to a special thermal anneal process at around 800°C to activate the dopants. This step releases electrons, and even more importantly, holes into the doped layers. Because *p*-type GaN is not very conductive so it is the usual practice to deposit a very thin (a few nanometers thick) layer of a suitable transparent, but electrically conducting, material at the very top to spread current uniformly throughout the p-GaN layer. This current spreading layer could be either indium tin oxide (ITO), aluminum-doped zinc oxide (AZO) or a very thin bi-layer of nickel and gold. DC magnetron sputtering is usually the method of choice for depositing this transparent conductive layer.

The peak wavelength emitted by such an LED structure is determined by two main factors. One is the band gap of the semiconductor material that forms the well regions of the LED. This can be adjusted, as needed, by selecting the proper stoichiometry for the InGaN layer. As the amount of indium is increased, the band gap shrinks and, consequently, the wavelength of emitted photons gets longer. The color of the emitted light then shifts toward the green region. Too much indium, however, can cause a well-known phase separation problem where instead of being distributed uniformly in stoichiometric proportion, indium begins to form small clusters within the GaN material. This problem prevented the manufacture of efficient true green LEDs using this technique for a long time. Techniques have been developed to overcome this phase segregation problem so that green GaN/InGaN LEDs are now produced commercially. Band gap tuning through InGaN composition adjustment can be used to produce GaN/InGaN LEDs that emit almost any wavelength from around 400 nm to nearly 520 nm. The other technique for adjusting the color of quantum well LEDs is through adjusting the quantum well thickness itself. As the well regions are made thinner, energy levels in the conduction and valence bands are effectively pushed apart. This increases their separations and results in the generation of shorter wavelength photons. As a result, the light shifts to shorter wavelengths, i.e. toward the UV end. In practice, both InGaN compositional tuning and quantum well width tuning are used together to produce commercial GaN/InGaN LEDs with emission wavelengths throughout the near-UV, violet, blue and green regions. Such a selection of LEDs is very useful because one can select the best peak emission wavelength to match with the peak absorption of a phosphor to make highly efficient white LEDs.

5.5.2 *Blue LED device fabrication*

Once LED wafers have been grown in an MOCVD reactor, they are used to make LED devices. To do this, some of the material is removed from the top, down to the *n*-type GaN layer. Electrical contacts are then made to the *n*- and *p*-type GaN which, together with the quantum wells, form the diode structure. If the contacts are connected to a power supply, then electrons and holes flow through the diode. These come together in the quantum wells where they merge with each other and in that process release light. On-wafer testing is carried out on several locations to make sure that proper LED functionality is present before the wafer is diced into individual chips. The individual LED chips may be further tested and 'binned' into groups of devices with similar peak emission wavelengths and output powers. The chips are finally packaged into suitable packages and go through a final round of testing. White LEDs and other 'color-converted' LEDs require a luminescent phosphor material which is applied on top of the LED chips during the device packaging process. This step is described in more detail, later in the chapter.

An alternative but more involved process for making HB LEDs is the so-called flip-chip technology. Here, as shown in Figure 5.30, after MOCVD growth, the LED wafer is first patterned to isolate individual LED chips, while these are still on the wafer (a). Next, a conformal silicon dioxide film is deposited on the top surface of the patterned wafer, followed by a platinum metal film and then a layer of photoresist (b). The platinum layer is to serve as a light reflector. The photoresist is used to open widows in the platinum/silicon dioxide layers to electroplate copper as a contact metal (c). The entire wafer is then flipped (inverted) to remove the sapphire used as the growth substrate by a technique called laser lift-off (LLO). This uses exposure to high power excimer laser pulses which cause material decomposition followed by delamination at the sapphire–gallium nitride interface (d). This step exposes the undoped GaN buffer layer which is removed and titanium/aluminum contacts are deposited on each LED chip to complete the fabrication process (e, f).

Flip-chip LEDs are significantly more efficient when compared to non-flip-chip devices. This is seen in Figure 5.31, and is due to a number of improvements. There is no current spreading layer in flip-chip LEDs as the platinum reflector acts as both a highly reflective mirror and a full-face electrical contact to the lower conductivity *p*-GaN layer. The top contact is a relatively small electrical contact which does not obstruct light too much. Furthermore, the bottom of the device is a layer of thick copper which acts as a very potent heat spreader, efficiently extracting heat from the device layers and transferring it to any external heat sink.

Low power LEDs make use of smaller chips (1 mm × 1 mm, or smaller) and are packaged in small through-hole or surface mount

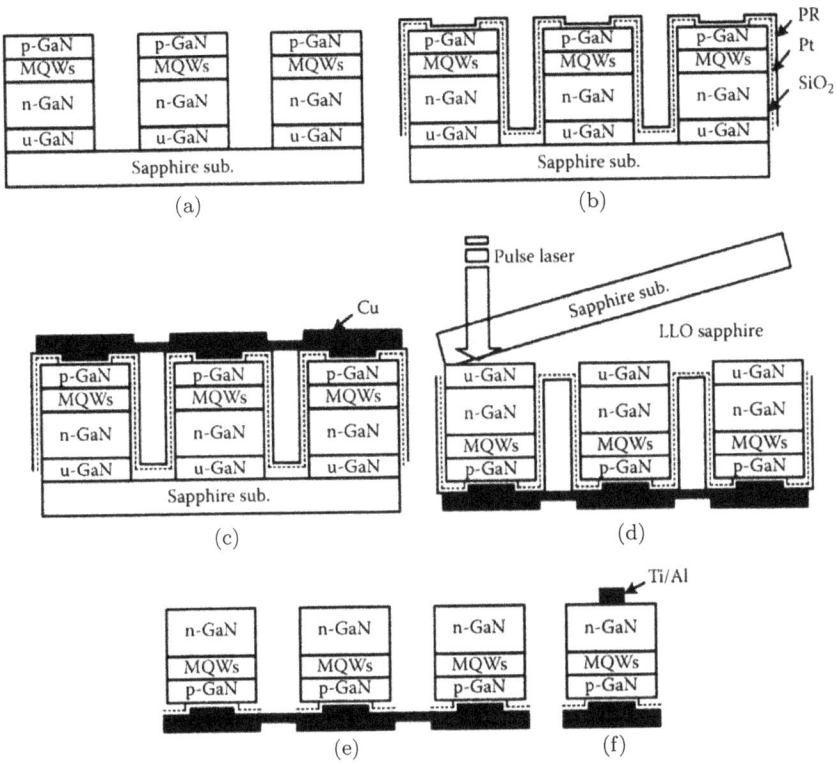

Fig. 5.30. Diagrams showing process steps used for fabricating flip chip LEDs.

packages. These devices are typically rated at up to 250 mA of maximum current flow and are used for electronic indication and other low luminosity applications. Low power LEDs are also used for various flat-panel display applications. Figure 5.32 shows low power LEDs in both through-hole and SMD packages.

Higher power LEDs are made from larger chips (3 mm × 3 mm, or larger) and are packaged in special packages with provision for adequate heat dissipation. Figure 5.33 shows a variety of LEDs with different SMD packaging styles.

5.6 Enhancing Light Output — Surface Texturing and Photonic Crystals

A particularly severe problem with all LEDs is that, due to the high refractive index of the semiconductor material, only a small amount of light

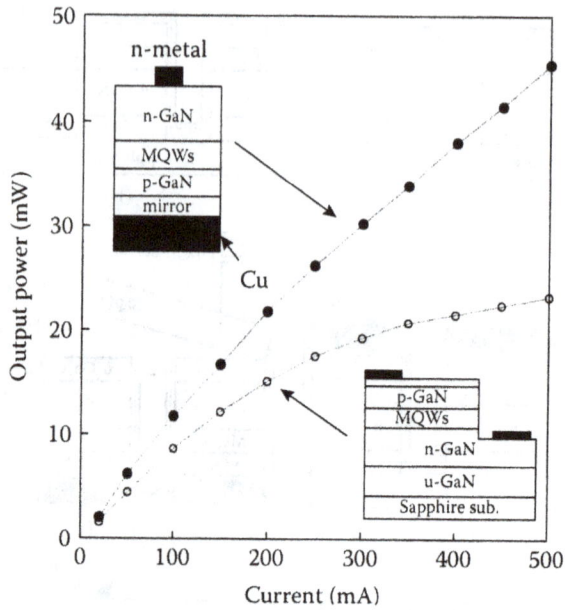

Fig. 5.31.   Comparison of optical output power versus drive current between 'ordinary' and flip chip LEDs.

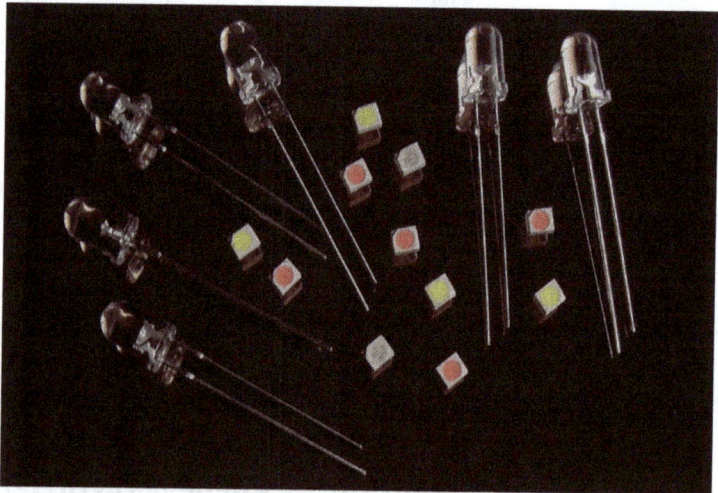

Fig. 5.32.   LEDs in through hole and surface mount packaging.

Fig. 5.33.   LEDs in different package styles.

generated inside the LED chips is able to escape as visible radiation. Most of the light remains trapped inside the confines of the chip due to total internal reflection at the chip's surfaces, as can be seen in Figure 5.34. It is estimated that if nothing is done about this problem then as much as 90% of the light can remain confined inside an LED chip. This phenomenon causes undesirable heating of the LED die and also lowers the efficacy (lumens output per watt electrical energy input) of LEDs as light sources. The high refractive index of GaN (2.4) makes light trapping a particularly serious problem. However, this phenomenon is not unique to GaN and arises in any transparent material of high refractive index. Diamond, for instance, admits light from all angles, but once inside, the escape cone is very narrow so that the trapped light can only escape through a few well-oriented facets, giving diamond its so-called 'fire'.

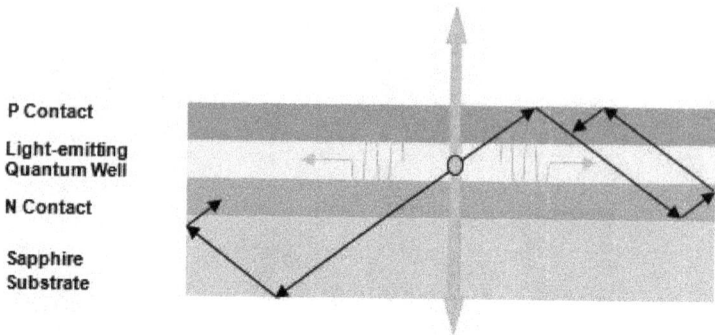

Fig. 5.34.   Light trapping due to total internal reflection inside a *pn*-junction LED.

Techniques have been developed to extract much of the trapped light from inside the LEDs — increasing their apparent brightness and efficacy. One widely used method is to simply roughen the top surface of LED chips through a short etch in warm aqueous potassium hydroxide solution. This process causes uneven etching, resulting in a rough surface with random topography. This roughness aids light to escape by making the chip geometry less favorable for total internal reflection to take place. Almost all of the LEDs commercially sold today have roughened emitting surfaces. Another technique relies on producing an ordered arrangement of tiny blind holes or shallow depressions on an LED's top surface (see Figure 5.35). Such a structure, with periodic spatial variation in refractive index, is called a photonic crystal.

Etched photonic crystals have a two-dimensional periodicity and act as a two-dimensional diffraction grating etched on the surface of the LED. With proper choice of the size of the holes and their spacings, the periodicity of the photonic crystal can interact with the periodicity of the confined light field in such a way that the light is steered outside of the chip. Typical sizes involved are around 200 nm for hole diameter and 500 nm for hole center-to-center spacing (hole pitch). The photonic crystal pattern is formed by a suitable lithography technique, such as nano-imprint lithography (NIL), and the structure is then etched into the surface using dry etching techniques. The etch depth is quite shallow and 50–100 nm is typical. A number of different designs are possible. Patterns using both regular symmetry and quasi-periodic symmetry have been used. Figure 5.36 shows blue light (from a laser beam) diffracting from an LED wafer patterned with a six-fold symmetry photonic crystal pattern.

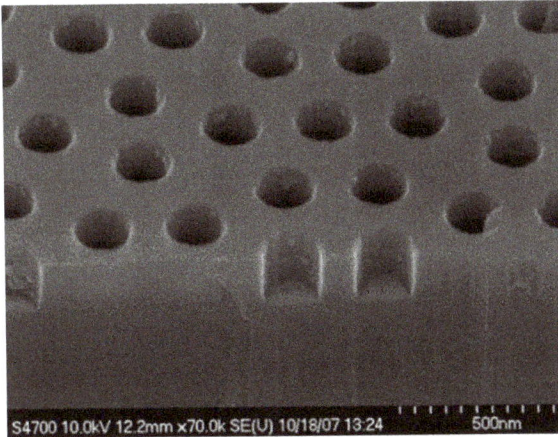

Fig. 5.35. Array of shallow holes, forming the photonic crystal structure on the top surface of a GaN LED wafer.

Fig. 5.36. Blue laser diode beam diffracting on transmission through a GaN LED wafer, patterned with a photonic crystal structure.

LEDs with photonic crystals are not only brighter than other types of LEDs but also emit light in a narrower, more collimated beam. This ordered structure is much harder to produce than a simple rough surface so photonic crystal LEDs are much more expensive compared to surface-roughened LEDs. Such LEDs have a limited market and are mainly used

Fig. 5.37.   Commercial photonic crystal LED from Luminus Devices Inc.
Courtesy: Luminus Devices Inc.

in some projectors and other display devices. Figure 5.37 shows a commercial photonic crystal LED from Luminus Devices Inc.

Finally, yet another technology that has been successfully used to make brighter LEDs is based on the use of patterned sapphire substrate (PSS). Instead of using a flat sapphire surface for MOCVD growth, use is made of sapphire wafers which have been etched to create tiny spherical or pyramidal patterns on the entire front wafer surface. A micrograph of one such wafer appears below in Figure 5.38, with the inset showing a close-up of the individual surface protrusions on the wafer surface.

PSS wafers offer two main advantages. During the MOCVD growth of the GaN buffer layer, the three-dimensional structure with slanted side walls causes threading dislocations to tend to change direction from vertical to horizontal. This reduces the number of such dislocations that propagate into the upper active device layers. Reduction in dislocation density results in less non-radiative recombination and, thus, higher photon yield during device operation. Reduced dislocation density also reduces LED heating by reducing overall device resistance and contributes to longer LED lifetime. The other prominent advantage of PSS comes from its ability to enable more light to be extracted from the device. This happens

Fig. 5.38. Surface of a patterned sapphire wafer for LED layer growth. Inset shows a close-up view of surface topography.

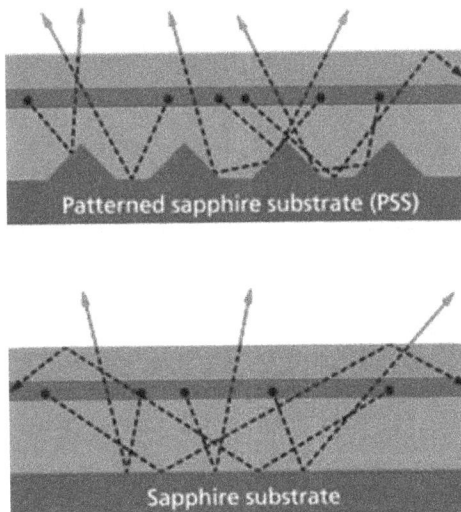

Fig. 5.39. Light extraction comparison between flat and patterned sapphire substrates for LED wafers.

because downward directed light rays encounter less-favorable angles for total internal reflection at sapphire-GaN boundaries, as seen below in Figure 5.39. The redirection of light considerably enhances the brightness of LEDs constructed on PSS wafers; making them a popular choice among commercial LED manufacturers.

5.7 Thermal Management of LEDs

Efficient removal of heat remains the most serious issue facing designers of high-power LEDs and LED-based luminaires. While LEDs are extremely efficient in converting electric power to visible light, they still generate a considerable amount of heat which must be quickly dissipated to ensure trouble-free operation. This is a particular problem with larger watt-class LEDs that make use of larger LED dies. Device temperature can easily exceed 100°C when more than an ampere of current is passing through a multi-watt LED. This can pose both short- and long-term reliability issues. The heat generated by LEDs is produced in a number of ways. These include, resistive heating inside the device, non-radiative electron–hole recombination inside the chip, self-absorption of radiation in the chip and absorption of light by phosphor and packaging material. In contrast to incandescent sources where much of the heat is produced as infrared radiation which is lost by radiation to the surroundings, no part of an LED gets so hot as to produce easily-radiated short wavelength infrared radiation. This makes cooling LEDs especially difficult and ingenious methods have to be employed to make sure that as much heat as possible is removed from the LED so that its temperature remains within safe bounds.

The first line of defense against thermal runaway in power LEDs is the use of appropriate low thermal resistance packages. Special LED packaging technologies have been developed where the LED die is located on top of a large metal base which can efficiently drain away heat produced by the chip. One such LED is seen in Figure 5.40. Through holes provided on the metal base plate, such LEDs can be attached to larger air- or water-cooled heat sinks or heat pipes for efficient heat removal.

For power LEDs mounted on printed circuit boards (PCBs), use is generally made of so-called metal core PCBs (MCPCBs) which use an aluminum plate as the PCB base. Such boards can properly diffuse away heat generated from the point where an LED is mounted. Furthermore, as LEDs are placed and soldered on to PCBs, it is generally advisable to use a very thin coating of a thermally conductive material between the LED's

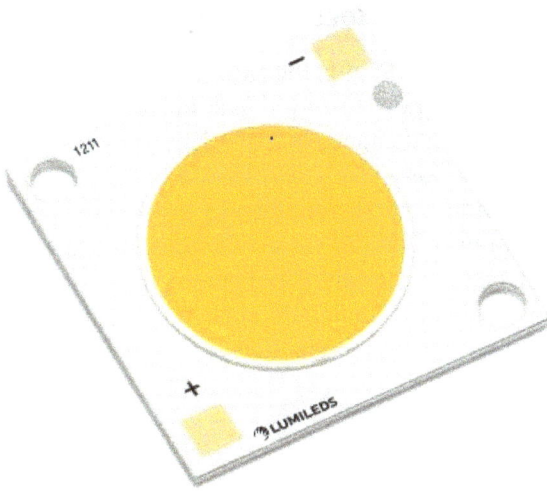

Fig. 5.40.   Power LED package on a metal core PCB.

Courtesy: Lumileds Inc.

bottom surface and the LED pad on the PCB. This fills up any air pockets between the two mating surfaces and allows heat to flow more efficiently between the LED and the PCB.

Other strategies have also been used for managing heat during LED operation. Devices made by Cree Inc. make use of SiC instead of sapphire as the substrate for the growth of LED structure. Silicon carbide has superior thermal conductivity when compared to sapphire and this helps in extracting heat from the LED layers. The ultimate design in this respect is one where the LED chip is inverted before being packaged. In this 'flip-chip' configuration, that has been mentioned previously, the top portion of the LED chip where the light (and heat) is generated is coated with a layer of metal to act both as a reflector and as the positive electrode and this side is then bonded to a metal pad which acts as a heat sink. The bottom part of the LED die is now exposed at the top, from where light can escape through the transparent substrate. As the heat no longer needs to escape through the substrate, so the chip can dissipate heat much more easily. This design prolongs LED life and enables these devices to run at higher currents, producing more light. Flip chip configuration is now standard for all moderate and high-power LEDs.

5.8 Luminescent Materials for LEDs

Due to the explosive growth of white LEDs, the majority of all blue LEDs produced annually are now used for making phosphor-converted white LEDs. Thus, the market demand for blue LEDs is, for the most part, being driven by the demand for white LEDs. However, blue LEDs, albeit essential, are only the pump sources for white LEDs. In order to produce white light, the blue emitters have to work with various kinds of luminescent wavelength conversion materials. Chief among these are the various phosphors but other luminescent materials, such as quantum dots, are also now beginning to be used in commercial devices.

5.8.1 *Phosphors*

While organic light converting materials exist, all commercial white LEDs use inorganic phosphors. A typical phosphor consists of a host crystal doped with a small amount of one or more 'activator' ions. The activator ions are usually (but not always) ions of one of the rare-earth atoms (see Figure 5.41). Electronic transitions within the rare-earth ions are responsible for color conversion. The unique structures of the atoms of the rare-earth series of elements (the lanthanides) bestow them with the rather special ability to absorb light at shorter wavelengths (typically, blue and

Fig. 5.41.   Rare-earth elements.

UV) and efficiently emit the absorbed energy at longer wavelengths. Because longer wavelength photons carry less energy than shorter wavelength ones so some energy is necessarily lost in this process. This is called the Stokes shift or Stokes loss, and it accounts for most of the energy loss in a phosphor-driven light emitter. This energy loss appears as heat in the phosphor.

It is important to understand that most phosphors are two-component systems, consisting of a host crystal matrix containing luminescent ionic centers. The role of the host matrix is crucial as it affects the energy levels of the dopant atoms and, thus, determines the absorption and emission wavelengths of the phosphor. The same ion, doped in different crystal hosts, will exhibit different light conversion characteristics. As an example, a typical phosphor used in many low-priced cool (slightly bluish) white LEDs is cerium-doped yttrium aluminum garnet (Ce:YAG). Here cerium ions are doped into a host crystal which is a mixed oxide of aluminum and yttrium. This was one of the first phosphors used for making white LEDs, and it is still widely used in the manufacture of LEDs for inexpensive LED lamps. Ce:YAG generates yellow light when excited with blue light. In combination with the residual blue light from the pump LED, the combined yellow and blue lights appear as a cool shade of white.

Figure 5.42 shows the absorption (excitation) and emission spectra of this phosphor. Such spectra are characteristic of all luminescent materials.

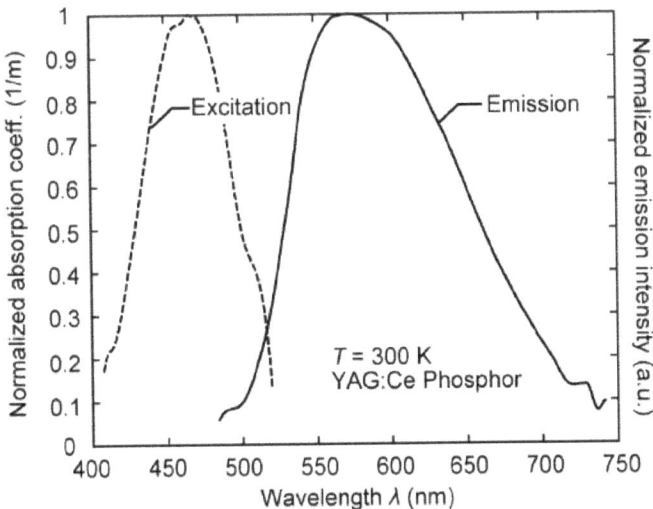

Fig. 5.42. Absorption and emission spectra of Ce:YAG phosphor.

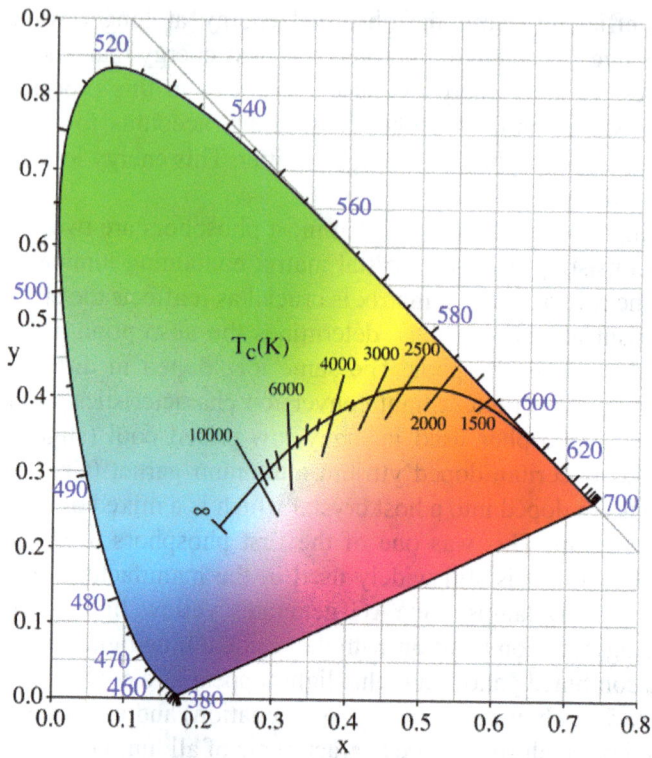

Fig. 5.43.   Chromaticity diagram with superimposed black body locus.

The difference in wavelengths between the excitation and emission peaks represents the Stokes loss in this case (when converted to energy unit).

The quality of white light emitted by LEDs is quantified by a number of parameters. These include its color or chromaticity point (x, y) in a chart of saturated and unsaturated colors, called a chromaticity diagram. As seen in Figure 5.43, the chart is bounded by monochromatic wavelengths. The curved black line inside the diagram is the blackbody locus which passes through the colors that would be emitted by a heated perfectly black object at different temperatures. The short black lines intersecting the locus show the various correlated color temperatures (CCTs) — temperatures of non-blackbody emitters emitting light similar to a blackbody. Light intensity itself is given in units called lumen, which quantifies how bright it appears to human eyes, whereas the efficacy of the device is measured in lumens per watt of electrical energy consumption. Finally, color rendering index (CRI) is stated to describe how closely the light from a white LED matches that of natural daylight. A perfect match would imply a CRI of 100, whereas lesser values stand for poorer approximations to natural white light.

Phosphors — of the kind used in GaN LEDs — are made by high temperature solid-state reactions. Usually, oxides of the various elements are finely ground and thoroughly mixed in stoichiometric proportions and then heated to temperatures that may exceed 1000°C. Phosphor firing furnaces can hold anywhere from 100 g to several kilograms of material at a time, that are heated under an atmosphere of nitrogen or similar inert gas. Once the reaction has completed, the material is removed, reground and may be washed in either water or organic solvents to remove any unreacted compounds. After mandatory testing, the phosphor is ready for use with blue pump LEDs. Most LED manufacturers prepare a slurry of phosphors mixed with a thermoplastic resin. Polycarbonate and silicones are widely used for this purpose — the polymer acting as a binder to keep the phosphor in place. The high viscosity liquid slurry is dispensed on to pump LED chips resting in their packages, using precision liquid dispensing machines (see Figure 5.44). The coating is cured with heat or UV radiation while the assembly is placed in vacuum to remove any air bubbles that might form. A further polymer molding step may be performed later to produce a dome lens on top of the phosphor-coated LED chip.

Rare-earth based phosphors owe their popularity to the ease with which almost any color in the visible spectrum can be created by choosing appropriate combinations of a host matrix and rare-earth ions. Much of

Fig. 5.44.   Phosphor slurry in silicone resin being dispensed on top of blue LEDs to make white light-emitting devices. Image on the right shows a close-up view.

current effort in phosphor development for high CRI and high efficacy light sources is directed toward developing phosphors with one or more rare-earth ions. Electronic excitation followed by de-excitation in these ions produce either narrow spectral line emissions (4f-4f transitions in triply-charged +3 rare-earth ions) or broad spectral band emissions (5d transitions in doubly-charged +2 rare-earth ions). Furthermore, their relatively slow luminescence kinetics in the range of 10^{-6}–10^{-3} s allows for effective energy transfer between optically-active rare-earth ions and the host matrix, relaxing pumping requirements and directly affecting device efficacy. Various emission properties of the phosphor can be tuned through proper selection of the host matrix, rare-earth ion(s) and doping concentration. The matrix material selection is a critical factor here since it has a profound effect on the active ion coordination chemistry. This allows phosphor makers to engineer the crystal field around the doping sites of the ions. It should be mentioned here that due to the cost and scarcity of rare-earths researchers have also been looking at other types of phosphors. These mostly resemble the traditional host + luminescent ion systems but are based on non-rare-earth elements, such as manganese. The table below lists three commercial phosphors with their relevant parameters.

Phosphor Color	Phosphor Composition	Phosphor Density (g/ml)	Peak Wavelength (nm)	Spectral Width (nm)	CIE Coordinates
Red	CaS:Eu	2.5	650	80	$x = 0.70$ $y = 0.30$
Green	$Lu_3Al_5O_{12}$: Ce	3.9	515	90	$x = 0.31$ $y = 0.58$
Yellow	$(Y,Gd)_3Al_5O_{12}$: Ce	4.4	570	120	$x = 0.48$ $y = 0.51$

Phosphors based on rare-earths have sometimes sparked discussions on the economics and scarcity of the rare-earths themselves. Apart from their use in phosphors, lanthanide elements have myriad other uses in such items as catalytic converters, powerful permanent magnets, batteries, various kinds of sensors, etc. Contrary to what their name would suggest, rare-earths are not truly rare and are in fact quite widely distributed in the earth's crust. However, due to their occurrence as secondary dilutions in the ores of other elements, and the close chemical similarities between different lanthanides, these elements are not easy to mine and separate from each other. And, while generally the rare-earth elements are quite equitably distributed over the world's land mass, substantial concentrations that are amenable to economic mining are found only in a few locations. The specter of a possible supply crunch is, however, quite unfounded because there is sufficient current and projected future availability of rare-earth compounds to ensure an un-interrupted supply for many years to come. Most applications of rare-earths use only small amounts of these elements, in percentage terms. For instance, in LED phosphors, elements such as europium, terbium or lanthanum are only used in minute amounts as optical dopants. For this reason, recycling of rare-earth compounds from LEDs is, currently, not economically feasible. From a long-term sustainability point-of-view, there do not appear to be any serious issues related to the availability of rare-earths, especially as there are good prospects for new sources, such as deep-sea mining, to be developed in the coming years.

5.8.2 *Quantum dots*

Phosphors, of the kind described above, are the most-used wavelength conversion materials in contemporary LEDs. These materials are now widely produced, are easy to obtain, and are reasonably inexpensive. Nevertheless, other luminescent materials have been investigated for this application. The most prominent of these are the quantum dots. Essentially, any pure solid substance prepared in a form such that it consists of nanometer-sized particles can be classified as quantum dots. In practice, very small crystallites of such materials as zinc sulfide (ZnS) or cadmium selenide (CdSe) are widely called quantum dots. Typical size range for quantum dots usually ranges from around 3 to 50 nm across. Because of their very small size, quantum dots possess a large surface-to-volume ratio which makes their surface properties of particular importance in whichever applications these are used in. However, for luminescence applications, their main utility arises out of their interesting and customizable energy level structure.

Recall from earlier discussion on energy levels and energy bands in this chapter, that when atoms come together, their energy levels form a system of closely spaced energy states. In bulk solids there are so many atoms present that the energy levels are extremely closely spaced — forming an almost continuous energy band. The situation is different in quantum dots where the far smaller number of atoms results in more widely spaced energy levels. In fact, the smaller the quantum dot, i.e. the smaller the number of atoms forming the dot, the larger is the spacing between energy levels. In actual practice, while energy levels still assemble to form a band structure, the band gaps are significantly larger than those observed in bulk matter. This feature gives rise to distinct across-the-band gap electronic transitions in the visible region. What is more, simply by controlling the size of the quantum dots, it is possible to control the band gaps and, thus, the colors resulting from electronic transitions. Therefore, by synthesizing quantum dots with a range of sizes, one can obtain different wavelength conversion materials from the same chemical substance. This size-dependency of the fluorescent emission wavelength of quantum dots can be seen here in Figure 5.45.

A distinguishing feature of quantum dot luminescence is the narrow width of spectral emission. Being an approximation to zero-dimensional entities, quantum dots have very sharply defined energy levels, as well as

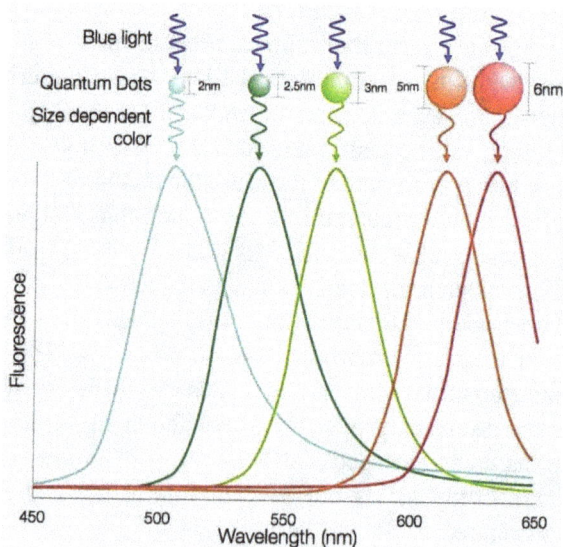

Fig. 5.45. Size dependency of the fluorescence emission wavelength from typical quantum dots.

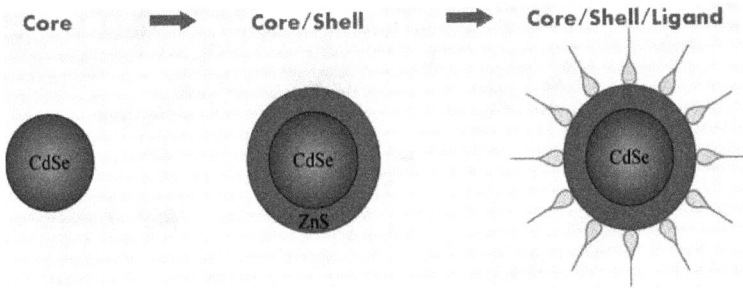

Fig. 5.46. Structure of core, core/shell and core/shell/ligand quantum dots.

density of energy states, and, thus, their emission widths are correspondingly narrower when compared to that of traditional phosphors. Thus, one can expose them to light with quite wide spectral distribution and get the energy converted to a much narrower range of longer wavelengths. This property of quantum dots has made them useful for several applications where quasi-monochromatic light generation is desired.

Most quantum dots, used in both research and commercial applications nowadays, consist of a so-called core–shell structure where one material is completely enveloped by another material. This is shown schematically in Figure 5.46. The shell provides physical protection to the quantum dot hosted inside it but, equally importantly, it also passivates the surface of the quantum dot. This is very important because, due to the large surface-to-volume ratio in quantum dots, they can have a very high density of surface defects. These structural imperfections can quench radiative transitions and, thus, cause loss of luminous efficiency. Passivation effectively seals off defect sites so that non-radiative transitions are greatly reduced. A further feature seen in modern synthesized quantum dots is the presence of long chain-like molecules anchored to the shell surface. These 'ligands' modify the surface properties of the core–shell quantum dots, making them compatible with their embedding environment in various ways. For instance, special ligands could be used to dissolve quantum dots in particular solvents, or to reduce their tendency to aggregate together. Quantum dot aggregation can be troublesome because it could result in concentration quenching where energy absorbed by one quantum dot is shared with other neighboring dots instead of causing well-defined luminous transitions. Thus, the presence of both the shell and the ligands can help increase the luminous efficiency of quantum dots.

Fig. 5.47.   Different size quantum dots in suspension, glowing under UV illumination.

A wide range of quantum dot material is now easily available commercially from different suppliers. Their price remains somewhat high because of the involved preparation process, but this is not of particular concern for many applications because only very small amount of material is required for most specialized light conversion applications. A major concern with the use of quantum dots is their extreme sensitivity to their environment. Oxygen, atmospheric contaminants, and heat, can all be harmful to them, and, thus, in any application the dots have to be properly embedded in a protective matrix to seal them from the deleterious effect of the surroundings. Figure 5.47 shows quantum dots of different sizes, suspended in a carrier liquid, glowing under excitation from UV radiation.

LEDs using quantum dots as wavelength conversion material have been demonstrated by several academic research groups but have not found a commercial market because of their high cost. For most ordinary lighting applications, phosphors make much more sense than quantum dots due to their much lower cost, lower toxicity, and long lifetime. However, there is one application where wavelength down-conversion by quantum dots is clearly superior to that done by phosphors. This application is generating white light for use in LCD TV backlights. LCD TV technology makes use of a flat, uniformly-illuminated, white surface called a backlight. In the early years of LCD TVs, small fluorescent tubes were used to illuminate TV backlights but this is now universally done using white LEDs positioned behind the backlight surface, as seen in Figure 5.48. The LCD panel sits just in front of the backlight.

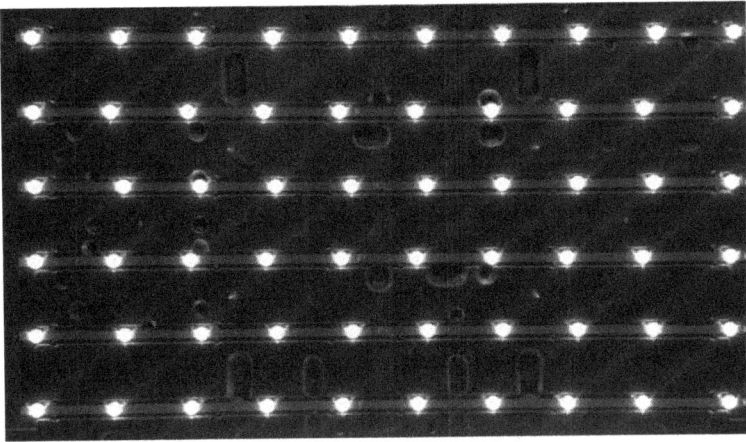

Fig. 5.48.   Full-array LCD TV backlight with white LEDs.

Individual pixels on the LCD panel act like tiny light valves which can be opened or shut to locally allow light from the backlight to shine through the screen or block it from view, respectively. Each color sub-pixel is a red, green, or transparent (for blue) dot of colored polymer, positioned in front of the LCD panel. In this way a trio of sub-pixels can be controlled by the LCD panel locally to generate a desired color. Most TV backlights use phosphor-converted white LEDs but TV manufacturers now also offer models that use quantum dots. In this case, blue-emitting, phosphor-free LEDs illuminate the backlight so that it glows with a uniform blue light. A polymer film containing red- and green-emitting quantum dots is positioned in front of the blue backlight. The emission from quantum dots contains sharp spikes of red and green wavelengths. Thus, the light arriving at the LCD panel consists of narrow-band emissions in the RGB regions instead of the broad spectrum produced by phosphor-based LEDs. The narrow spectral widths result in highly saturated colors, which enable a better color gamut, i.e. allow for the reproduction of a wider range of more accurate life-like colors. TVs with backlights lit by quantum dot films are popularly marketed as QLED TVs.

Even more advanced TVs make use of a printed screen of red- and green-emitting quantum dots placed in front of an LCD panel. Blue light from a blue LED-lit backlight selectively shines through the LCD panel and gets converted to red or green colors, as desired. No filters are used in this arrangement and thus the pictures can be brighter and the power consumption smaller. Other displays employing quantum dots with GaN micro-LED arrays have also been commercialized. More futuristic displays where quantum dot emission is driven by electrically injecting carriers directly into the dots have also been demonstrated, but not commercialized.

Research on quantum dots keeps on introducing new materials as well as new application possibilities. Particular attention has been paid in recent years on creating dots with non-toxic elements. Indium phosphide quantum dots, for instance, have received much interest due to this reason. Another research frontier has been dedicated to developing quantum dots from Perovskite materials. Perovskites have attracted much attention in the context of solar cells where these materials have enabled rapid enhancements in photovoltaic conversion efficiency. Perovskite quantum dots have been developed from such materials as caesium lead halides ($CsPbX_3$, with $X = Cl$, Br or I) and are receiving much interest because of their non-toxicity, very narrow emission widths and short luminescence lifetimes, which result in significantly brighter light emission.

A significant amount of work has also been done on developing organic wavelength conversion materials. Most such materials are light-emitting polymers which emit over a wide wavelength band. Their advantages over inorganic phosphors include easy integration with pump LEDs through application from a liquid solution and the ability to emit a wide range of wavelengths for better CRI of light source. Their disadvantages include higher cost, lower thermal endurance and susceptibility to degradation from atmospheric oxygen. These drawbacks have held back their use in commercial LEDs, although several experimental devices have been demonstrated.

5.8.3 *Phosphor degradation*

While white LEDs are made by coating appropriate phosphors on top of blue LEDs, solid-state light bulbs can employ either phosphor-coated white LEDs or blue LEDs that project their light on a phosphor-coated envelope. This latter approach is termed 'remote phosphor', and can result in higher brightness and longer bulb lifetimes because phosphor heating is greatly reduced in such an arrangement. As temperature rises, the wavelength conversion efficiency of phosphors comes down. Above a certain temperature, called quenching temperature, there is a steep fall in conversion efficiency. Luckily, most phosphors quench at temperatures that are much higher than room temperature so ordinary heating during normal operation is not of much concern. However, inadequate heat removal does cause a perceptible change in the color of LEDs and LED-based luminaires because decrease in phosphor efficiency alters the spectrum of light emission.

Phosphors are extremely robust materials but continued heating over long periods of time leads to permanent reduction in their wavelength conversion efficiency. For this reason, LED bulbs progressively get dimmer over thousands of hours of usage rather than failing suddenly, as is the case with incandescent lamps. Their end-of-life is reached when their light output falls to around 50% of the initial flux.

Phosphor degradation is caused by chemical changes in the phosphor as well as the appearance of crystal defects over a long period of time. Certain phosphors, such as red-emitting Eu-doped CaS phosphor, are prone to degradation from atmospheric moisture and must be carefully encapsulated with water-resistant barriers. Others are susceptible to oxidation at high temperatures. Some Eu-containing phosphors show a change in ionic valency, i.e. Eu^{2+} ions changing to Eu^{3+} ions which leads to a precipitous fall in light converting efficiency. Phosphor crystals also exhibit dislocation multiplication at elevated temperatures where existing dislocations spawn new ones, above a certain threshold temperature.

Both chemical and physical degradation effects get accelerated at higher temperatures, as can be seen in the 'luminance maintenance' plot shown in Figure 5.49, where reduction in light output from initial brightness is shown as a function of continuous use time. Behavior over three

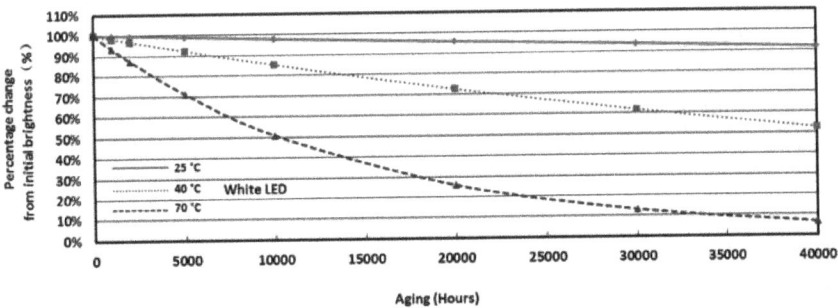

Fig. 5.49.   Decrease over time of output luminous flux from white LEDs maintained at different temperatures.

different operating temperatures is shown, and it is clear that higher temperatures cause enhanced degradation. This makes it clear that proper heat management considerations are important all the way from LED device design and packaging to the design of bulb and luminaire housing. Another factor, besides phosphor degradation, that also contributes to the reduction in LED light output, is increased opacity of the lens and phosphor binding material due to photochemical reactions in their polymer matrix. But even with these life-limiting factors, LED bulbs far outlive their more traditional counterparts.

5.8.4 *Other advances in LED phosphors*

Over the past several years, much progress has been made in improving the quality of light emitted by LEDs. The very first white LEDs were made using cerium-doped yttrium aluminum garnet (Ce:YAG) phosphor. Those devices emitted a harsh blue-white light. Consumers prefer a rich golden-yellow color tone to white light — close to that produced by conventional tungsten incandescent lamps. Cool white light produced by early white LEDs was considered of poor quality, incapable of bringing out the natural color of illuminated objects. The color rendering can be improved by the use of better phosphors that generate a more balanced spectrum. Modern white LEDs utilize phosphors that are actually mixtures of two or three phosphors for optimizing the spectral distribution of emitted light. Thus, for instance, warm white LEDs are made from phosphors that have a small amount of europium-containing red-emitting phosphor mixed with more traditional yellow-emitting phosphor. By properly compounding phosphor mixtures, it is possible to generate very high quality full-spectrum (broadband) white light from LED sources. Wide spectrum light contains all wavelengths of visible light (from 400 to 700 nm) in varying proportions. Such light is essential for bringing out the true colors of illuminated objects, and is, thus, valuable for applications such as museum and retail lighting. The absence of infrared (heat) and UV radiation in light from LEDs is also beneficial for preserving art works and museum displays.

Researchers are also working on developing phosphors where a single host matrix contains multiple rare-earth ions for generating several colors through the same phosphor. Such a phosphor may contain a mixture of europium, terbium and cerium for producing red, green and blue lights, respectively. This approach promises an even better lighting quality than has been achieved so far. Figures 5.50(a) and 5.50(b) show the spectrum

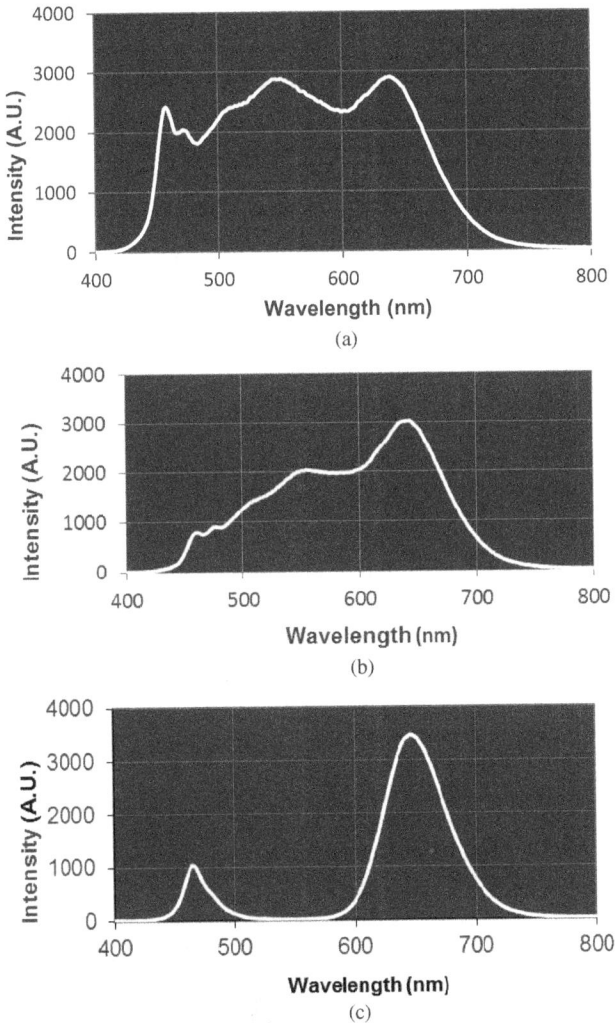

Fig. 5.50.   Spectrum of (a) broadband flat-white LED, (b) tungsten lamp simulation LED and (c) plant growth LED.

from a broadband white LED and that from an LED designed to mimic light from tungsten halogen lamps. It is remarkable that present day LEDs can generate light that is completely indistinguishable from that emitted by thermal sources. Spectral engineering has also been extended to applications such as lighting for indoor plant cultivation. Horticultural LED lamps are now available that can generate spectrum closely matched to what is best absorbed by vegetation. Most plants appear green because they do not absorb green light; instead reflecting it back to our eyes. Photosynthesis in plants requires mostly red and deep blue light. Figure 5.50(c) shows the spectrum from a broadband red LED that is used as a plant grow light. Notice that the spectrum contains all shades of red light and, in addition, also contains some blue light.

Specialized LED lighting, targeted toward horticultural applications, is now commonplace in the indoor cultivation industry. Their use is enabling the cultivation of unseasonable fruits and vegetables, as well as farming in inclement weather, and in remote high altitude and/or high latitude locations where sufficient sunlight may not be readily available all year round. Figure 5.51 shows a horticultural LED lamp for use in indoor farms. Such lamps contain red, blue and white LEDs with individual intensity controls for each color. These controls can be adjusted to produce spectra that would suit almost any crop under cultivation.

While solid-state white light illumination is now predominantly carried out with white-emitting phosphor-converted LEDs, there is considerable advantage in using color-mixed RGB LED sources for certain applications. As has been pointed out before, properly mixing red, green and blue lights to create white light is a difficult task. Light is inevitably lost in the mixing process, severely reducing the efficiency of mixed RGB

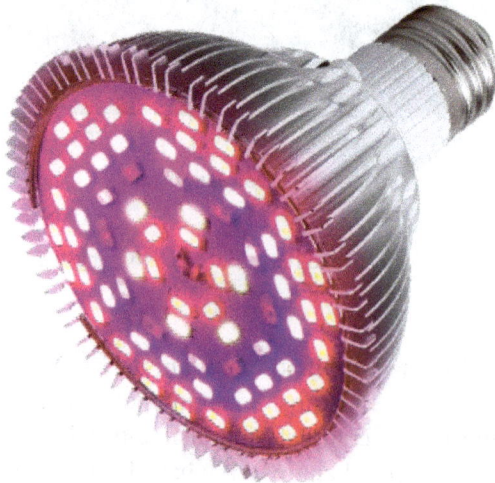

Fig. 5.51.   LED horticultural lamp.

Fig. 5.52. Alteration in perceived scenic details due to changes in the spectral content of illuminating radiation.

white light sources. However, such systems offer on-the-fly customization of illumination spectrum by varying the amounts of red, green, and blue lights in the mix. Object lighting in museums, art galleries, and retail displays, can all benefit from adjustable spectrum lighting, where the color balance could be adjusted to best bring out desired attributes of illuminated objects. An example of this can be seen in Figure 5.52, which shows two views of the same oil-color portrait seen under light with enhanced red bias (left) and blue bias (right). In this picture of a girl in a field of flowers, reddish white light clearly tends to emphasize the flowers in the scene, whereas bluish white light de-emphasizes the flowers but picks out the girl's dress and the sky. In this way, the observer's gaze could be selectively drawn to desired features through the proper selection of spectral components in the illumination source.

5.9 LED Metrics

Performance comparisons between LEDs are made using several measurable attributes that characterize various aspects of device operation. These metrics are usually listed on LED datasheets and are used in device selection for particular LED applications. Electrical engineers and lighting designers rely on these metrics to specify devices of choice. Although many operational metrics are in use, below are some of the most commonly encountered ones.

- **Maximum drive current** — Largest allowable current through the device with the LED consistently maintained at room temperature. If the LED temperature is higher during operation, then the value has to be appropriately 'de-rated' to a lower value.
- **Maximum operating temperature** — LED temperature has to remain below this value for continued operation without adverse effects due to thermal degradation. Adequate heat sinking must be provided so that device temperature never comes too close to this value.
- **Maximum power dissipation** — This is simply the product of the maximum current rating and the forward voltage drop of the diode at that current.
- **Peak emission wavelength** — Wavelength at which the LED's spectral output is at its maximum.
- **Spectral width** — Wavelength interval at the mid-point of the spectral curve, i.e. the FWHM of the spectral curve.
- **Output intensity** — Perceived optical brightness, usually given in milli candela or lumens.
- **Efficacy** — This is the effective light output for a given electrical power input. This is a much better measure than device efficiency for LEDs. It is given in lumens/watt.
- **Emission angle** — Angular width of the cone in which most of the LED light is emitted. This is generally twice the angle from the normal to the device to the point where the light intensity becomes half of the value from the normal.
- **Correlated color temperature (CCT)** — This is the temperature of a black body that has the same color appearance as that of the LED emission. 'Cool' white LEDs have higher CCTs when compared with 'warm' white LEDs.
- **Color rendering index (CRI)** — The quality of white light for illuminating colored objects is specified by this metric. A perfect illuminant (such as sunlight), capable of accurately rendering colors of illuminated objects, is given a CRI of 100. Smaller CRIs are less desirable when accurate color visibility is required. High quality white LEDs possess CRIs of 90 or more.

- **Chromaticity coordinates** — The x and y coordinates of the location of the chromaticity point of the LED light on a chromaticity chart is usually specified, so that accurate color matching between different LEDs could be performed.
- **Luminance maintenance lifetime** — The light output from LEDs gradually diminishes over time. This metric gives the average expected useful lifetime of an LED. It is the time by which a continuously operating LED will lose half of its starting brightness.

5.10 'Droop' in Gallium Nitride LEDs

Multiple quantum well GaN/InGaN blue LEDs are now the fastest developing class of optoelectronic devices, being the mainstay of the fast-developing solid-state lighting industry. Several of their operational parameters such as light extraction efficiency, and device resistance require further improvements. However, the largest single concern with them is due to their relative inefficiency when driven at high current levels. The decrease of electrical-to-optical energy conversion efficiency at high drive currents in GaN LEDs is a well-known effect. If this were not the case then the light flux from a blue LED could simply be increased by pumping more current into the diode (provided heat was efficiently extracted from the device). However, in actual practice, increase in drive current beyond a certain value results in decreasing internal efficiency, so that the dependence of light intensity on drive current becomes strongly sub-linear. This 'droop' effect has been widely debated in recent literature and, while progress has been made, an unequivocal explanation of its origin is yet to emerge. There is overall consensus, however, that droop will likely result in a fundamental limit to the operating efficiency of LED-based light sources. Despite much work devoted to understanding the origin of loss of LED efficiency at high drive currents there is still considerable uncertainty about the exact or dominant cause of its origin. Among the likely candidates causing droop are effects such as Auger recombination (where an energetic electron shares its energy with an electron bound to an atom, rather than undergoing radiative recombination), over-the-barrier leakage and various polarization-induced effects. If indeed deleterious effects on LED performance that give rise to droop cannot be overcome then LEDs will face a serious obstacle where their applications in HB sources are concerned. For instance, creating

LED-based replacements for 100-W incandescent bulbs will be a monumental task if droop cannot be eliminated. Such pressing concerns are driving the quest to better understand the causes of droop in GaN LEDs.

5.11 Commercial LED Luminaires

Going to a supermarket, looking for LED lightbulbs, one finds that while 40 and 60 W equivalent bulbs are readily available, it is difficult or impossible to get 100 W equivalent LED bulbs. A major reason for this is the difficulty in removing excess heat generated by LEDs. Although LEDs are very efficient when converting electrical energy to light, they still generate very significant amounts of heat, especially when running at high current levels. Tungsten bulbs generate even more heat but the heat they generate is mostly radiated away in the form of near- to mid-infrared radiation, radiatively cooling the bulb in that process. As a result, incandescent bulbs radiate both light and heat. LEDs are different in that they produce less intense heat, which cannot be easily removed. Good thermal engineering and use of substantial heat sinks are needed to safely dissipate heat from LEDs, and this makes the design of LED lightbulbs somewhat complicated. Generally speaking, proper thermal management is the most challenging aspect of designing any HB LED-based illumination system. As discussed earlier, power LEDs (devices with power ratings of 1 W or more) come packaged in special power packages, where the LED chip is mounted directly on top of a metal slug. Such packages are mounted directly on metal heat sinks such that heat can flow through a low thermal impedance path from the chip to the heat sink. The metal slug is highly polished so use of a thermal interface compound is usually not required, and in some cases may impede the flow of heat rather than aid it.

Simply making an LED die larger is not a viable solution for making brighter LEDs through increase in luminous area, because as the LED size gets bigger its efficiency begins to decrease markedly. This happens because it is increasingly more difficult to extract heat from an LED chip as its size increases, beyond about 3 mm × 3 mm die size. A practical alternative is to package several separate LED dies together on the same metal substrate. If phosphor needs to be used then it is applied to the entire LED array as a single uniform coating. This produces the larger multi-watt LED light engines, with ratings that can go above 100 W. Smaller LED arrays with ratings around 10 W also make use of this technology, which

Fig. 5.53. Lumileds Luxeon LED in a COB package.

Courtesy: Lumiled Inc.

usually goes by the name of chip-on-board (COB) package. Figure 5.53 shows a Luxeon LED from Lumileds that comes in a COB package.

In many designs, multiple power LEDs are first mounted on a MCPCB made of electrically-insulated aluminum sheet. The mount serves as a heat spreader, and is attached to an aluminum or copper heat sink for convective cooling. This method of mounting power LEDs is now widespread and many PCB makers offer custom MCPCB manufacturing services. Electrically insulating but thermally conductive materials used in commercial technologies such as SinkPAD™ MCPCB's allow thermal conductivities as high as 385 w/mK to be achieved. For more demanding applications, for example in LED searchlights, water cooled mounting substrates, such as that used with high power dissipation microprocessors, can be used. For low-cost retrofit LED bulbs, several small power LEDs are usually mounted on an MCPCB (see Figure 5.54), which is thermally connected to an external body-conformable heat sink. A small switch-mode power supply is contained in the base of the lamp to provide a pulsed DC drive for operating the LEDs.

Driving LEDs seems straightforward enough, but care is needed in the design of long-life lighting systems. Being diodes, LEDs are best driven using constant current sources. Both bipolar transistors and MOSFETS can be used for this purpose. Variable voltage at the base or gate of a transistor can then be used to adjust the brightness of LEDs. Like all semiconductor devices, LEDs too become electrically more conducting as their

Fig. 5.54.   Internal construction (left) and outer appearance (right) of a retrofit LED bulb.

Fig. 5.55.   Seoul Semiconductor's direct mains-driven power LED.
Courtesy: Seoul Semiconductor Corporation.

temperature increases. This can lead to failure through thermal runaway, if their temperature is allowed to rise unchecked. Thermistors can be used to sense LED temperature, and adjust their drive current, to maintain a constant current and, therefore, constant brightness.

Simple AC mains-driven LED lighting systems can be designed with a number of different techniques. Seoul Semiconductor produces LED modules which can be driven directly from 120- or 240-V AC mains. These modules (see Figure 5.55) contain a chain of series connected LEDs together with a rectifier diode. Although somewhat pricier than

Fig. 5.56. Infineon's LED driver IC ICL5102.
Courtesy: Infineon Technologies AG.

simple DC-driven LED modules of the same brightness, Seoul Semiconductor LEDs make it very easy to develop mains-powered solid-state lighting systems. Yet another approach for driving LEDs from mains power is to implement a switch-mode buck converter with constant current drive. ICs, such as SSL5031BTS from NXP Semiconductors, allow designers to develop a simple mains-driven LED lighting system of up to 30 W power with the addition of a few external components.

Specialized ICs are also available from several different companies to build complete mains-powered LED lighting systems. These devices can be used to build highly power-efficient dimmable LED drive systems with the addition of few external components. ICL5102 from Infineon (Figure 5.56) is an example of such an LED driver IC.

Figure 5.57 shows an application circuit for driving an LED string (series connection of several LEDs) from mains power, using an LED driver IC and a few discrete components. This shows how easy it is to build practical circuits for powering LEDs from the mains power supply.

Solid-state lighting is making rapid strides with developments taking place on many fronts. With on-going advances, LED-based lighting sources are becoming increasingly common with applications such as automobile headlights being the latest making the transition from incandescent to solid-state lighting (see Figure 5.58). Other challenging applications where LEDs have been replacing traditional light sources, include street lights and outdoor architectural lighting.

Even aircraft interior lighting has migrated to the use of LEDs. Figure 5.59 shows the passenger cabin of Boeing's 787 Dreamliner

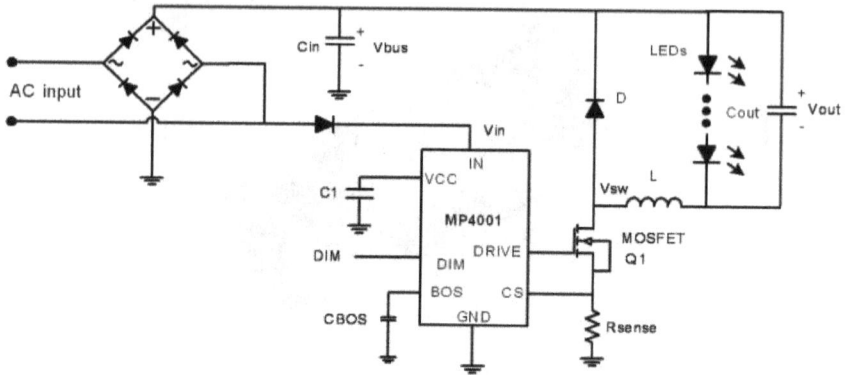

Fig. 5.57. Typical application circuit for driving an LED string from mains power.

Fig. 5.58. LED headlights on a BMW automobile.
Courtesy: BMW AG.

commercial jet which uses only LED-based illumination. Lighting engineers designed the cabin lighting for this series of planes to satisfy both decorative sensibilities and the circadian rhythms of air passengers. The result not only looks aesthetically pleasing but also serves the utilitarian purpose of keeping passengers from getting tired over long journeys. This is accomplished by adjusting the color and brightness of interior lighting to mimic normal light variation over the course of a day, as the travelers complete their journey. The result, with other refinements to the cabin

Fig. 5.59. Interior view of the passenger cabin of a Boeing 787 Dreamliner aircraft, showing overhead LED lighting.

Courtesy: Boeing Corporation.

environment, such as slightly higher humidity and pressurization, is that people arrive at their destination fresher and less fatigued than was usual before the advent of such technologies.

5.12 Further Advances in LED Technology

GaN LEDs made from epitaxial GaN/InGaN epilayers grown through MOCVD on sapphire substrates remain the main commercial technology for blue LED wafer and device manufacturing. Other material technologies have been investigated to fully or partially replace GaN LEDs with even better performing devices in the future. These include the use of silicon in place of sapphire substrates for GaN/InGaN MOCVD epitaxial growth, use of ZnO as an alternative LED material, and even the use of organic semiconductors for making light-emitting devices. Next, we take a look at these technologies to see where they might lead to, in the future.

5.12.1 *Gallium nitride LEDs on silicon substrates*

In order to reduce the manufacturing cost of blue LEDs, manufacturers have developed a new technology that makes use of silicon instead of sapphire as the substrate for the LED structure. This approach offers several advantages. Silicon is a much better thermal conductor compared to sapphire so heat can be removed much more easily with this so-called GaN-on-Si scheme. Use of silicon rather than the optically-transparent sapphire also enables GaN-on-Si LEDs to have better light output geometry (see Figure 5.60). Silicon can also be easily separated from the LED structure formed on top of it. This makes it possible to fabricate thin-film substrate-less LEDs with much superior characteristics, when compared with ordinary LEDs. Perhaps the most important argument in favor of GaN-on-Si LEDs is that silicon is available economically in wafer sizes as large as 12 inches in diameter whereas sapphire wafers are mostly available in only 2- to 4-inch diameters — making possible great economies of scale. It costs about the same to process a 2-inch diameter wafer as a 12-inch diameter one, so using large silicon wafers as LED substrates becomes very attractive. There are also other incentives with this approach, such as the availability of established silicon IC processing tools for processing GaN-on-Si LED wafers. As these tools have been proven in production settings for a long time so no new process-related developments need to be undertaken for GaN-on-Si material processed on such equipment.

It took a long time to develop GaN-on-Si technology, because it is very difficult to grow high quality III-nitride epilayers on silicon substrates. Not only the lattice constants are different but the crystal structures are also different, with silicon having a cubic structure and GaN having a hexagonal structure. Furthermore, the problem is further exacerbated by the fact that the thermal expansion coefficients of GaN and silicon are also very different. The difference in thermal expansion coefficient between GaN and silicon, 5.59×10^{-6} K^{-1} and 3.59×10^{-6} K^{-1}, respectively, at room temperature, leads to tensile stress in GaN epilayers and, thus, to crack formation upon cooling down from high growth temperatures. The lattice parameter and thermal expansion coefficient mismatch between Si and GaN also cause wafer bowing. In order to avoid wafer cracking and bowing, special proprietary growth techniques and buffer layers have to be used for III-nitride epitaxy on silicon. AlN interlayer is a common buffer layer that is grown on clean oxide-free silicon surface before GaN growth is started, although other materials have also been

BETTER THERMAL PERFORMANCE

Sapphire

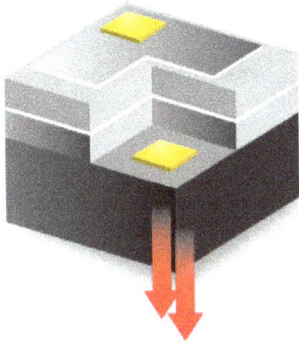

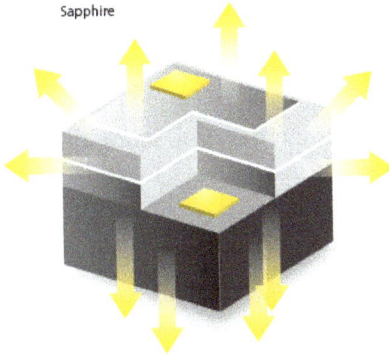

GaN-on-Silicon

HEAT EXPELLED
Sapphire 27w/mk & Silicon 149 w/mk

FOCUSED LIGHT-EMITTING SURFACE

Sapphire

GaN-on-Silicon

LIGHT EMITTED

Fig. 5.60.   Thermal and optical performance comparison between LED devices fabricated on a sapphire (left) and silicon (right) substrate.

Fig. 5.61.   Power LED made from GaN-on-Si material.

successfully used as buffer layers. The AlN layer is usually followed by a graded AlGaN layer in which the amount of Ga gradually increases while the amount of Al decreases in going from the bottom AlN/AlGaN interface to the top AlGaN/GaN interface. Incorporation of this strain compensation layer enables up to 1200 nm thick crack-free GaN epilayer to be grown on top. GaN-on-Si wafers with a top GaN epilayer and with complete GaN/InGaN LED structures are both commercially available.

After years of intense research, this technology is now well-developed and GaN-on-Si blue LED wafer material is readily available. Companies, such as Plessey Semiconductors of the UK, are currently manufacturing thin film LEDs and micro-LED arrays based on this technology. Available in different package options, these LEDs are targeted at a range of applications from general lighting and plant grow lights to emissive microdisplays. Power LEDs have also been constructed from this material (see Figure 5.61).

5.12.2 *Zinc oxide LEDs*

While GaN has become the standard semiconductor for making blue LEDs that serve as the engines for all white LEDs, a different semiconductor has also received significant interest as another contender for this role. This semiconductor is zinc oxide (ZnO) — a wide band gap II–VI oxide semiconductor with a 3.3-eV wide direct band gap. The potential of zinc oxide as a blue LED material has been recognized for many years but there continue to be formidable challenges in making efficient and stable LEDs out of this transparent oxide. Whereas GaN also had to go through a phase of intense material and processing developments, the challenges with ZnO seem to be even more formidable. However, its several advantages, compared to those offered by GaN, are such that work on ZnO keeps going in many academic and government labs around the world.

Zinc oxide's higher exciton binding energy (60 meV) makes it a, potentially, better light emitter compared to GaN (26 meV). In simple terms, higher exciton binding energy reduces the likelihood of non-radiative recombination by encouraging bound electron–hole pairs (excitons) to form and then decay into a photon. In lower exciton binding energy materials, in contrast, fewer excitons get formed at room temperature. While individual electrons and holes can still recombine and result in photon emission, as separate unbound entities, these charge carriers can more easily recombine non-radiatively at, for example, defect centers. This results in reduced photon emission and, thus, less bright light sources.

Another major advantage of ZnO over GaN — the current mainstream LED material — is its environmental sustainability. Whereas gallium, for making GaN, is an expensive and hard-to-get element, zinc is cheap and abundant. Furthermore, until recently, GaN was not available as a free-standing wafer and, thus, epitaxial films of GaN had to be prepared on 'foreign' substrates such as sapphire. Commercial production of GaN LED wafers is still carried out on sapphire substrates due to their much lower cost. Pure, mono-crystalline wafers of ZnO are already widely available (see Figure 5.62), making it possible to grow device-quality epitaxial layers on the native substrate. Currently, hydrothermally-grown ZnO wafers are mostly available in small sizes but if their demand increases then they can also be produced in larger sizes, and in much larger quantities. The availability of free-standing bulk material not only makes ZnO epitaxial growth much easier to accomplish but also produces material of much higher quality with fewer threading dislocations and other growth-induced defects. The relative scarcity of defects, together with the high exciton binding energy, promise to produce LEDs with much higher performance metrics than existing GaN-based devices.

As an LED material, ZnO offers a number of other benefits. The refractive index of ZnO is 16% lower than that of GaN (2.15 versus 2.56, at 400 nm). Consequently, light can be extracted more easily from the confines of a ZnO-based light emitting device. Photonic crystal light extraction structures on ZnO devices can be shallower and easier to fabricate than those on GaN devices. This should again lead to brighter devices

Fig. 5.62.   Hydrothermally-grown ZnO wafers.

that have a convincing external quantum efficiency advantage over GaN-based devices.

With ZnO, not only bulk wafer material is available but it is also easier to process it for making device structures. The chemical nature of ZnO is such that it can be easily etched through wet chemical processes. While dry etch chemistries are available for ultra-small features, where needed, many devices could be fabricated by simpler wet etching processes. GaN, in contrast, is very difficult to etch through wet chemistries.

Superior epitaxial optoelectronic material is possible with ZnO because a ZnO light emitter can be constructed entirely from the same material system with the substrate, buffer, charge transport layers, active layers and current-spreading layers all epitaxially grown in one MOCVD reactor with an integrated process flow. This improves both material quality and reduces growth cost and complexity.

ZnO can challenge the might of GaN if and when the technology for making good, stable p-type ZnO becomes available. Despite persistent efforts, over many years, device quality p-ZnO still remains elusive. The situation is, interestingly, reminiscent of the problem with GaN before it became a mainstream optoelectronic material. Just like ZnO, *n*-type GaN was easy to obtain but repeatable *p*-doping remained a tough challenge until it was finally demonstrated, thus opening the doors to all kinds of GaN-based devices — both electronic and optoelectronic. Doping, and especially *p*-type doping, is generally difficult with wide band gap semiconductors but ZnO presents its own challenges. The material is naturally *n*-type. Even when doped *p*-type the hole concentration achieved is not as high as desired because of the presence of natural background *n*-type doping. More troublesome still, the *p*-type conductivity often falls off over time so that the material is electrically unstable and, thus, not suitable for making commercial devices. There have been many reports of *p*-type doping of ZnO but making robust *p*-type material is still a research goal.

Given the difficulty of making *p*-type ZnO, most ZnO LEDs demonstrated so far have been hybrid LEDs, utilizing *n*-type ZnO epitaxial layer paired with *p*-type GaN epitaxial layer. In the context of ZnO LEDs, such devices are called heterojunction LEDs. The growth of GaN on ZnO, or vice-versa, is especially interesting because the two materials have a rather small lattice mismatch of around 2.2%. In recent years, homojunction ZnO LEDs, where both *n*- and *p*-type material are ZnO epilayers, have also been demonstrated. These device emit a bright violet or blue light (see Figure 5.63), but have short lifetimes and low operating efficiencies. While we are still far from the goal of having a practical, low-cost, efficient ZnO LED, the progress that is being made is an indication of the importance that the worldwide scientific community attaches to the prospect of realizing LEDs from this II-VI oxide semiconductor.

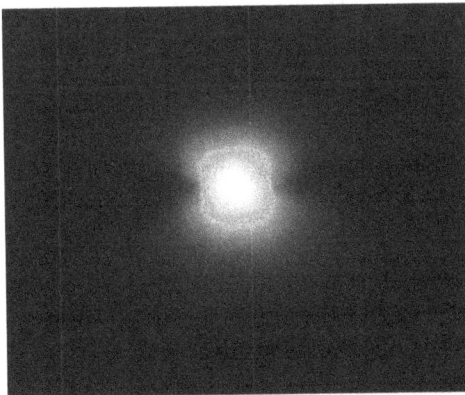

Fig. 5.63. Light emission from a homojunction ZnO LED.

5.12.3 *Organic light-emitting diodes (OLEDs)*

Diodes that emit light can also be constructed from organic materials. Many organic materials not only conduct electrical charge but also possess distinct semiconductor-like properties. These include the existence of a band gap and the presence of either free electrons, holes, or both. Some of these compounds are polymers whereas others are non-polymer organic molecules. Almost all of them contain a system of conjugated bonds, i.e. an alternating arrangement of single and double bonds. Perhaps the most well-known of such carbon-based conductors is graphite which consists of sheets of carbon atom hexagons separated from each other by weak Van der Wall bonds. Each carbon hexagon, and thus also each extended sheet, contains a system of free electrons located both above and below the planes. These itinerant electrons can carry heat and electricity in much the same way as free electrons do in metals.

While graphite has been known since antiquity and its electrical properties have been known for a long time, it was the discovery of the polymer polyacetylene and its ability to be doped to produce a highly conducting variety that launched the field of organic conductors. Alan Heeger, Alan MacDiarmid and Hideki Shirikawa were awarded the Nobel Prize in Chemistry in the year 2000 for the discovery and development of conducting polymers. Whilst theirs is a modern discovery, way back in 1862, the English chemist Henry Letheby had also synthesized a somewhat conductive material by the anodic oxidation of aniline in sulfuric acid. Most likely, this compound was polyaniline.

Almost all organic conductors are characterized by the presence of sp^2 molecular orbitals, where three of the four valence electrons of carbon atom take part in covalent bonding. The remaining π electron becomes delocalized and, thus, is available to carry electric current under the influence of an applied potential difference. All conventional organic conductors possess an energy band structure derived from molecular orbital states, which is somewhat similar to that found in conventional inorganic semiconductors. Thus, there is a band of states, each belonging to the so-called 'bonding' and 'anti-bonding' molecular orbital states. The former states are occupied by electrons while the latter are all empty. There is also a distinct energy band gap which separates the highest occupied molecular orbital (HOMO) and the lowest unoccupied molecular orbital (LUMO), as seen in the figure here. This system is reminiscent of the valence and conduction band system encountered with inorganic insulators and

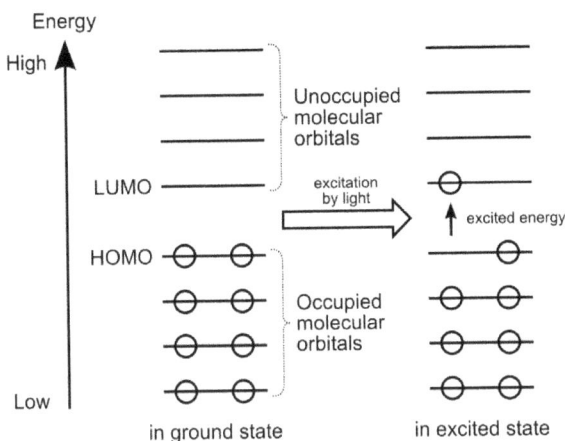

Fig. 5.64.   Electron excitation from HOMO to LUMO band in an organic conductor.

semiconductors, as discussed earlier in this chapter. Also, in an analogous manner, a photon can be absorbed and result in the excitation of an electron from normally occupied orbital states to normally unoccupied orbital states. Excited states decay after a while, either radiatively or non-radiatively to the normal ground state configuration. Electrons and holes can be created in organic conductors through charge injection from electrodes, doping or photo-excitation. Figure 5.64 also shows the excitation of an electron from HOMO to LUMO band under illumination with light of appropriate wavelength.

Given the similarities between the energy level systems of inorganic semiconductors and organic conductors, it is no surprise that similar device structures can be constructed with organic conductors as have been traditional with inorganic semiconductors. Thus, both diodes and field-effect transistors can be fabricated with schemes that closely parallel those used with traditional inorganic semiconductors. Many organic semiconductors possess band gaps that are suitable for the absorption and emission of visible light, and this has been utilized for making photo-sensors, solar cells, and LEDs. Organic LEDs, or OLEDs, behave in a similar manner to their inorganic counterparts. The schematic structure of an OLED is shown in Figure 5.65.

An emission layer is sandwiched between layers that can inject electrons and holes. Usually, each of these consists of a carrier injection layer and a carrier transport layer. Thus, on the cathode side, there is an electron

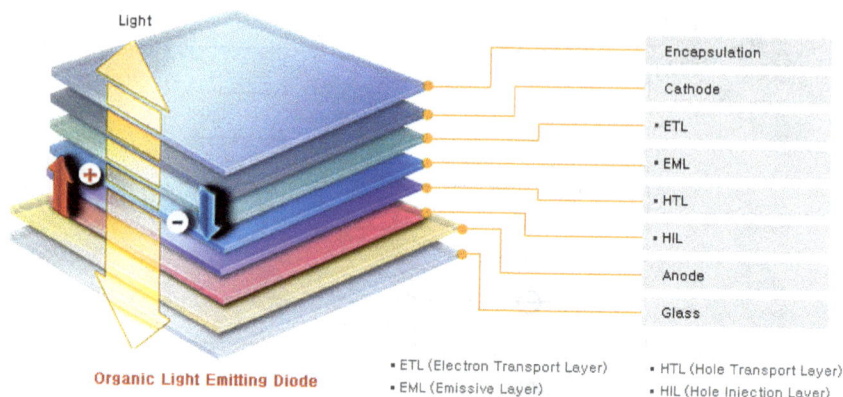

Light

Encapsulation
Cathode
• ETL
• EML
• HTL
• HIL
Anode
Glass

Organic Light Emitting Diode

• ETL (Electron Transport Layer)
• EML (Emissive Layer)

• HTL (Hole Transport Layer)
• HIL (Hole Injection Layer)

Fig. 5.65.   Schematic structure of an OLED device.

injector — usually a low work function metal film, such as calcium or lithium, covered by a protective metal film, generally aluminum. On the anode side, usually a transparent conductive film, such as ITO or AZO is used. These films could be deposited on either a rigid substrate such as glass or on a flexible plastic sheet — depending on the kind of device desired. Transparent conductive films have high work functions that suit their function as hole injectors. Again, a layer of hole transport film is placed between the hole injector and the emissive layer to ease the flow of holes into the emission medium. Electrons moving into the emissive layer get injected into the LUMO band of the emitting material whereas holes get injected into the HOMO band. There, the two usually bind together through Coulombic interaction to form an exciton. This electrostatically-bound electron–hole pair decays through recombination to yield either a photon (radiative recombination) or just heat (non-radiative recombination).

There are two broad classes of OLEDs. One is based on non-polymeric light-emitting organic semiconductors, such as tris(8-hydroxyquinolinato) aluminum ($Al(C_9H_6NO)_3$); often abbreviated as Alq_3. The other uses polymeric materials, such as poly(p-phenylene vinylene) and its derivatives. The former, also known as small molecule OLEDs, are made by vacuum deposition of material layers on a suitable substrate. This type of OLEDs exhibits good performance with HB and respectable efficiency figures, but are expensive to manufacture and challenging to make on large substrates. Polymer OLEDs, on the other hand, are made using solution-based coating techniques which make them relatively in-expensive and easy-to-produce on large substrates using various printing techniques. However, these

OLEDs, generally, show inferior performance when compared with the small molecule variety. Both small molecule and polymer OLEDs require effective encapsulation to protect the active layers degrading from the moisture and oxygen present in ambient air. In the absence of protective barriers, OLEDs will degrade very quickly, so sealing off these devices from the surrounding environment is of paramount importance. Much research effort has been devoted over the years on developing practical and effective OLED encapsulation technologies. Advances in this area have resulted in the commercialization of many OLED products. Materials for making both types of OLEDs are now commercially available from a number of suppliers, such as Merck, TCI, Materion, American Dye Source, etc. Major manufacturers of OLED products, such as Samsung and LG also produce their own OLED ingredients.

Despite their potential, OLEDs have not been very successful in penetrating the general illumination market. This is because their cost is still too high while their longevity and efficiency are still low, when compared with gallium nitride-based LEDs. However, OLED lighting offers some features that conventional LEDs cannot match. These include, glare-free lighting as well as the availability of large area thin and flat lighting panels that can be easily mounted on walls in various decorative arrangements (see Figure 5.66). The large area glare-free lighting also means that

Fig. 5.66.   OLED overhead lighting panels.

OLED lighting produces only very soft shadows. Furthermore, OLEDs can be produced on flexible substrates which makes possible OLED luminaires in a variety of curved formats — something that cannot be achieved with inorganic LEDs. The promise of such consumer-friendly features has kept OLED lighting research going, and we expect to see gradually increasing penetration of OLED panel lighting in the years to come.

While OLEDs have lagged behind gallium nitride-based white-emitting LEDs when it comes to space lighting, OLED-based displays have been immensely successful. From TVs and mobile phones to tablets and even laptop computers, OLED displays are now widely available for most consumer devices that require visual output. LCD displays, backlit with LED-based backlights, are gradually giving way to OLED display screens, which offer a number of benefits. All modern OLED display panels consist of a dense two-dimensional array of closely-spaced individual OLED elements (pixels), each controlled by dedicated switching transistors on a separate backplane. This active matrix approach enables separate access to each OLED pixel, greatly increasing the speed (responsivity) and intensity control for the entire display. Consequently, OLED displays do not show motion blurs and other streaky artefacts that early generations of LCD displays suffered from. Individual pixel-level control of OLED arrays provides another outstanding benefit. Unlike backlit LCD displays, OLED displays have self-emissive pixels, so that each pixel can be turned on to emit light or turned off to remain completely dark. This enables enormous light-to-dark contrast, producing exceptional dark levels, which cannot be matched by LCD displays where black areas remain, at best, a dark shade of gray. The almost perfect black levels, combined with rich color reproduction, wide viewing angles, high frame rate capability and the possibility of making curved screens, all contribute to the widespread popularity of OLED displays (Figure 5.67). Quantum dots are also being integrated with OLED displays in order to further enhance color reproduction accuracy. Q-OLED displays could be made from a single-color active matrix OLED array with red-, green- and blue-emitting quantum dots printed on adjacent pixels to provide full-color capability.

5.13 The Road Ahead

When considered alongside the development of highly efficient and affordable solar photovoltaic devices, the solar cell/LED combination is remarkable in its scope for energy conservation and safeguarding our

Fig. 5.67.   OLED TV.

environment. Together these devices hold great promise for reducing damage to the earth's climate through greenhouse gas emissions and reducing our reliance on limited reserves of fossil fuel. Their fortuitous combination has also been a boon to developing countries with less reliable mains power grids, where even remote villages now have electric lighting thanks to these devices (Figure 5.68).

LEDs themselves are benefiting from a diversity of research directions. As discussed earlier, native GaN substrates have been developed, and with the passage of time their cost will come down so that freestanding GaN wafers could, eventually, be used for the fabrication of LEDs. This would lead to further increases in LED efficiency because of the curtailment of non-radiative recombination through growth-induced epilayer defects. Together with better surface texturing, this will enable the manufacture of brighter LEDs. Non-sapphire substrates and use of high-performance thermal interface materials will make it possible to drive LEDs with higher current drives than is currently possible — again obtaining brighter emissions from smaller packages. Use of quantum dots — alone or in combination with metallic nanoparticles — will also result in brighter, high color purity LEDs, which are especially sought after for display applications. No doubt, other strategies will also be developed in the coming years that will see LEDs proliferate in all kinds of lighting applications.

Fig. 5.68.   Use of LED-based lights for illumination in off-grid locations.

The main argument in favor of LEDs is, of course, their significantly superior electricity-to-light conversion efficiency when compared to older electric lighting technologies. This has the potential to transform the way we generate, distribute and use electric power, given that a large fraction of electricity generated by various means (renewable and non-renewable) is used for illumination purposes. According to the US Energy Information Administration (EIA), 12% of all electricity usage in the United States was for lighting in the year 2012. The figures for other developed industrialized countries are similar. As the transition from tungsten filament bulbs to LED luminaires is completed in the next decade or two, lighting will represent a much smaller share of electricity usage worldwide. It is projected that by the year 2030, lighting will account for an estimated 8% of electricity consumption in the US.

The shift from conventional tungsten bulbs to LED lighting is a great success story of our times, and continuing developments will ensure that incandescent bulbs will become a relic of the past, much as vacuum tubes were relegated to museums with the advent of solid-state electronics. Since its advent, LED technology has been advancing at a rapid pace.

Prominent ongoing developments include the use of laser diodes instead of LEDs as pump sources for color-converting phosphors (discussed in Chapter 7) and the use of quantum wire structures for making LEDs of almost any desired color without the use of phosphors. Both LEDs and OLEDs also keep advancing the frontiers of display technology — making our world bright, and display screens ever more vibrant.

6 Lasers

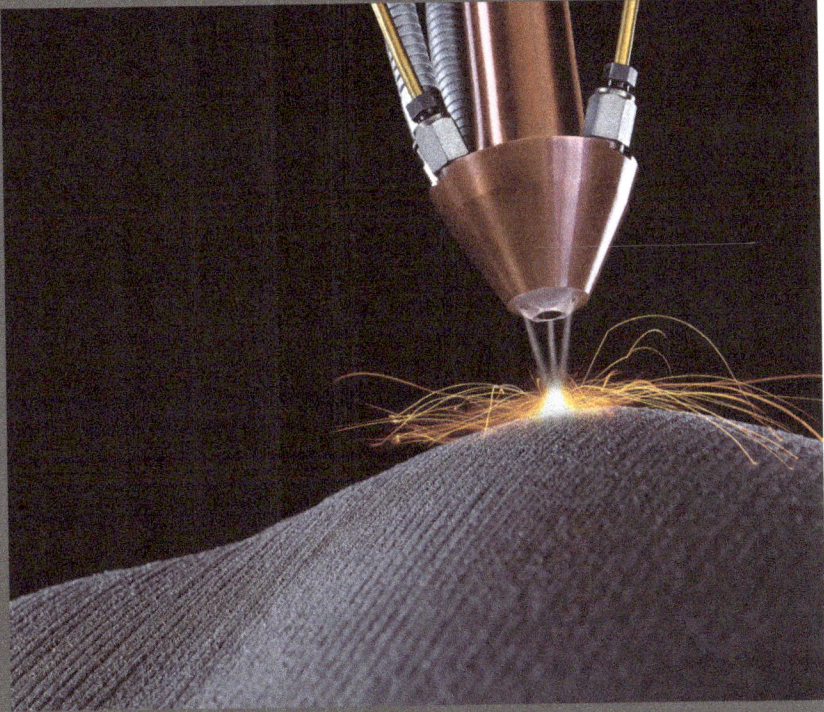

6.1 Introduction

The previous chapters have described 'ordinary' light sources. This chapter, in contrast, looks at a class of 'extraordinary' sources of light. If one could describe light as having some kind of order then ordinary light sources display very little order. The sources discussed in this chapter, however, emit light which is highly ordered. This concept of order, as it relates to light waves, obviously needs explanation so we start with that.

As detailed in the very first chapter, in free space, light exists as a sinusoidal variation of electric and magnetic fields that sustain each other. The electric field variation creates the magnetic field, as described by Ampere's law while, at the same time, variation of the magnetic field generates the electric field, as described by Faraday's law. Each field is the time derivative of the other, and this intertwined process works successfully because each field has a sinusoidal functional dependence on both space and time. As the derivative of a sine wave is another sine wave (with a phase difference), so the sinusoidal electric and magnetic fields keep each other alive. Every light source gives out a sine wave variation of electric field strength, in accordance with the above picture (the magnetic field variation is understood to be of the same form and to accompany the electric field wave). If one could look at such a sine wave in its schematic form then it would be noticed that the waves are not continuous but are of some finite length. Each 'wave train' contains a finite number of cycles of the sine wave. These wave trains are also of random lengths and bear no relationship to each other. It is this chaotic emission of different cycle length wave trains that gives the impression of light having no specific order. Most ordinary light sources generate light of this form. Since the middle of the 20th century, however, it has been possible to generate 'ordered' light where the wave trains can be very long and, more importantly, are emitted in phase with each other, i.e. any two simultaneously emitted wave trains have sinusoidal variations that are in step with each other, or in other words they go through the same phase angle at the same time. Light composed of such synchronized wave trains is technically called 'coherent'. This character endows light with a great deal of order, as all wave trains that are emitted together are in lock step with each other. Optical coherence is such a distinctive feature of a light source that all such sources fall in a class of their own, and this chapter is devoted to their description. For historical reasons, coherent light sources are called lasers. We shall see later in this chapter, how this name came about.

Besides coherence, the light emitted from lasers shows other interesting characteristics, such as a very high degree of monochromaticity (very small wavelength spread) and high directionality.

This chapter is divided into two distinct halves. The first deals with the history of and the science behind the laser. The next half describes a number of important lasers covering the region from the infrared to the ultraviolet part of the electromagnetic spectrum. The first part mainly focuses on the developments until the 1960s and chronologically describes the emergence of ideas that form the basis of the laser. The latter part looks at some prominent types of lasers that are useful as light sources for a wide range of applications.

6.2 The Foundations

Before we delve into the specifics of several different types of lasers, it is essential to look at the scientific foundations behind laser action. Because, on a fundamental level, these light sources are very different from all the others we have encountered so far in this book, so a proper understanding of their operation requires a good grounding in the basic principles of laser operation. We begin by looking at the concept of the photon and how it can induce electronic transitions in atoms and ions. Next, we will explore some historical developments that were key to the appearance of this new light source. We'll also examine some engineering principles that are essential to the operation of almost any laser system available today.

6.2.1 *Particles of light — photons*

The true nature of light — whether it consists of waves or particles — has been the subject of debate for hundreds of years. While Isaac Newton favored a corpuscular picture, where light was thought of as a swarm of tiny luminous particles, later experiments by Thomas Young, Augustin-Jean Fresnel and others could best be interpreted by assuming a wave-like nature for light. Gradually, enough experimental evidence accumulated so that by the beginning of the 19th century, light was firmly believed to be a type of wave disturbance in a somewhat mysterious medium called the 'ether'. Optical experiments on interference, diffraction and polarization of light had conclusively established light as a transverse sinusoidal wave and by the middle of the 19th century this became regarded as a proven

Fig. 6.1.   Max Planck.

fact with enormous experimental backing. The development of the electromagnetic theory of radiation, based on James Clerk Maxwell's celebrated equations, only furthered this model by explaining light waves as sinusoidal variations of electric and magnetic fields.

Against this backdrop, it came as quite a surprise when the German physicist Max Planck (Figure 6.1) developed a theory where light was supposed to be emitted and absorbed as discrete bundles of energy. Planck was compelled to arrive at this description while trying to explain the spectrum of light emitted by hot black bodies. The characteristic spectrum of radiation coming out of a small hole drilled in a metal cavity that is heated to high temperatures is similar to that from an ideal completely black body. In the late 19th century Germany, Otto Lummer and Ernst Pringsheim teamed up to measure the distribution of energies at different wavelengths, emitted from the inside of a heated cavity. A simplified version of the spectrum of the so-called 'cavity radiation' obtained by them appears in Figure 6.2.

It is clear from the black body radiation spectral plot that the radiant energy is not equally distributed among the wavelengths. There is a very characteristic distribution with a distinct peak which moves to shorter wavelengths with increase in temperature. There is also a short wavelength cut-off — the shortest wavelength emitted at a given temperature. This cut-off also shifts to shorter wavelengths with increase in the cavity temperature.

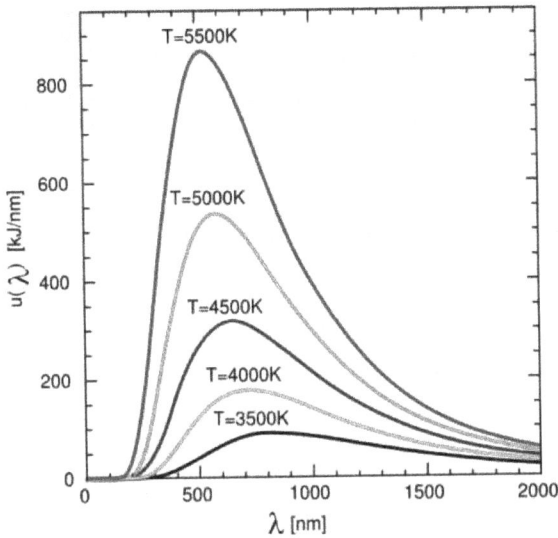

Fig. 6.2.   Spectra of black body (cavity) radiation at different temperatures.

Several physicists of the time tried explaining the spectral distribution of cavity radiation using physical principles that were known at that time, but only had partial success. A number of different 'radiation laws' were obtained which were successful in describing certain features of the black-body spectra but none could fully describe the entire range of observations. This was the state of affairs when Planck started to take an interest in this problem. At first, he tried approaches that were similar to his contemporaries, but had no better luck. After some time, he tried a bold speculative hypothesis by postulating that energy exchange between the internal walls of the cavity and the radiation field present in the cavity takes place in discrete quantities or 'quanta'. At any given wavelength λ, the quantum of energy was given by hc/λ (or equivalently, $h\nu$). Here, c is the velocity of light in free space, and ν is the frequency of light. The constant h that appears here was introduced by Planck and is called the Planck's constant. It has the value, $h = 6.626 \times 10^{-34}$ m^{2}kg/s. With this assumption, Planck was able to develop a theory which, while it built to some extent on those of his other contemporaries, was able to finally explain the entire blackbody spectral curve at any temperature. This was not only a great success as far as the explanation of the radiation emission characteristics was concerned, but it also proved to be a turning point in our understanding of the nature of light. According to Planck's radiation law the spectral energy density ρ(ν), i.e. the energy density per unit volume at a given radiation frequency ν and absolute temperature of the radiator T, is given by the equation:

$$\rho(\nu) = [8\pi h\nu^3/c^3] \cdot 1/(e^{h\nu/kT} - 1) \tag{6.1}$$

Here, k is Boltzmann constant, with the value, $k = 1.38 \times 10^{-23}$ m²kg/s²K and c is the speed of light.

A few years later, Albert Einstein (Figure 6.3) too explained the photoelectric effect by invoking energy exchange taking place in quantized amounts. It was known that light falling on clean surfaces of some metals could release free electrons (see Figure 6.4). Experiments showed that there was a certain metal-dependent cut-off wavelength for this effect to be observed. When the wavelength of light was above a certain value, which differed from metal to metal, no photoelectric effect was observed; no matter how high the intensity of light was. For wavelengths below the cut-off value, however, photoelectric electron emission was easily detected. Moreover, the number of photoelectrons emitted per unit time, i.e. the photoelectric current was seen to depend only on the intensity of light. Once a clean surface of an appropriate metal, such as sodium, potassium etc., was

Fig. 6.3.   Albert Einstein.

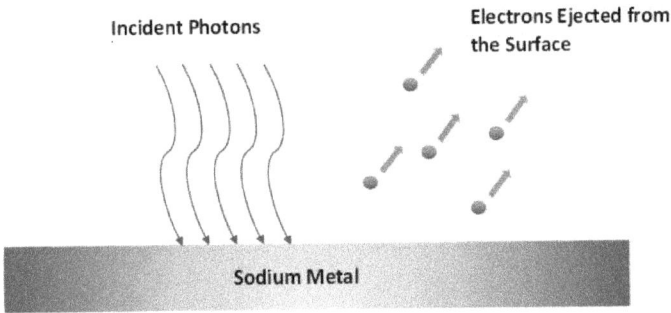

Fig. 6.4.   Schematic Illustration of photoelectric effect.

exposed to visible light below the cut-off wavelength, photoelectrons started coming out of the metal and their flux increased as the intensity of light was increased. Einstein explained these observations by following Max Planck in postulating that energy was delivered in discrete quanta, each carrying energy equal to hv, as explained above. This energy had to be above the energy required to liberate a free electron from the pull of the positively charged metal ions, called the 'work function' of the metal. Thus, for wavelengths shorter than a certain value or, equivalently, frequencies above a certain value, there was enough energy available in the light quanta to overcome the work function and liberate a free electron from the surface of the metal. Any energy in excess of the work function simply appeared as the kinetic energy of the photoelectron. This hypothesis neatly explained all observations related to the photoelectric effect.

Albert Einstein was awarded a Nobel Prize in physics in the year 1925 for his explanation of the photoelectric effect. By that time, scientists had generally accepted the notion of light being composed of discrete quantized energy packets. In 1926, the chemist Gilbert N. Lewis coined the term 'photon' for the quanta of light and this name was quickly adopted by the worldwide scientific community. A year later, in 1927, Arthur Holly Compton (Figure 6.5) was awarded the Nobel Prize in physics for his experiments on the inelastic scattering of electromagnetic radiation by free electrons, which also demonstrated that radiation is made up of particle-like entities. Here, inelastic simply means that the incident photon transfers some of its energy to the electron which, thus, acquires some kinetic energy while the scattered photon ends up with less energy,

Fig. 6.5.   Arthur H. Compton.

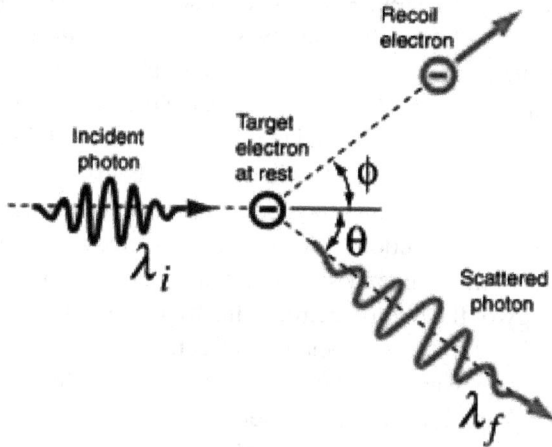

Fig. 6.6.   Schematic illustration of Compton scattering.

as can be seen in the schematic illustration in Figure 6.6. These energy exchanges clearly show that photons carry quantized amounts of energy. Thus, by the end of the 1920s, the notion of photons as quantized particles of electromagnetic radiation was firmly established.

Given all the experimental evidence and theoretical insights, light can be best understood as spatially-localized packets of electromagnetic waves. The amount of localization (roughly speaking, the inverse of the size of the wave packet) increases with increase in frequency, i.e. with decrease in wavelength. Thus, photons are more sharply localized in space as their energy increases. The somewhat diffuse nature of photons is in sharp contrast with the extremely localized, point particle-like character of elementary particles, such as electrons. This feature also explains the centuries long contention between light as waves or particles. The wave packet picture satisfactorily explains this dichotomy and has greatly helped in further advancing the science of light.

6.2.2 *Photons and electronic transitions in atoms*

Electrons bound to atomic nuclei can change their energy by absorbing and emitting photons, as has been described in the previous chapter. Electrons are arranged in a shell-like structure in atoms. Electrons in the inner shells (closer to the atomic nucleus) generally do not take part in physical or chemical interactions but those in the outermost (valence) shell are affected by their environment. If a photon is absorbed by an atom, then it disappears from existence but its energy is utilized by one of the outer electrons which becomes 'excited' or, in other words, transitions to a higher energy state. The entire atom is then said to be in an excited state and is sometimes indicated by an asterisk, as for example, $Ca \rightarrow Ca^*$, for such a process with a calcium atom. Excited states do not last forever. After some time, which is usually (but not always) of the order of a few nanoseconds, the atom 'de-excites' with the excited electron transitioning back to its original energy level. In this original state, the atom is said to be in its lowest energy or ground state. The energy held by the excited atom is released through the formation of a photon with energy exactly equal to the difference between the excited state and ground state energies. This mode of atomic de-excitation is called spontaneous emission because this is a statistically random process with no apparent external influence. If the potential energies associated with the lower and upper levels are denoted by E_1 and E_2, respectively, then the energy of the absorbed and emitted photon is simply given by the relation: $E_2 - E_1 = h\nu = hc/\lambda$. Due to the energy-time uncertainty principle, the transition actually produces photons within a narrow range of frequencies called the spectral linewidth.

The concepts of atomic excitation and de-excitation through photon absorption and spontaneous photon emission had become established facts quite early on and are depicted here in Figure 6.7.

In 1916, Albert Einstein examined these processes on statistical grounds and put forward an additional mechanism for atomic

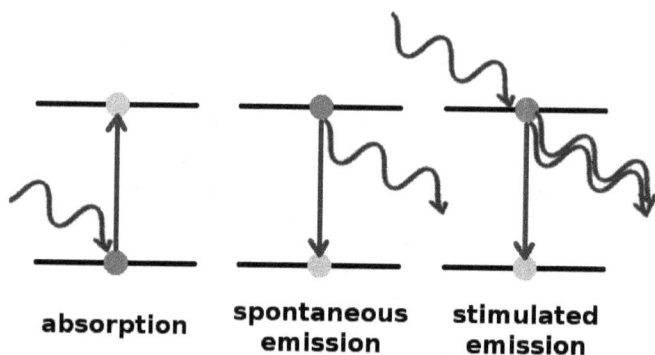

absorption   **spontaneous emission**   **stimulated emission**

Fig. 6.7.   Diagram illustrating absorption, spontaneous emission and stimulated emission processes.

de-excitation that he called stimulated emission. This is also shown in Figure 6.7. Stimulated emission takes place when in the vicinity of an excited atom there is a photon with exactly the same wavelength as that which will be spontaneously emitted by the atom in due course of time. In the presence of a 'matching' photon, the atom is coaxed into releasing the impending photon so that after de-excitation to the ground state there are two photons of exactly the same wavelength and direction of travel. The process is aptly named because the first photon stimulates the emission of the second photon. The two photons are exactly alike and this gives rise to interesting effects that we'll examine later. While the spontaneous emission process could linger on, stimulated emission is abrupt because the stimulated photon is produced almost as soon as the stimulating photon starts interacting with the excited atom. The statistical delay involved in spontaneous emission arises from the fact that a spontaneous transition is actually stimulated by fluctuations in the background electromagnetic field or, in other words, by virtual photons. These are extremely short-lived photons that are generated momentarily, and at random, from the electromagnetic background and just as quickly merge back into the background. Because their appearance and disappearance are completely random so this accounts for the random nature of spontaneous emission.

The dynamics of photon absorption and emission in any atomic, molecular or ionic system in the presence of a radiation field can be described by using Einstein coefficients: A_{21}, B_{21} and B_{12}. If n_1 and n_2, respectively, denote the number density (number per unit volume) of electrons in energy levels E_1 (lower) and E_2 (upper) then the rate of spontaneous emission, i.e. the rate at which the number density of upper-level electrons is depleted with time, can be written as:

$$dn_2/dt = -A_{21}n_2 \tag{6.2}$$

Thus, the rate of upper-level depletion is proportional to the number of electrons in the upper level at any given time. This is reasonable, because if there are more excited electrons then there will be proportionally more spontaneous emission transitions from level 2 to level 1. The constant of proportionality, A_{21}, here is one of the Einstein coefficients. This one describes spontaneous emission.

Spontaneous emission obviously also causes the population of ground state electrons to increase at a rate given by

$$dn_1/dt = A_{21}n_2 \tag{6.3}$$

For stimulated emission there is a similar relation, giving the rate of increase of n_1 as

$$dn_1/dt = B_{21}n_2\rho(v) \tag{6.4}$$

Here, B_{21} is the Einstein coefficient for stimulated emission and $\rho(v)$ is the spectral energy density of the radiation field around the excited atoms, molecules or ions. This is simply the energy present in the field per unit volume and per unit frequency, and is given by the Planck radiation law, mentioned above. Its unit is J.s/m³.

Finally, photon absorption is described by

$$dn_1/dt = -B_{12}n_1\rho(v) \tag{6.5}$$

Note that the Einstein coefficients are simply the probabilities of respective transitions taking place. Their units are 1/s for A_{21} and m³/Js² for B_{21} and B_{12}.

Photon absorption and emission dynamics is such that there is no net change in the number of electrons with time that populate energy level E_1. Setting $dn_1/dt = 0$ in the equations above gives us the equation for detailed balance as:

$$A_{21}n_2 + B_{21}n_2\rho(v) - B_{12}n_1\rho(v) = 0 \tag{6.6}$$

On using the equation for Planck's radiation law, given earlier, and the Maxwell–Boltzmann distribution function:

$$n_i = [n\, g_i\, e^{-E_i/kT}]/Z \tag{6.7}$$

where n_i is the number of electrons at an energy E_i, g_i is the multiplicity of this energy level, i.e. the number of actual energy levels at energy E_i and Z is the partition function, we can show the following relations between the Einstein coefficients:

$$B_{21}/B_{12} = g_1/g_2 \tag{6.8}$$

and

$$A_{21}/B_{21} = 8\pi h\nu^3/c^3 \tag{6.9}$$

In most cases, $g1 = g2$ and then the first equation here tells us that in such cases $B_{21} = B_{12}$, i.e. the probabilities of absorption and of stimulated emission are the same, which is reasonable because from a thermodynamic point-of-view, these processes are inverse of each other. The second equation here (6.9) tells us that the probability of spontaneous emission increases over that of stimulated emission as the frequency of radiation increases, i.e. as the wavelength gets shorter.

6.2.3 *Radiation amplification through stimulated emission*

The stimulated emission process provides a route for increasing the number of photons, i.e. for amplifying radiation to increase its intensity. This is very easy to understand because any stimulated transition effectively produces a cloned photon which can go on to stimulate other transitions, producing ever more photons. Thus, even a single photon can generate an enormous cascade of identical photons, if it travels through a suitable medium containing excited atoms, molecules or ions. There will, of course, be competition with absorptive transitions as well, but if the number of electrons in the excited state is more than that in the ground state then there will be a net gain in the number of photons, resulting in radiation amplification. In normal circumstances, the Maxwell–Boltzmann distribution ensures that the number of ground state species (atoms, molecules or ions) far exceed those in any excited state. However, for radiation amplification, as outlined here to take place, the number of excited centers must be more than the ground state population. This necessary condition is termed 'population inversion'.

Population inversion is hard to achieve in most cases because it goes against normal thermodynamic equilibrium. This configuration is transitory, i.e. can only exist for a very short time, unless a continuous energy source maintains it against relaxation to thermal equilibrium. In order to obtain population inversion, a suitable medium has to be excited or 'pumped' so as to populate more of an upper energy level. A suitable combination of an adequately long-lived excited state and an intense energy source can be used to obtain inverted electron populations. The energy can be supplied to the active medium through one of several possible routes, such as an electric discharge, exposure to suitable electromagnetic radiation, excess electron-hole injection in a semiconductor etc.

6.3 Early History of Lasers

The laser — a coherent source of monochromatic visible radiation — was developed in the 1960s. To trace some of its developmental history, it is best to start with an understanding of the basic concepts. The fundamental scientific concepts behind laser operation have already been described in the previous section. Next, we examine scientific developments during the 1950s and 60s that led to the experimental realization of the first lasers. As it happened, this train of developments was initiated not by optical scientists but by practitioners of microwave engineering — a field that had quickly developed during the Second World War, and was witnessing rapid advances in the years following the war. In going with the historical order, we start by examining the microwave cousin of the laser, as it paved the way for the later development of the laser. It is no accident that stimulated emission-based devices were first developed in the microwave region. Equation (6.9) shows that the fraction of stimulated emission increases at lower frequencies and, thus, compared to optical frequencies, microwave frequencies are more suitable for observing stimulated emission. In practical terms, population inversion is comparatively easy to achieve with modest pumping powers in this region. Thus, stimulated emission-based devices were first demonstrated in the giga hertz frequency region — opening the doors to an entirely new field, of profound importance to both science and technology.

6.3.1 *Masers*

The birth of the laser was preceded by the invention of its microwave counterpart, called the maser (an acronym for microwave amplification by stimulated emission of radiation). This was the first application of the stimulated emission mechanism in a device of practical utility. Joseph Weber of the University of Maryland is credited with describing the operating principles of masers in 1952. At almost the same time, Nikolay Basov and Alexander Prokhorov from the Lebedev Institute of Physics in the USSR also independently came up with the same scheme. Exactly a year after that, the first maser was demonstrated at Columbia University by Charles H. Townes, James P. Gordon and Herbert J. Zeiger. They used ammonia gas as the working medium that provided the necessary quantum states for population inversion and stimulated emission to be achieved.

6.3.1.1 *The ammonia maser*

An ammonia molecule, NH_3, is shaped like a trigonal pyramid with three hydrogen atoms forming a nearly planar triangular base and a nitrogen atom forming the apex of the pyramid, as shown here in Figure 6.8.

Due to the symmetry inherent in this structure, the nitrogen atom can be located at any one of two positions relative to the hydrogen triad. In fact, the nitrogen atom keeps incessantly flipping between the two configurations. The energy required for such an inversion is very small (about 10^{-4} eV). This is equivalent to a frequency of 23.79 GHz, which corresponds to microwave radiation of wavelength 1.26 cm. Nitrogen inversion

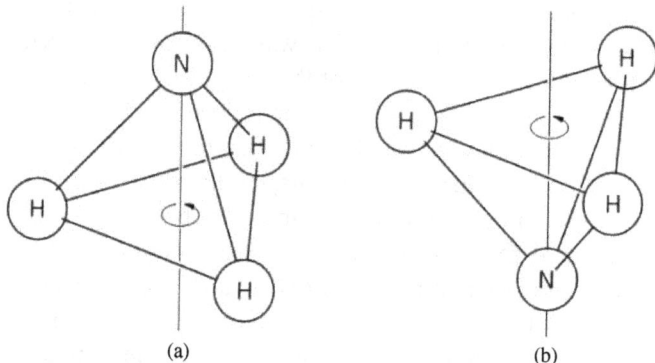

(a)   (b)

Fig. 6.8.   Schematic illustration of nitrogen atom flipping between two equivalent positions (a & b) through the hydrogen basal plane in an ammonia molecule.

in ammonia molecules was first observed in 1934 at the University of Michigan through microwave spectroscopy by C. E. Cleeton and N. H. Williams. The flexional motion of the nitrogen atom provides a two-level system between which energy transitions can take place. The very small energy barrier to the transition makes it very easy to energize and invert the population in a collection of ammonia molecules. In fact, a simple way of producing a population inversion in this system is to simply mechanically remove one configuration from ammonia gas, which normally contains equal number of molecules in each configuration at any given temperature. Gas that is enriched with one configuration is in a non-equilibrium state which will tend to achieve thermodynamic equilibrium by inter-conversion of some of the molecules to the opposite configuration so that an equipartitioned population is achieved once more. Left to itself, this will be accompanied by the spontaneous emission of photons at 23.79 GHz. However, the spontaneously emitted photons can stimulate other ammonia molecules to emit photons if the gas is confined to an appropriate resonant cavity.

The mechanism described above was the basis of Townes' apparatus, which consisted of a source of ammonia gas, a molecular configuration filter and a metal resonant cavity, as seen in Figure 6.9. The front of the metal enclosure has been opened up to show the inner workings of the equipment. A thin stream of ammonia gas at room temperature entered the apparatus at the extreme left. The speed of the ammonia jet was regulated by a leak valve. The gas stream traveled through a short tube for collimation and then traversed a region of highly non-uniform static electric field. This high gradient field was established by an electrical quadrupole, consisting of an arrangement of four copper rods visible in the center. The quadrupole field served to filter out one molecular configuration such that those molecules left the beam while the molecules in the opposite configuration were focused on to the quadrupole's axis. This was possible because ammonia molecules possess an electrical dipole moment owing to the presence of a lone pair of electrons. The filtered ammonia beam then entered a microwave resonant cavity fitted with a vertical waveguide and a beam dump. White parts seen in the photograph are electrically-insulating supports. A weak microwave radiation field (~10 nW) was generated in the cavity through stimulated emission from ammonia molecules. The radiation was weak because the filtering of the ammonia gas stream was highly inefficient. Nevertheless, the signal was strong enough to be easily measured. In Figure 6.9, Charles Townes is seen adjusting the

Fig. 6.9.   Charles Townes working on his ammonia maser.

waveguide through which the microwave radiation was extracted. This apparatus could be used to both amplify a weak microwave signal or to generate self-sustained oscillations and, thus, could work as both an amplifier and an oscillator. Three oil diffusion pumps for maintaining vacuum in the equipment chamber are seen in the lower part of the picture.

To name their system, Towne's research group came up with the acronym MASER, which stood for microwave amplification by stimulated emission of radiation. Many of his colleagues were initially highly skeptical of the possibility of building such a device; to the extent of even deriding him by saying that MASER stood for Means of Acquiring Support for Expensive Research!

Towne, Gordon and Zeiger published their paper on the maser in the journal *Physical Review* in 1954. A decade later, in 1964, Townes, Basov and Prokhorov were jointly awarded the Nobel Prize in physics for their work on both masers and lasers. Masers quickly became objects of great fascination and many researchers started working on different types of masers. A flood of papers on maser-related topics started to appear in prominent physics journals of the time. This became such a problem that for some time the *Physical Review* declared a moratorium on publishing any papers that dealt with masers! As maser technology started to proliferate and become a field on its own, Townes and Harold Lyons — a collaborator on atomic time keeping at the US National Bureau of Standards (now the US National Institute of Standards and Technology, NIST) — came up with the term 'quantum electronics' to describe the new discipline. Later, laser technology also got included under this nomenclature.

An interesting observation that was made with masers as soon as the ammonia maser was demonstrated was the sharpness of its output frequency. The later development of lasers also showed that their light was extremely monochromatic. This is a characteristic property of such stimulated emission-based oscillators and is one reason why masers and lasers have found widespread applications in science and technology. While this property was theoretically envisaged even before a maser was physically realized, it could be properly tested only after the first ammonia maser became available. Here is an account, in Charles Towne's own words, of the effort that was involved in verifying the narrowness and constancy of emission from the very first masers:

We started building a second maser almost immediately after the first one worked, in order to check the frequency of one against the other. We were joined by Tien Chuan Wang, a student from China with considerable engineering experience, and we had the second one operating in about 6 months. Each used the 1.25-centimeter transition in ammonia, with a frequency of about 24 billion cycles per second. Although they were essentially identical, they were not expected to have exactly the same frequencies. Slight differences in the dimensions of their resonant cavities could displace the two signals from each other by a tiny amount — by 1 part in 100 million or so. To test their constancy, we overlapped the outputs of the two masers so that they "beat" together. The signals came in and out of phase with each other at an audio frequency of a few hundred cycles per second. What resulted thus resembled, somewhat, the warble of a twin-propellered airplane, in which one engine is running just slightly faster than the other — the drone of one propeller alternately reinforcing, and then damping, the noise of the other. With our masers, the beat signal was very steady. Its pure sinusoidal form told us immediately that, indeed, both masers were operating at precise, nearly unvarying frequencies. If either of the maser's wavelength varied appreciably, the beat would have been noisy or irregular, but it was not. With data from this demonstration and other tests, we published in August 1955, a longer and more detailed paper on the maser in The Physical Review, which gave more complete information to other physicists on its intriguing properties.

Owing to such phenomenal accuracy in the frequency domain, masers now form the basis of some atomic clocks and extremely low-noise microwave amplifiers in radio telescopes and deep space spacecraft communication ground stations.

6.3.1.2 *The hydrogen maser*

Another maser device with origin in the 1950s is the hydrogen maser. It uses transitions in the hydrogen atom as the basis for its operation and is, thus, an atomic rather than a molecular maser. In a hydrogen atom the electron and the proton both have intrinsic spin but their spin orientations may be dissimilar. The spins may be parallel (aligned in the same direction) or anti-parallel (pointing in opposite directions). The former configuration has a higher energy compared with the latter. The difference in energy between these hyperfine levels is 5.87×10^{-6} eV. This is a much smaller energy separation than even that encountered with the operation of the ammonia maser. This is reasonable because, in this case, only a spin flip is involved, whereas in the case of nitrogen inversion in ammonia, an entire atom re-positions itself in the molecular framework. Using the Planck relation, $E = h\nu$, the frequency of this spin-flip transition can be calculated to be very close to 1.42 GHz. The corresponding wavelength in free space is 21 cm. Figure 6.10 provides a schematic illustration of a hydrogen atom's spin-flip transition. The spontaneous transition probability A_{21} of this spin-flip is extremely low at 2.9×10^{-15} s^{-1}. This makes it more amenable to stimulated emission, if the right conditions are created.

A maser working on the spin-flip hyperfine transition of atomic hydrogen was developed at Harvard University by D. Kleppner, H. M. Goldenberg and N. F. Ramsey in 1960. The general design of hydrogen masers has remained the same from that time and is shown as a schematic diagram in Figure 6.11.

Hydrogen gas is bled into the system through a micro leak valve and passes through a glass ampoule surrounded by a wire coil. High frequency RF energy is fed to the coil which generates an RF discharge in the

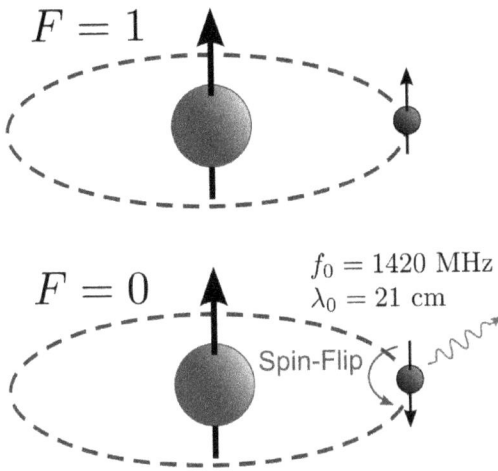

Fig. 6.10.   Schematic illustration of electron spin flip in a hydrogen atom.

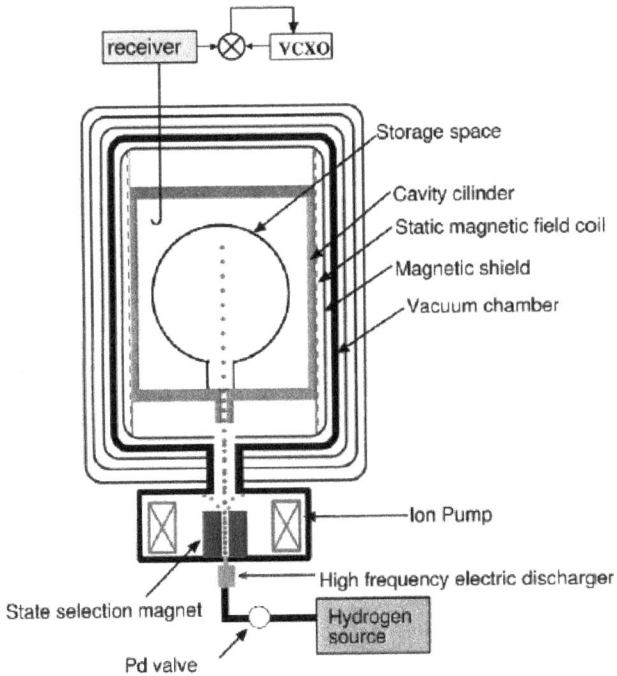

Fig. 6.11.   Schematic diagram of a hydrogen maser.

hydrogen flowing through the ampoule. Molecular hydrogen is dissociated into separate hydrogen atoms which are collimated into a narrow beam by going through a pair of slits into a highly evacuated drift tube. The drift tube is surrounded by a twelve-pole magnet which creates a strong highly non-uniform magnetic field through the length of the tube. This radial field is exactly zero at the central axis. Hydrogen atoms with parallel electron-proton spins keep traveling straight through while those with anti-parallel spins get deflected out of the beam. The parallel spin hydrogen atoms then enter a quartz envelope coated inside with a thin highly uniform Teflon coating. The Teflon serves to reduce atomic recombination to molecular hydrogen as hydrogen atoms have repeated collisions with its surface. This lengthens the residence time of atomic hydrogen. The quartz-Teflon bulb is surrounded by a precision- machined cylindrical copper cavity which functions as a microwave resonant cavity with dimensions tuned to the 21 cm hydrogen transition. A weak magnetic field is also applied along the hydrogen storage bulb. The cavity assembly is surrounded by several mu-metal magnetic shields to avoid perturbations from external magnetic fields. Hydrogen atoms undergo spin-flip transition in the storage bulb which gets amplified through stimulated emission. Some of this energy is syphoned out of the cavity through a wire loop and taken to external electronics such as a phase locked loop (PLL) which gets locked to the maser frequency. This extremely precise and stable frequency can be used for precision time keeping and other applications that require a well-defined frequency standard of very small spectral width (frequency spread). Two identical hydrogen masers would only differ by about one thirty-billionth of a second over an hour's operation time.

While the ammonia maser is no longer as widely used as was the case up to three decades ago, hydrogen masers are still in wide use in radio astronomy, for navigation on earth and in space, and in several precision spectroscopy applications. Figure 6.12 here shows a commercial hydrogen maser system with all hardware and electronics inside a single compact cabinet.

6.3.2 *From masers to lasers*

Lasers are conceptually the same devices as masers, except that their frequency of operation is much higher. It is customary for resonant stimulated emission devices operating in the far-infrared region, or shorter

Fig. 6.12. A commercial MHM 2010 active hydrogen maser system, developed by Microsemi Corporation.

Courtesy: Microchip Technology Inc.

wavelengths, to be called lasers. The term itself, an acronym standing for Light Amplification by Stimulated Emission of Radiation, was coined by Gordon Gould in 1957. From a historical perspective, lasers were a natural development stemming from the desire to extend maser operation to higher frequencies, i.e. shorter wavelengths. The maser itself, as the forerunner of the laser, owes much to the field of microwave spectroscopy for its invention. It was the detailed knowledge of rotational and vibrational energy levels in gases and solids, obtained through microwave spectroscopy that made the maser possible.

After the invention of the maser — and even before that — there was considerable speculation if stimulated emission could be used for amplifying light. The physics of stimulated emission seemed to favor the possibility of photon multiplication in an appropriate population-inverted medium. Shortly, after Charles Townes and his students had demonstrated a working ammonia maser, he started giving serious consideration to amplifying light by a similar technique. However, initially it seemed a daunting prospect because there appeared no way to gradually move from the microwave to the optical region, as the space in-between — terahertz and infrared — were largely scientifically undeveloped. Good emitters and detectors had not yet been developed for those regions, so that creating population inversion and detecting radiation emission looked formidable. The near-infrared and visible regions were a different story because radiation generation and detection technologies for them were highly developed. Thus, the initial enthusiasm quickly waned and most researchers in the field diverted their attentions to the further development of masers and their applications. The idea, however, remained at the back of Townes' mind, and by 1957 he began to give the prospect of photon multiplication through stimulated emission fresh scrutiny.   Townes also started discussing the idea with his colleagues at Columbia University. He understood the importance of creating a substantial population inversion in a suitable medium for stimulated emission to be observed at optical frequencies. One of his Columbia University colleagues, Polykarp Kusch, was working, among other things, on optically pumping metal vapors. Optical pumping had been developed in the early 1950s by Alfred Kastler at the École Normale Supérieure in Paris. It can be used to directly excite specific transitions. Exciting various media by shining visible or infrared radiation on them, for spectroscopy purposes, was an idea that was beginning to develop during that time. Several prominent physicists, including I. I. Rabi, also at Columbia, were interested in this technique. In October of 1957, Townes held a meeting with Gordon Gould — a graduate student of Kusch who had been working on optical pumping of thallium vapor. Gould immediately became very interested in the idea and started independently working on it. Both Townes and Gould developed and refined the idea for such a device, with Gould also coming up with the term 'laser' in his original notebook that detailed possible operating principles of such a device (see Figure 6.13). At that time, neither Townes nor Gould had fully appreciated the fact that optical pumping likely will not work for creating population inversion in gases or metal vapors because of their low density.

Townes later collaborated with Arthur Schawlow (see Figure 6.14) to write a detailed physics paper on the proposed operating principles of such a device. They called it an optical maser and their paper, titled: Infrared and Optical Masers, appeared in the December 1958 issue of the *Physical Review*. Schawlow's contribution was equally important to this joint effort as he proposed the idea of using two flat highly reflecting mirrors to build an optical resonant cavity for maser action to be observed at optical frequencies. This structure, also called a Fabry-Perot etalon, allows photons to repeatedly reflect between the mirrors, draining energy

Some rough calculations on the feasibility of a LASER: Light Amplification by Stimulated Emission of Radiation.

Conceive a tube terminated by optically flat

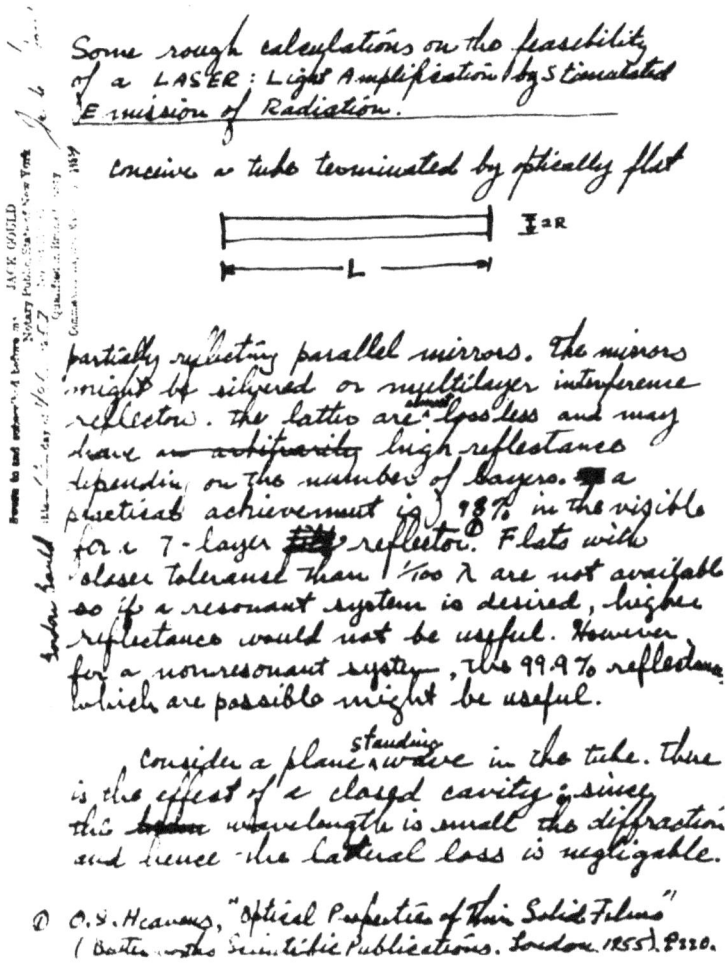

partially reflecting parallel mirrors. The mirrors might be silvered or multilayer interference reflectors. The latter are lossless and may have an arbitrarily high reflectance depending on the number of layers. A practical achievement is 98% in the visible for a 7-layer reflector. Flats with closer tolerance than 1/100 λ are not available so if a resonant system is desired, higher reflectance would not be useful. However for a nonresonant system, the 99.9% reflectance which are possible might be useful.

Consider a plane standing wave in the tube. There is the effect of a closed cavity; since the wavelength is small the diffraction and hence the lateral loss is negligible.

① O.S. Heavens, "Optical Properties of Thin Solid Films" (Butterworths Scientific Publications. London 1955). P.220.

Fig. 6.13. A page from Gordon Gould's notebook, showing the description of the operating concepts of a laser device.

from the excited gain medium by causing its atoms, ions or molecules to undergo stimulated emission. It appears that Gordon Gould also had this idea and, thus, the general consensus seems to be that the ideas underlying a practical laser device were developed by all three scientists through their mutual interactions between Columbia University and the Bell Labs. Schawlow went on to share the 1981 Nobel Prize in Physics with Nicolaas Bloembergen and Kai Siegbahn for his work on using lasers to determine

Fig. 6.14.   Arthur Schawlow.

atomic energy levels with great precision. Townes and Schawlow's paper became the landmark paper which laid the theoretical foundation of the laser. While not comprehensive from a modern perspective, it provided enough details and requirements to enable experimental work toward a working device.

Once the physical foundations and proposed engineering details had been published, interest in the optical analog of the maser quickly escalated. Several research groups started attempts at building an optical frequency maser (the term 'laser' was not in use at that time). As it turned out, the first demonstration of laser action not only borrowed conceptually from the earlier development of the maser, it also relied on experience gained from a very special kind of maser.

Of the several academic research groups, companies and government labs that became interested in the maser during the 1950s, NASA's Jet Propulsion Laboratory (JPL) in Pasadena, California was especially active. This was because of the realization that stimulated emission offered the possibility of amplifying very small microwave signals with unprecedented low noise. All other technologies for microwave amplification during that era were based either on the use of legacy vacuum tubes, such as klystrons and traveling wave tubes, or solid-state transistors. By their nature, they had noise figures that limited their signal-to-noise ratio, so that extremely weak signals could not be amplified. But with mankind venturing into outer space, there was an urgent need to find the means to perform low noise amplification of extremely week radio signals. Deep space interplanetary flybys and radar mapping of planets, like Mars and Venus, all required superbly sensitive high frequency radio receivers and this explains JPL's interest in maser technology. They sent Walter Higa to Columbia to study masers from its inventors. Soon, however, it became clear that gas masers, like the molecular ammonia and atomic hydrogen masers were not suitable for use in sensitive RF receiver installations. For best performance, the maser device, used as a signal amplifier, had to be installed at the prime focus of large steerable radio dishes and gas masers were far too bulky to be easily used in such manner. A less unwieldy maser device was required, and along came the ruby maser.

A ruby maser, developed during the late 1950s, is a completely solid-state device. It is small and compact enough (although it does require cryogenic cooling) to be used at radio telescope installations. Research teams at JPL had a pioneering role in developing ruby masers and deploying it at the sites of several of NASA's deep space network (DSN) dish antennas. This maser relies on the use of synthetic pink ruby crystals which consist of aluminum oxide (Al_2O_3) doped with a small amount (~0.05%) of chromium oxide (Cr_2O_3). Aluminum and chromium are isovalent in these compounds — both having a valency of 3. When chromium oxide is doped in aluminum oxide, the chromium atoms occupy sites that are normally occupied by aluminum atoms. It is the presence of substitutional chromium atoms in ruby that gives the, otherwise colorless, aluminum oxide its pink color. Chromium atoms exhibit paramagnetism from the three unpaired electrons in their outer (valence) shell. When a chromium atom is placed in a moderate magnetic field, its ground state energy level splits into four separate levels due to the Zeeman effect. These Zeeman levels differ in energy, with energy separations corresponding to frequencies from sub-1 GHz to several GHz. Energy transitions among the Zeeman levels can be used for stimulated emission. This system then acts as a suitable medium for maser action in the solid state. An additional, and very important, advantage offered by this device arises due the fact that the Zeeman splitting is dependent on the strength of the applied magnetic field with larger fields causing larger splitting. This allows the energy level separations, and, thus, the operating frequency of the maser to be tuned to a certain extent, so as to match a particular communication interlink frequency. Generally, the ruby crystal is placed inside a resonant cavity which is coupled to a coaxial feed or wave guide

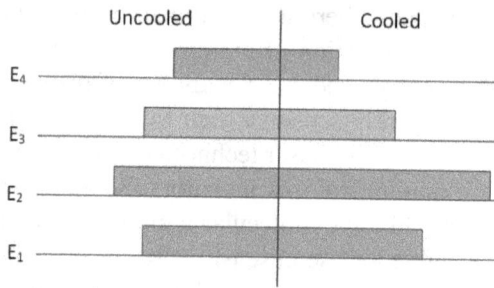

Fig. 6.15.   Schematic diagram showing energy level occupancies of Zeeman-split energy levels in uncooled (left) and cooled (right) ruby crystal.

through which pump power and input signal could be coupled to the ruby crystal. A permanent magnet supplemented with an electromagnet (for field strength control) provides the magnetic field bias for the ruby crystal. The entire cavity is enclosed inside a liquid helium bath which itself is contained inside a liquid nitrogen jacket. Cooling the maser active medium to extremely low temperatures (invariably using liquid helium) has remained a major limitation for the widespread use and miniaturization of solid-state maser oscillators and amplifiers. Cryogenic cooling is necessitated because the very small energy separations between the Zeeman-split energy levels causes all four levels to be nearly equally occupied. By cooling the ruby crystal, the occupation of electrons in the higher energy levels is reduced (unless the crystal is pumped). This is seen in Figure 6.15 where the energy levels and their occupancies of a pumped crystal before cooling (left of vertical line) are compared with that of a cooled crystal (right).

As explained earlier, application of a magnetic field to a ruby crystal causes the ground state energy level of trivalent chromium ions to split into four distinct energy levels, as seen in the figure here. The Maxwell Boltzmann distribution function dictates that, under thermodynamic equilibrium, at any given temperature, higher energy levels will have fewer electrons in them whereas the sates at lower energies will have higher occupancies. In the case of Zeeman-split energy levels in chromium ions, the inter-level energy differences are so small that at room temperature electrons possess enough energy to occupy all four levels to almost the same extent. This makes creating population inversion difficult. When ruby, immersed in an appropriate magnetic field, is pumped with microwave radiation at a frequency ν corresponding to the energy difference $E_3 - E_1$ then some electrons from level 1 are promoted to level 3. These then decay to level 2 through non-radiative transitions so that the occupancy of level 2 rises significantly. This creates the desired population inversion between levels 1 and 2 with the potential for stimulated emission between these levels. However, the high occupancies of levels 2, 3 and 4 do not let large population inversions to be created. This leads to noise generation through processes such as amplified spontaneous emission and also limits the amount of amplification that can be obtained if a signal corresponding to the energy difference $E_2 - E_1$ is injected into the system. Because three distinct energy levels: 1, 2 and 3 are involved in the stimulated emission process, so the ruby maser is called a three-level maser.

Matters can be greatly improved, if the ruby crystal is cooled down to 4.2 K — the temperature of liquid helium. At that temperature, the ground state has markedly higher electron occupancy than any of the other Zeeman levels. The higher occupancy of the lowest levels means correspondingly lower occupancy of upper levels which makes inversion pumping much more energy efficient. This makes the maser, in turn, more efficient and, thus, all ruby masers are operated at liquid helium temperature. The only major obstacle in operating ruby masers comes from the elaborate cryogenic systems needed to maintain the active ruby medium at such low temperature. The first ruby masers were very large and required extensive support, but within a couple of years of their invention, these devices were re-engineered to a much more practical size. This improvement not only made masers practical as ultra-low noise microwave amplifiers but, historically, as we see below, also indirectly led to the invention of the laser.

Ruby masers of different designs were built and operated by JPL and other labs during the late 1950s. Radio astronomy communities pioneered the use of masers as extremely sensitive ultra-low noise preamplifiers. Figure 6.16 shows Harvard University's 60-feet parabolic dish antenna with a maser installed at the prime focus. It can be readily appreciated that replenishing the liquid cryogens needed for maser operation with the device located at such height from the ground is not an easy operation, and this remained a problem for many years. Nowadays, cryogen-free closed cycle cryo-coolers have eliminated this service problem.

The US military also became interested in using this maser for space-to-ground communications and several defense contractors became active

Fig. 6.16.   Harvard University's 60-feet parabolic dish antenna with a maser installed at the prime focus.

in further developing and improving the ruby maser. One of these companies was the Hughes Aircraft Company, then being relocated to Malibu in Los Angeles, California. There, Theodore H. Maiman led a ruby maser redesign project for the US Army Signal Corps. Maiman was a skilled experimenter, with very good grounding in optics and general engineering. Over a short period of time, he and his colleagues had managed to greatly reduce the size of ruby masers; making them much more practical for low-noise X-band microwave amplification. It was this familiarity with the ruby maser, and especially with the optical and spectroscopic properties of ruby, that led Maiman on to his next big project — demonstrating stimulated emission at optical frequencies. This objective had become the goal for many scientists after the appearance of the famous 1954 paper by Schawlow and Townes. For Maiman, in many ways, it was his pet project. After some lobbying, he succeeded in convincing his company's management to undertake this project and even obtained $50,000 for necessary expenses.

For his effort toward developing a working 'optical maser', Maiman brought to bear all of his knowledge of optical physics, atomic spectroscopy, then-recent literature on stimulated emission and, of course, proposals on how to construct such a device. Because he had had recent experience with the ruby maser so it is not surprising that he chose that medium for his initial work. Instead of the low energy magnetically-split ground state he had in mind higher energy transitions in the chromium ion that had energies in the optical region. Many of his colleagues were skeptical of his work and told him that he was unlikely to succeed in making a working optical maser with ruby crystals. Maiman, nevertheless, persisted, and made several devices with optical resonant cavities. Finally, on May 16, 1960, from a newly-constructed device, he was able to observe the emission of red light that had all the characteristics predicted for the output from an optical maser. The very first laser was born!

Laser development took off as soon as Maiman demonstrated his device, and many research groups with interest in optics started looking at developing either new types of lasers or trying to better understand their function and develop new applications. Gradually, the term 'optical maser' fell by the way and 'laser' came to be used for all stimulated emission amplifiers and oscillators that worked at frequencies beyond the microwave region. Lasers got so much press soon after their invention that the Time magazine raved about laser as the 'hottest topic in solid state physics since the transistor'. In the sections that follow, we look at a number of different prominent lasers that have been developed over the decades, since that fateful day in May 1960.

6.4 Some Prominent Laser Systems

The stimulated emission process — as explained by Albert Einstein — is the foundational concept behind the laser. However, one needs three essential material components to realize a working laser. First, a gain medium (also called working medium and active medium) with the right kind of atoms, ions or molecules for producing a thermodynamic population inversion. Based on the state of this medium, lasers are often classified as solid-state, liquid or gas lasers. Over the years, thousands of materials have been found with energy level schemes suitable for building lasers. Next, a means for energizing the active laser species in the lasing medium must be provided. This supplies the pump power to excite electrons to a higher energy level. Depending on the medium used, the pump energy may be imparted optically, electrically, thermally, or even through chemical means. Lastly, a specialized enclosure — a resonant cavity — needs to be in place so that photons can make multiple passes through the gain medium, releasing the energy in excited electrons to create ever more photons. The resonant cavity's shape and dimensions are carefully chosen to reinforce the optical field inside the gain medium. Without such a cavity, we can get stimulated emission and even limited light amplification but not self-sustained oscillations. Thus, a resonant cavity is essential if the laser is to be a source of light.

Hundreds of types of lasers are commercially available — capable of producing radiation from the far infrared all the way to the X-ray region. These wonderful devices serve in so many applications that a proper count is impossible to provide. In what follows, we look at only a few prominent laser systems, in roughly a chronological order of their development.

6.4.1 *The ruby laser*

The distinction of being the very first laser to be invented goes to the ruby laser. It was Theodore Maiman's deep familiarity with the energy levels available with ruby crystals, owing to his work on the ruby maser that made him to consider ruby as the gain medium for a possible laser. Ruby crystals for such applications are artificially grown from aluminum oxide melt containing a precise amount of chromium oxide. Trace amounts of chromium oxide (~0.01% w/w) give the crystal a slight pink color. With increasing amount of chromium oxide, the color deepens toward darker shades of pink. Once grown, synthetic ruby crystals can be cut and polished to a required size and shape. Due to the sapphire (Al_2O_3) matrix, ruby crystals have good physical and chemical properties: excellent hardness and durability, high thermal conductivity and good chemical stability.

Whereas the ruby maser uses magnetic Zeeman levels with very small energy separations for its operation, Maiman (see Figure 6.17) relied on electronic shell states with much larger separations for constructing a laser.

The energy level system of interest for ruby laser operation is shown in Figure 6.18. The ground state is at the very bottom, shown as the long dark line. With optical pumping, electrons, initially residing in the ground

Fig. 6.17.   Theodore Maiman.

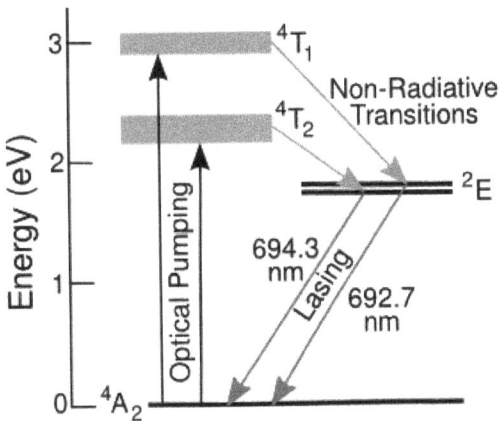

Fig. 6.18.   Schematic diagram showing the energy level system involved in the operation of a ruby laser.

state, could be sent up to two bands of very closely-spaced energy states that appear as pink bands in the illustration. These bands are most efficiently pumped at wavelengths around 404 nm (violet) and 554 nm (green). Both are plentiful in the light from xenon flash discharge lamps. From each of these bands, excited electrons transition down to a pair of closely-spaced energy levels (dark lines in the figure). These transitions are both quick and non-radiative. The transition energy is simply dissi-pated as heat. The spin-split ^2E energy levels are called metastable levels because of their relatively long lifetime (~3 ms at room temperature). This feature is very important for this so-called three-level laser system. The presence of a metastable state below an upper excited state allows elec-trons to accumulate into the metastable state through quick decay from the short lifetime upper state. This allows a significant electron occupancy to build up in the metastable state, providing a large population inversion relative to the ground state. Lasing then takes place through stimulated emission between the 2E levels and the ground state. Even a single spon-taneously emitted photon due to electronic transition between these levels can start a cascade of stimulated photons to appear — all moving in the same direction, having the same wavelength and the same oscillation phase. The ruby laser can lase at both 694.3 and 692.7 nm wavelengths, depending on which metastable level participates in the lasing transition. Which wavelength actually appears as the output is determined by the size of the resonant cavity, i.e. the optical distance between the two resonator mirrors. An integral number of half wavelengths has to fit between the two mirrors, so by carefully aligning and positioning the two mirrors, rela-tive to the ruby rod, it is possible to select one of the two possible lasing wavelengths. This technique can be used with any laser with multiple possible lasing transitions, to select a specific lasing wavelength. Most ruby lasers are operated at the 694.3 nm wavelength which is in the deep red region of the visible spectrum.

In order to provide optical excitation to the ruby rod, the choice ini-tially fell on cine projector lamps. The large size and awkward geometry of those mercury arc lamps, however, were not suited to this application, so Maiman started thinking of other possibilities. His lab assistant, sug-gested using photographic flash lamps. The tubular lamps, shaped as a spiral, could fit snugly over a cylindrical ruby rod. Furthermore, these xenon filled flash lamps could be operated in a flash mode with high volt-age pulses, generating short intense bursts of wide spectrum light. This way, higher peak powers could be obtained for more intense pumping than would be possible with continuous operation of the flash lamp. This idea was suggested by Gordon Gould while attending a conference in 1959 with Theodore Maiman. A commercial GE FT-506 xenon flash lamp was finally used for this purpose.

To make the ruby rod serve as an optical cavity, both ends of the ruby cylinder (1 cm in diameter and 2 cm long) were ground, polished flat and

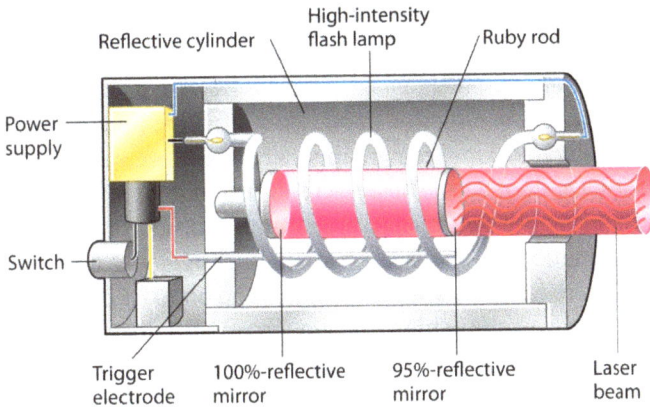

Fig. 6.19.   Schematic illustration of a ruby laser.

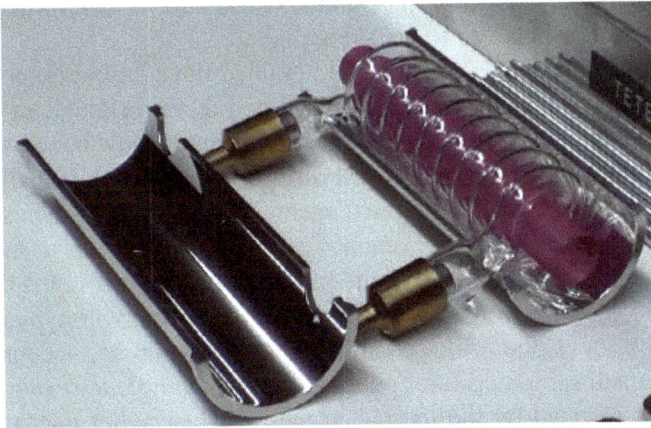

Fig. 6.20.   The disassembled parts of a working ruby laser.

parallel, and were coated with evaporated silver. A 1 mm diameter hole in one of the silver coatings was provided to extract the laser radiation. Some designs had one end of the ruby rod made fully silvered (100% reflecting) whereas the other end was made somewhat less reflective for radiation to escape through. Maiman enclosed this arrangement inside a polished aluminum cylinder whose internal walls served to reflect even more light on to the ruby rod. A model of a similar laser appears in Figure 6.19, whereas Figure 6.20 shows the disassembled parts of a working ruby laser.

After putting together his first experimental ruby laser system, Maiman connected the flash lamp to a pulsed high voltage supply and started driving the lamp with gradually increasing voltages. Beyond a certain value of the lamp drive voltage, the radiation coming out from the hole-bearing silvered end of the ruby rod became remarkably narrow in spectral width as well as getting sharply collimated. These were unmistakable signs of successful stimulated emission-based radiation generation inside the ruby rod. More detailed examination also revealed that the laser light was coherent, i.e. had photons that were in phase with each other. The onset of these changes was very sudden, once a certain high voltage was applied to the flash tube. This was the first observation of what is now known as a 'lasing threshold' — the establishment of laser oscillations once the pump energy, in any form, goes above a certain value so that population inversion is attained. This was also the first observation of spectral narrowing, as spontaneous emission gave way to stimulated emission. The sudden collapse of spectral line width is now one of the most important features looked for in cases where stimulated emission is believed to be taking place.

By mid-July 1960, Maiman had obtained three new ruby crystals of much improved optical quality. When tested in the original set-up, they provided even more dramatic spectral and beam narrowing, as well as a sharp threshold for laser action. All of these systems were small enough to not require special cooling arrangements but larger systems needed water cooling because the xenon flash lamp generated a significant amount of heat at high pulse repetition rates. Unlike other lasers that were developed later, the ruby laser is a pulsed laser system with typical pulses lasting for about one millisecond. Maiman wrote up his account of the ruby laser and submitted it for publication in the *Physical Review Letters*, but it was rejected by the journal in the mistaken belief that it was yet another manuscript on masers! Maiman then sent it to the journal *Nature* which published the short (two pages long) landmark paper, titled: Stimulated Optical Radiation in Ruby. Figure 6.21 (left) shows one of the original ruby crystals used by Maiman for his demonstration of laser action. Figure 6.21 (right) shows an original cylindrical enclosure that housed a ruby rod for flashlight pumping.

With Maiman's demonstration of the ruby laser, the Hughes Aircraft Company won the race to build the first working laser. Maiman, understandably, received much publicity for his invention. He later moved from Hughes to establish companies of his own for manufacturing and selling ruby lasers.

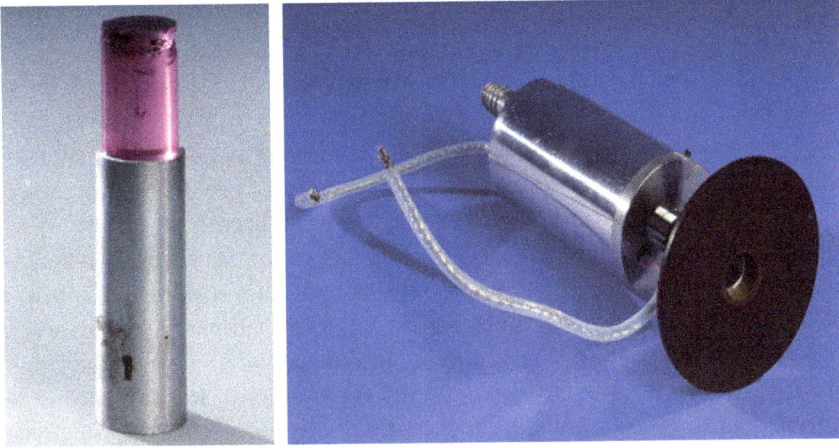

Fig. 6.21.   (Left) A ruby rod used in one of Maiman's original ruby lasers. (Right) An original cylindrical enclosure that housed a ruby rod for flashlight pumping in a ruby laser.

After its invention, the ruby laser was pressed into several applications, ranging from studies of light–matter interaction to ophthalmic surgery. Famously, it was also used, with retro-reflectors left on the lunar surface by Apollo astronauts, to accurately measure the earth-to-moon distance. But, here on Earth, one of its first applications was to explore the behavior of matter under irradiation with intense light beams. In 1961 — just a year after its invention — the ruby laser was used by Peter Franken, A. E. Hill, C. W. Peters, and G. Weinreich at the University of Michigan in Ann Arbor to demonstrate the intriguing phenomena of second harmonic generation. Intense, coherent, light beams passing through certain 'nonlinear' materials can result in photons mixing together to produce new photons at twice the frequency (and thus, half the wavelength) of the original photons. To practically observe this effect requires very high light intensities that were not available before the invention of the laser. In their experiment, the four collaborators tightly focused 694 nm pulses of deep red light from a ruby laser into a quartz crystal. The light coming out of the crystal was analyzed by a spectrometer which clearly showed the generation of light at 347 nm in the UV region. The spectrum was recorded on photographic paper and formed part of the manuscript that the team submitted to *Physical Review Letters* for publication. The copy editor of the journal misunderstood the faint line at 347 nm for a speck of dust and removed it from the publication! This error was found and acknowledged later but became a case in point for the utility of newly-developed lasers for uncovering subtle physical phenomena.

Ruby lasers, for long, have been extensively used for producing large holograms, in sizes up to a meter square. Many non-destructive testing labs also use ruby lasers to create holograms of large objects, such as aircraft tires, to look for weaknesses in the lining. Ruby lasers were used extensively in tattoo and hair removal, but have been replaced by other lasers in such cosmetic applications. The ruby laser has several outstanding characteristics, such as simple construction, a very narrow spectral line width of about 0.5 nm and operation at room temperature. However, it is no longer widely used because of its low conversion efficiency, and its applications have been taken over by other much better lasers. As the very first laser to be invented, though, it will always retain a special place in the pantheon of laser systems.

6.4.2 *The helium-neon laser*

Laser action in a gas, or rather a mixture of two gases, was first demonstrated in 1960 at the Bell Labs, shortly after Theodor Maiman's demonstration of the ruby laser. Not only was this the first gas laser to be developed, it was also the first continuously operating (continuous wave or CW) laser — distinct from its predecessor, the pulsed ruby laser. In contrast to the ruby laser, the helium-neon laser is still widely manufactured and used in a variety of applications.

Lasing in gases, with pumping provided by an electrical discharge, was considered by several early maser and laser pioneers. The principal issue in building such a laser is finding a suitable gas with an energy level scheme that will allow for efficient population inversion through pumping with ion collisions. Due to the low density of gases, optical pumping is not practical for gas lasers. Neon was proposed quite early on as a prospective gaseous lasing medium. It is easy to ionize, as evidenced by its use in neon signs, and as a starter gas in sodium vapor lamps. It also has a set of energy levels which are easy to populate with favorable state lifetimes. With the ruby laser having demonstrated the operation of a laser, several academic and industrial groups started looking into the possibility of building a gas laser.

The first gas discharge laser was built and operated at AT&T Bell Labs by Ali Javan (see Figure 6.22) in 1960, in collaboration with William Bennett, Jr. and Donald Herriott. Javan was Townes' PhD

Fig. 6.22. Ali Javan.

student at Columbia University. In the summer of 1958, he was inter-
viewed for a job at Bell Labs, where Arthur Schawlow told him about the
laser, which at that time had not yet been demonstrated. Intrigued, Javan
rushed back to Columbia that afternoon and began investigating laser
concepts. He was soon convinced that electrical discharge was the only
viable route toward producing population inversion in gases. To this end,
he favored a two-step excitation process in a dilute mixture of neon gas
in helium. In a two-step excitation process, electrons would first collide
with helium atoms, exciting them to a higher energy level. Energized
helium atoms would then transfer their extra energy to the less-abundant
neon atoms, exciting them to metastable states with energies close to
those of the excited helium atoms. Javan had worked out that this scheme
could produce a population inversion in neon atoms such that an infrared
transition could be excited. He persuaded Bell Labs to hire William
Bennett Jr. to help with his experiments. Bennett had recently finished
his Ph.D. research on just this type of electrical discharge. Javan and
Bennett teamed up with Donald Herriott — then an optics specialist at
Bell Labs — to develop a suitable optical resonator for their laser.
Initially, they did not meet with much success, but after several months
of painstaking work, one day while Herriott was adjusting the mirrors at
the ends of the resonator, Javan noticed the first signs of laser oscillations
in the laser plasma tube. The helium-neon laser had been born.

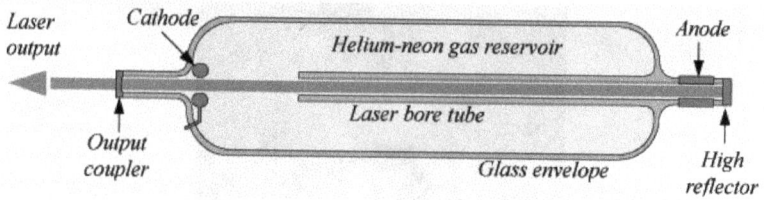

Fig. 6.23.   A schematic diagram of a helium-neon laser.

Fig. 6.24.   Ali Javan's original helium-neon laser.

This laser used a mixture of helium and neon gases in roughly a 10:1 ratio at a pressure of about 1 Torr, inside a sealed glass tube. A fully reflecting and a partially reflecting mirror were placed at the two ends of the discharge tube. Electrodes fitted at both ends were used to pass a high voltage DC discharge through the tube. Later, use of RF discharge at a few milli amperes became the norm. Figure 6.23 shows a schematic diagram of a helium-neon laser whereas Figure 6.24 shows Javan's original helium-neon laser unit which is now housed in the Smithsonian Institution of Washington DC.

The laser emitted an invisible infrared beam at 1.15 μm which was detected by semiconductor infrared radiation detecting devices. This was the first demonstration of not only a gas laser but also a continuously-operating laser and an infrared-emitting laser.

Electric discharge in low pressure neon gas is quite efficient but is not effective enough to easily create a population inversion. Adding helium to neon makes for a much more energy-efficient lasing medium. Accelerated electrons in the discharge, as mentioned above, can transfer energy to

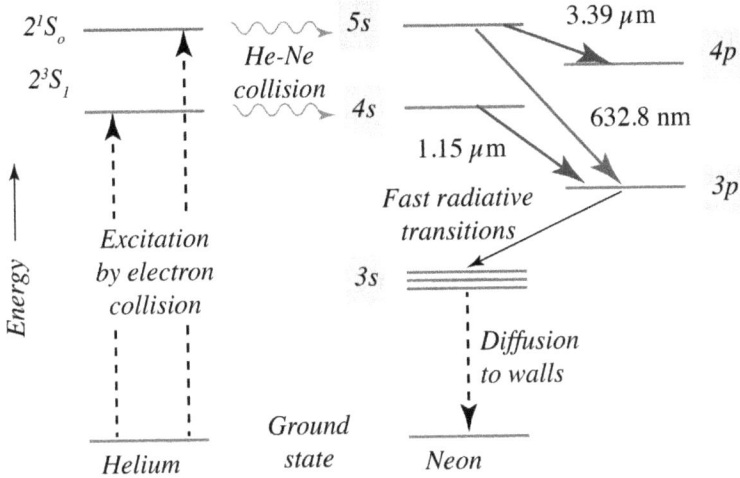

Fig. 6.25. Schematic diagram showing the energy level system scheme involved in the operation of helium-neon lasers.

Courtesy: XuPanda.

helium atoms which can then transfer this energy to neon atoms during atom-atom collisions. This method of exciting one kind of atoms (or molecules) in a gaseous medium by collisions with a different species was an early development in experimental spectroscopy. It was initially called 'excitation of the second kind', to distinguish it from direct collisional excitation. The energy transfer is particularly effective in a mixture of helium and neon gases because of good alignment between the excited energy levels of helium and neon atoms, as seen in Figure 6.25. The two electrons in a helium atom normally occupy the 1S level in the ground state. Once excited, they move up to one of two possible 2S levels — one with parallel spins, 2^3S_1, and one with anti-parallel spins, 2^1S_0. These energy levels are almost perfectly aligned with the 4S and 5S levels, respectively, of neon atoms. Thus, electronic excitation energy can be readily transferred from excited helium atoms to neon atoms. The excited neon atoms can then decay radiatively through one of several available channels, as seen in Figure 6.25. All final states in such transitions can then quickly decay to the 3S excited state of neon which eventually de-excites to the L-shell ground state through atomic collisions with the walls of the discharge tube. Fast decay to neon's 3S level is important in this scheme because it empties the second level of all laser transitions. Thus, easy population of the upper laser levels through collisional energy transfer from excited helium atoms, combined with quick depopulation of the lower laser levels, makes this gas mixture especially suited for laser action.

The development of the helium-neon laser caused considerable excitement. The US Army Signal Corps requested the Bell Labs to make an infrared helium-neon laser for their use. It also got extensive media coverage during the time of its first demonstration. Nowadays, Helium-neon lasers are widely used for many general purpose as well as specialized applications. Their wide availability, relatively low cost and high optical quality monochromatic beam make them a staple of both teaching and research labs, all over the world. Several companies manufacture helium-neon lasers of different sizes. Figure 6.26 shows a low power commercial helium-neon laser unit.

W. R. Bennett, Jr., who worked with the original team that developed the helium-neon laser, continued further investigations of the lasing mechanism in neon-containing gas mixtures. In collaboration with another Bell Labs physicist, C. Kumar N. Patel, he succeeded in showing lasing in pure neon gas by gradually reducing helium gas pressure. Their team was later joined by Walter Faust and Ross McFarlane, and together they looked for lasing transitions in several promising gas mixtures. This work resulted in the discovery of laser action in multiple gases and gas

Fig. 6.26.   A commercial intensity and frequency stabilized helium-neon laser with its power supply.

Courtesy: Excelitas Technologies Corporation.

mixtures, such as, neon-oxygen, argon-oxygen, as well as pure argon, krypton and xenon. Due to reasons having to do with power conversion efficiency and the difficulty of making such gases lase, it is only the helium-neon laser from that era which went into commercial production and remains in use to this day.

6.4.3 *The carbon dioxide laser*

Carbon dioxide was the first molecular gas in which laser action was successfully demonstrated. Later, lasing was achieved in several other molecular gases, such as nitrogen and carbon monoxide. The carbon dioxide laser, however, remains the most important molecular gas laser and has found widespread applications in materials processing. It is the most widely used industrial laser; almost universally used for cutting, drilling and machining operations with a host of materials, ranging from metals and ceramics to wood, glasses and plastics. The combined share of all other laser types used for such materials processing applications is easily dwarfed by this laser. The carbon dioxide laser emits CW radiation principally at 9.4 μm and 10.6 μm with the latter wavelength used more frequently. Many natural and synthetic materials have strong characteristic absorptions in the 9–12 μm spectral range in which CO_2 lasers emit. This results in numerous opportunities in materials processing and spectral analysis where CO_2 lasers are ideally suited as radiation sources.

CO_2 lasers make use of electronic transitions among vibrational energy levels in tri-atomic carbon dioxide molecules. Because these energy levels are close to the ground state so carbon dioxide lasers are able to convert a large fraction of input electrical energy into useful laser output. Their energy conversion efficiency exceeds 20% and, correspondingly, these lasers are the highest power CW molecular gas lasers available. Units capable of delivering several kilowatts of optical power are readily available.

The carbon dioxide laser was invented in 1964 by C. Kumar N. Patel (see Figure 6.27) at the Bell Labs. Patel was hired by the Lab in 1961 and put to work on discharge-based laser systems. He was closely familiar with the work of Ali Javan and his collaborators, developing the helium-neon laser, and wondered whether molecular gases could also be made to lase. This was achieved only three years later when he showed laser action in CO_2 gas. His was not an accidental invention as Patel had carefully worked out the operating principles beforehand and expected high

Fig. 6.27.   C. Kumar N. Patel.

efficiency before any experimental work took place. Theoretical work on possible laser action in carbon dioxide had commenced in 1963, and by 1964 Patel had become confident enough to assemble the required equipment to demonstrate laser operation. In an interview, many years after the first demonstration of the carbon dioxide laser, he recalled his experience: 'It did the first time we tried. It worked marvelously well. We got tens of milliwatts on the first shot." Later, realizing that diatomic gases should also lase, he tried carbon monoxide (CO) which also lased well. Continuing with the development of CO_2 lasers, he added nitrogen and then also helium to carbon dioxide and succeeded in raising the output power to 200 W, by the middle of 1965. This gas mixture is used with few changes in present-day commercial CO_2 lasers.

Lasing in carbon dioxide lasers takes place through energy transitions between the vibrational and rotational levels of the carbon dioxide molecule. These energy levels are close to the ground state and are separated from each other by relatively low energies. Thus, high energy discharges are not required for laser excitation and the emitted photons have low energies (long wavelengths). The gas mixture in CO_2 laser tubes typically consists of 10–20% of CO_2, 10–20% of N_2, ~5% of H_2 and the rest 65–75% of He. CO_2 is the active lasing gas whereas N_2 and He serve to

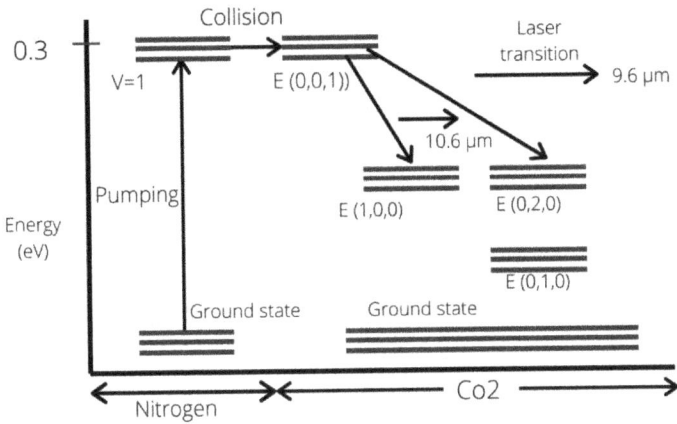

Fig. 6.28. Schematic diagram illustrating the electronic and vibrational energy levels involved in the operation of a CO_2 laser.

populate and depopulate the upper and lower lasing levels, respectively. The lasing process is depicted in Figure 6.28.

Electric discharge in the CO_2 laser tube excites N_2 molecules through collisional excitations with fast moving electrons. Being homonuclear, N_2 molecules cannot easily lose energy through photon emission — making their excited state fairly long-lived. Energy is transferred from the metastable excited state of N_2 molecules to asymmetric vibrational state of CO_2 molecules. These two states in N_2 and CO_2 molecules are very closely matched in energy, which greatly facilitates inter-molecular collisional energy transfer, and is the prime reason for the high efficiency of CO_2 lasers. Laser transitions can take place from the excited asymmetric vibration modes of CO_2 molecules to one of two symmetric vibration modes. The one emitting 10.6 μm photons have a higher transition probability and, thus, emission at this wavelength is more intense, compared with emission at 9.4 μm. It is possible to select any of these wavelengths for lasing by the use of suitable laser cavity optical elements. Once a laser transition has taken place, the symmetric vibration mode population has to be depleted quickly to ensure CW operation. This is achieved by collisional de-excitation with He atoms. Energy is, thus, transferred from CO_2 molecules to He atoms which, in turn, lose their thermal energy by

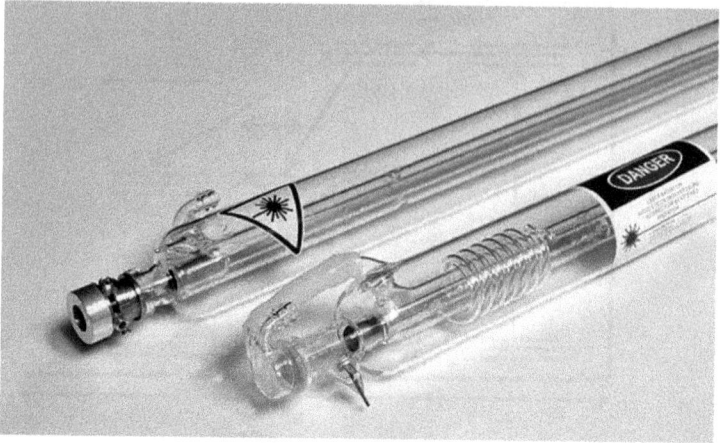

Fig. 6.29. CO_2 laser tubes.

colliding with the wall of the discharge tube. This makes the discharge tube run hot and it has to be constructed from temperature-resistant materials, such as quartz or refractory ceramics (see Figure 6.29). For low power outputs (a few watts) the laser tube can be cooled by forced air circulation but higher power lasers require efficient water cooling of laser tubes to keep them operating. This is essential, because if the tube is not cooled then not only it will suffer thermal breakdown, hot helium will be unable to deplete electrons from the lower lasing level of carbon dioxide molecules without which population inversion, and thus lasing, cannot be sustained. To achieve even higher output powers (multi kilowatt) continuous streaming (flow-through) CO_2 lasers have been developed where the lasing gas mixture is continuously fed to the discharge tube and the hot gas is removed at the other end by exhaust pumps.

The CO_2 laser offers a certain amount of limited wavelength tuneability. The vibrational levels are associated with bands of closely spaced rotational energy levels which allow tuning of emission wavelength by using a diffraction grating inside the optical resonator. Furthermore, CO_2 molecules have three atoms and a variety of isotopic compositions are possible by altering the C and O isotopes that form the molecules. Possible isotopes are ^{12}C, ^{13}C and ^{14}C for carbon and ^{16}O, ^{17}O and ^{18}O for oxygen. With increase in isotopic mass, the laser emission shifts toward longer wavelengths. By using a combination of both grating tuning and isotopic tuning, the output wavelength of CO_2 lasers can be set to one of several possible values in the range of 8.98 μm to 10.6 μm.

CO_2 lasers are, generally, CW devices but can be made to output a train of high intensity pulses through a technique called Q-switching. This involves the use of a fast-rotating mirror, in place of the fixed fully reflecting mirror. While the mirror is positioned off the optical axis (straight line through the center of the laser tube) the laser cavity cannot oscillate and the energy input from the electrical discharge simply builds up the population of the upper lasing level. When the rotating mirror gets aligned with the beam axis, laser light can bounce back and forth between the two ends, extracting power from the CO_2 molecules in the form of a high peak power, short duration, laser pulse. Q-switched CO_2 lasers are often used in machining operations that require high peak power pulses.

Carbon dioxide lasers are widely available due to their popularity in many materials processing and medical applications. Figure 6.30 shows a typical commercial CO_2 laser cutting and engraving system. Equipment like this is extremely popular for manufacturing signage, decorative items

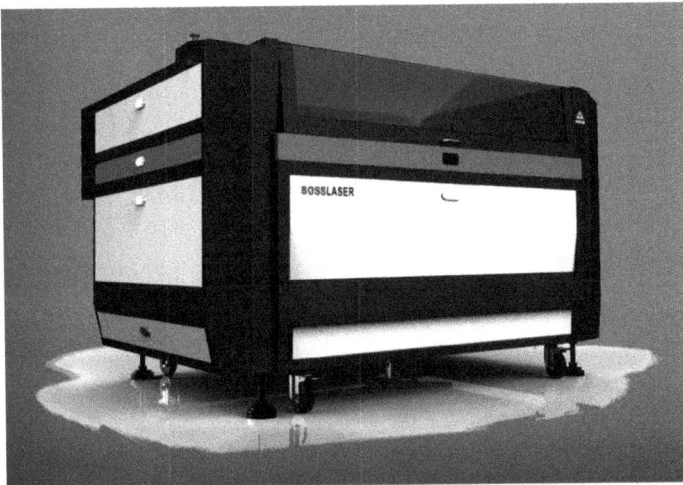

Fig. 6.30.   A commercial CO_2 laser cutting and engraving system.
Courtesy: Boss Laser LLC.

and multi-layer fabric processing. The work piece is usually held fixed in position while the laser head moves in an X-Y pattern, determined by a design file, to cut or engrave material in a desired shape. Similar laser units are also available for laser sintering, marking and welding applications for industrial manufacturing and assembly lines.

CO_2 lasers are also used for cosmetic surgery procedures, such as tattoo, mole and skin blemish removals. A system marketed for these and other 'skin resurfacing' treatments is shown in Figure 6.31. Note the articulated arm for delivering laser radiation to desired locations. Multi-joint arms containing beam steering mirrors are common in machines like this one, because fiber delivery of CO_2 laser radiation is not efficient due to high absorption in most materials. If fiber coupling is a must for some application, then a carbon monoxide laser may be used. CO lasers emit around 5 μm where radiation can be delivered over short distances using

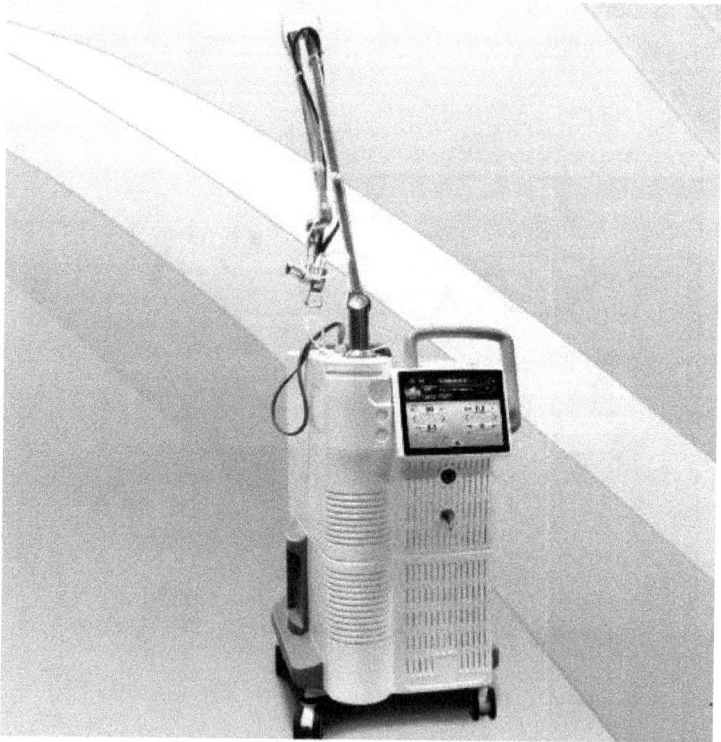

Fig. 6.31.   A commercial CO_2 laser system for medical applications.

special optical fibers. Another advantage of CO lasers is that because of their shorter emission wavelength, radiation can be focused more tightly into a smaller spot size than is possible with CO_2 laser radiation.

CO_2 laser systems of a similar design are also found in some dental practices. Radiation around 9.6 mm is strongly absorbed by hard and soft tissues in the oral cavity, which makes carbon dioxide laser especially valuable for a number of dental procedures.

Apart from its many uses in medicine and manufacturing, carbon dioxide lasers are also an enduring research tool. High atmospheric transparency in the 8–12 μm range makes CO_2 laser radiation particularly suitable for atmospheric remote sounding and laser detection and ranging (LIDAR) applications. Their heating capabilities have also been utilized to great effect in vaporizing molten tin droplets in extreme UV light sources, as discussed in Chapter 4. Even the laser interferometer gravitational wave observatory (LIGO) makes use of carbon dioxide lasers for thermal compensation of optical distortions in its sensitive optics chain. The separation of isotopes by laser excitation (SILEX) process — a technology originally developed in Australia — uses CO_2 lasers for enriching uranium isotope ^{235}U in natural uranium for making fuel for nuclear reactors. Thus, while the carbon dioxide laser may be one of the oldest types of lasers in widespread use, it keeps finding new applications that fit its special capabilities.

6.4.4 *The argon ion laser*

Laser development picked up immediately after the demonstration of the first ruby laser. All kinds of laser media and pumping schemes were suggested in the aftermath of the first successful laser operation. 1960s was a year of particularly rapid developments in laser technology. Several new lasers, that remain very important, more than half a century after their first demonstration, were invented in that decade. Among them is the venerable argon ion laser — still very widely used for a variety of uses, just like the helium-neon and carbon dioxide lasers, described above. This laser is made by several major laser manufacturers in North America, Europe and Japan.

The argon ion laser belongs to a special category of gas lasers where the active luminescent species are gas ions, rather than neutral atoms or molecules. Lasing in ionized gases was first reported in 1964 from singly ionized mercury atoms (Hg^+), by William E. Bell of Spectra-Physics Inc.

He stumbled on this discovery while trying to improve the short lifetime of early helium-neon lasers by adding mercury vapor to the gas mixture. A green glow appeared near the cathode which Bell attributed to spontaneous emission from mercury. It occurred to him that this might lead to a mercury laser and so he started earnestly trying to build such a laser. After some work he found that high current discharges through mercury-helium vapor were capable of laser emission and also that the energy levels involved were that of singly-ionized mercury rather than neutral mercury atoms. This was a surprise because ions hadn't been considered suitable for lasers because their energy levels were high above the neutral ground state. But it was also encouraging, because ions tend to have higher transition energies than neutral atoms, offering the potential for shorter-wavelength lasers. After this first observation, several gases were investigated for lasing in singly- and multiply-ionized species, and within the year no less than eleven gases were found capable of lasing in their ionized states. The energy level systems in ions are completely different from that in corresponding neutral atoms. Thus, lasing is generally observed at distinctly different wavelengths when compared with lasing in the corresponding neutral gas. Bell was able to observe lasing at two wavelengths, 615 nm (orange) and 567.8 nm (yellow-green) in the visible region, as well as at one infrared wavelength, using a pulsed discharge in a mixture of helium gas and mercury vapor. The gas mixture was contained in a 3-meter-long glass tube with 3.5 mm internal diameter. Pulsed electrical discharge was generated using a high voltage (10 kV) ceramic capacitor bank which was charged with a high voltage DC power supply. Bell measured the characteristics of laser oscillations at the aforementioned two wavelengths and surmised that with higher discharge currents CW lasing action may be possible.

This development was received with interest at several labs with active laser research programs. One of them was the Hughes Aircraft Company — the birthplace of the ruby laser, a mere four years earlier. William Bridges (see Figure 6.32), at Hughes, constructed his own pulsed mercury-helium laser and while studying various modifications to this laser, accidentally discovered laser action in ionized argon. This was truly an accidental development because Bridges was simply investigating the effect of replacing helium in mercury-helium laser with various other noble gases. He first replaced helium, as the buffer gas, with neon and then with argon. To his surprise, he found that with argon, he not only observed the orange and yellow-green mercury lines but an entirely new

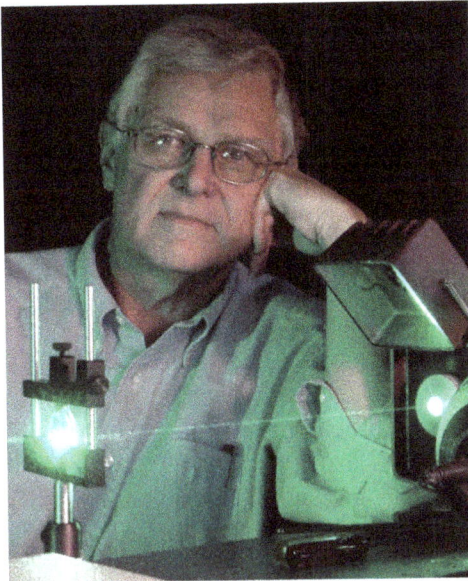

Fig. 6.32.   William Bridges.

blue line at 488 nm. Using spectral line tables in the lab's library, Bridges quickly found that the line most likely came from ionized argon. He tested this by using a new tube filled only with high purity argon gas and observed as many as ten lines. Soon afterward, he extended this work to ionized krypton and xenon. These two gases yielded laser emission at many wavelengths that spanned the entire visible spectrum. In his words: "Lines just tumbled out all over the place". In particular, with krypton, laser action could be simultaneously sustained at multiple wavelengths such that the emitted light appeared white in color. Xenon was notable in that more lasing-capable lines were found in the spectrum of ionized xenon than in almost any other element. Shortly afterward, Gene Gordon of Bell Labs demonstrated CW lasing in all three noble gases. Later, several other elements including iodine, oxygen, nitrogen, chlorine, zinc, cadmium, sulfur and phosphorus were found capable of lasing in singly and multiply ionized states. Because of relatively low gain, however, practical ion lasers are generally not constructed using these elements. It is notable that due to the often-larger energy level separations in ionized gases, lasing at ultraviolet wavelengths is usually easier to observe in ion lasers. For instance, lasing can be obtained at 275.4 nm from doubly-ionized argon.

The brightness of argon ion lasers at visible wavelengths earned ion lasers some important early applications, but their tough cooling requirements caused problems. At Hughes, Bridges developed argon ion lasers for an Air Force night reconnaissance system. The system performed well but it never went into production because the cooling system didn't meet the requirements for installation in military planes.

The energy level scheme involved in the operation of argon ion lasers is shown in Figure 6.33. An excessive electron population is created in a set of closely-spaced argon ion energy levels about 35 eV above the ground state energy. While several transitions are possible, two have very high radiative emission probabilities and dominate lasing action in argon ion medium. The lower lasing level is quickly emptied through a fast radiative transition to the ground state of argon ion which, in turn, decays to the ground state of argon atom after capturing an electron. These downward transitions are the cause of high heat production in argon ion laser tubes.

To produce laser action in ionized gases takes considerable amount of electrical energy. This energy is expended in an electric discharge which has to be powerful enough to create a large population of ionized atoms

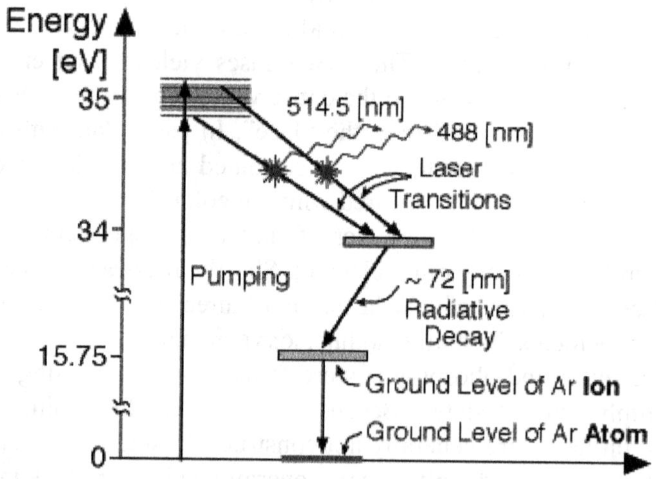

Fig. 6.33.   Schematic diagram showing the energy level system involved in the operation of an argon ion laser.

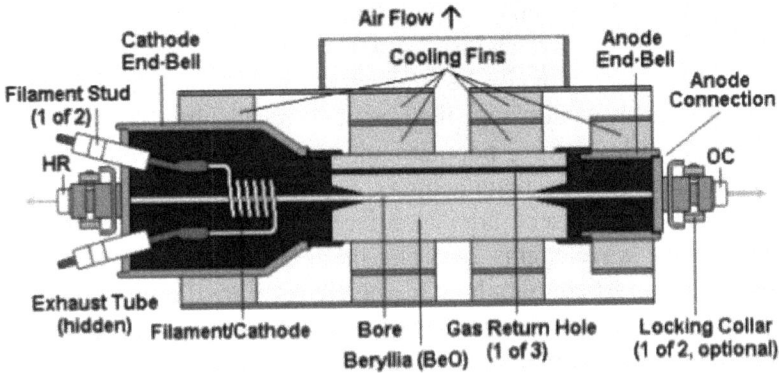

Fig. 6.34.   Schematic diagram of an argon ion laser.

in a rarefied gaseous medium. The higher the charge on the ion, the larger the required energy input. Fortunately, almost all ionized gas lasers make use of transitions in singly charged ions, keeping the energy needed within practical limits. The electrical-to-optical energy conversion efficiency of ion-based lasers is generally quite low (typically, significantly less than 1%). This means that most of the energy provided to generate and sustain a population of energy-inverted ions goes to waste as heat. This heat needs to be quickly extracted out of the system if catastrophic thermal failure is to be avoided. Thus, all ionized gas lasers make use of high temperature-resistant silica, beryllia or other similar ceramic tubes for containing the laser medium (see Figure 6.34). These tubes, and the electrodes fitted to them, require efficient cooling with chilled water. For this reason, all argon ion lasers, except for the smallest units, producing less than 10 mW of output power, require a cooling water connection. The narrow lasing tube is contained inside a much wider outer tube which serves as a reservoir of argon gas. Large cap-like anode and cathode structures are fitted to each end of the wider tube. A heated filament is also provided to generate electrons and make it easier to strike an arc through the tube. Often one or more magnets are also used to longitudinally 'pinch' the plasma in the discharge capillary. This confines the ions more strongly, resulting in somewhat higher gain and operating efficiency. Finally, the optical resonator is formed by two dielectric mirrors at either end of the outer envelope, with spectral reflectance optimized for the operating wavelength.

While ion lasers have very low operating efficiency, they also have several outstanding characteristics. Among these are their very high optical gain (in the region of 20% per meter) and the capability for lasing at many different wavelengths — with some very deep into the UV region. These lasers can also be engineered easily to produce kilowatts of optical power. Together with their good beam quality, these attributes have made ion lasers very popular and several types are routinely available from various laser manufacturers.

Like other ion lasers, argon ion lasers can also emit at any one of many possible wavelengths, covering the region from the UV to the IR. However, the most commonly used wavelengths are those that fall in the blue-green portion of the visible spectrum, specifically at 488 nm and 514.5 nm (see Figure 6.35). As a coherent source of blue and green light, argon ion lasers are often used in spectroscopy and for pumping luminescent materials and other laser media. A typical commercial argon ion laser is capable of operating at one of several distinct wavelengths in the 454–514 nm range. A specific lasing wavelength is chosen by the proper selection of optical cavity mirrors. Beam diameter is about 0.6 mm with low beam divergence of about 1 milli radian. The radiation has a Gaussian profile with very high beam quality and low noise, making argon ion lasers ideal for metrology and spectroscopy applications. Other noble gas

Fig. 6.35.   An operating argon ion laser.

ion lasers are also commercially available, with the krypton ion laser having the largest market share after argon ion lasers. Mixed gas argon-krypton ion lasers are also available for applications where selection from a very large number of possible lasing wavelengths may be required.

Argon ion laser's blue and green emissions penetrate deep into water, and this makes the laser suitable for several surgical applications that require intense focused radiation to be sent deep inside tissues. Opthalmic procedures are prime examples of such surgeries. For the same reason, argon ion lasers were also proposed for underwater communication systems. Other applications include recording of holograms, laser light shows, DNA sequencing, semiconductor manufacturing, and as a pump source for dye lasers.

6.4.5 *The neodymium YAG laser*

Another very important laser, which is somewhat similar to the ruby laser, uses an yttrium aluminum garnet (YAG) rod doped (around 1% by weight) with neodymium (Nd). YAG is a double oxide of yttrium and aluminum ($Y_3Al_5O_{12}$). When doped with neodymium, Nd^{3+} ions substitute for some Y^{3+} ions in the host crystal. Due to the similarity in charge (isovalency) and ionic radius, this substitution occurs readily. The resulting material is symbolized by the chemical notation: $Nd:Y_3Al_5O_{12}$. Incorporation of Nd^{3+} ions produce a pale to dark lilac coloration in the YAG crystal (see Figure 6.36), depending on the Nd concentration. Like the ruby laser, it is the dopant ions, Nd^{3+} in this case, which behave as luminance centers and provide the lasing action in a Nd:YAG laser.

The first report of lasing in this material appeared in 1964 from a research team based at the Bell Labs. J. E. Guesic, H. M. Marcos and L. G. van Uitert had been exploring the possibility of lasing action in the

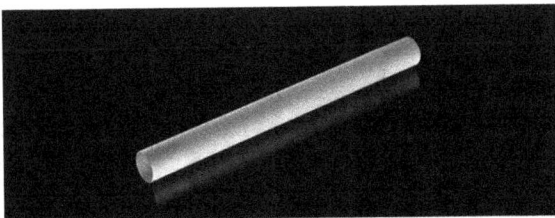

Fig. 6.36. A Nd:YAG laser rod.

energy manifold of triply-charged Nd ions for some time. Trying to build a working solid-state laser utilizing energy transitions in this ion, they examined various mixed oxide crystals as hosts. A number of different materials were available to them — some sourced from other companies and some grown in-house by crystal growers at Bell Labs. They found that mixed oxides containing any two of yttrium, aluminum, gallium and gadolinium had favorable properties for use as host materials for Nd^{3+} ions. These, so-called garnets, are hard isotropic crystals that show high transparency with low optical loss and are easy to dope with both transition metals and rare earths during the crystal growth process. Their work led to the creation of an infrared-emitting laser based on Nd-doped YAG crystal. The first device was constructed from a 2.5 mm diameter cylindrical Nd:YAG crystal with flat parallel end surfaces. Both ends were coated with multi-layer dielectric coatings that were 99% reflecting at 1.06 μm. High power tungsten lamps were used as the pump source, with the laser operating at room temperature in both CW and pulsed modes. This laser could operate at several wavelengths, with the most intense emission at 1064 nm in the infrared.

Figure 6.37 shows the energy levels involved in the operation of Nd:YAG lasers. The principal transition is the $^4F_{3/2} \rightarrow ^4I_{11/2}$ transition which emits, and thus lases, at 1064 nm. The lasing upper state lifetime is 230 μs.

Nd:YAG laser rods are made from single crystal material that is pulled from a melt of the mixed oxide and doped with a controlled amount of neodymium. The exact amount of neodymium depends on the desired mode of laser operation. Due to its high thermal conductivity, good mechanical properties, and the ability to incorporate a variety of dopant ions in its crystal, YAG is one of the most popular materials for laser gain media. Besides Nd, YAG-based lasers are also made using other rare-earths, such as holmium, dysprosium, erbium and samarium. Nd:YAG CW lasers use rods with lower Nd concentration, compared with pulsed lasers. With increase in Nd concentration, the crystal changes from pale lilac to a bluish-pink color. To make a laser rod, Nd:YAG boule pulled from the melt is cut to size, ground at the circumference, and fashioned into long cylindrical rods with accurate parallel end faces. After polishing with increasingly fine grades of cerium oxide and other abrasives, a suitable multi-layer dielectric coating is evaporated on to the end faces. The coating is carefully chosen to be highly reflective at the laser's chosen operating wavelength. The laser rod is mounted within a secure

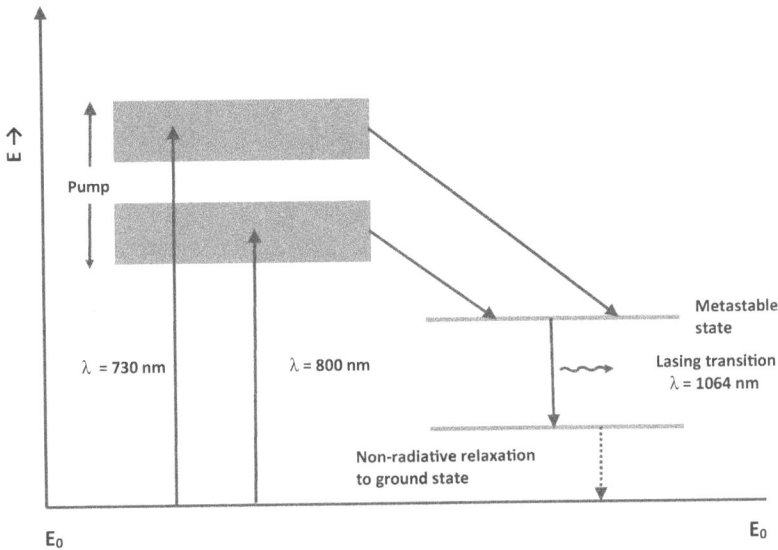

Fig. 6.37.   Energy level scheme of a Nd:YAG laser.

mechanical holder with provision for adequate cooling during laser operation. Long tubular flash-lamps, enclosed inside elliptical mirrors, sur-
round the Nd:YAG rod to complete the optical part of the laser assembly,
as shown in Figure 6.38.

Nd:YAG lasers can be pumped by a variety of light sources, including
krypton and xenon flash lamps and near-infrared semiconductor diode
lasers. The versatility of pump sources is a prominent advantage of
Nd:YAG lasers that can convert either coherent or incoherent radiation to
longer wavelength coherent radiation. Although the most common output
wavelength for Nd:YAG lasers is 1064 nm, radiation at 946 nm, 1120 nm,
1320 nm and 1440 nm can also be obtained with proper choice of laser
cavity optics. With pulsed operation, higher power can be generated than
with CW operation. Often the high-power pulses are used to pump other
crystals for generating frequency doubled output. Thus, pulsed Nd:YAG
lasers are very commonly used to produce green light at 532 nm and, less
commonly, other higher frequency harmonics at 355, 266 and 213 nm.
Nd:YAG lasers are also commonly used to generate very high peak power
pulses through Q-switching. In this technique, as described earlier, a laser
is prevented from lasing by spoiling its end-to-end reflectivity until sufficient population inversion builds up from pumping so that a very large

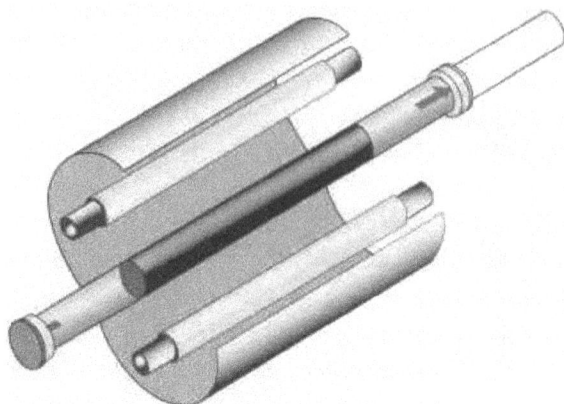

Fig. 6.38.   Schematic illustration showing the construction of a Nd:YAG laser.

optical output could be extracted, as a short ultra-high-power pulse. The long fluorescence lifetime of 230 ms enables the Nd:YAG gain medium to be easily used in Q-switched laser systems. The peak power in Q-switched laser pulses, lasting just a few nano seconds, can reach a good fraction of a giga watt.

While YAG is an excellent host for Nd, other widely used hosts include yttrium lithium fluoride (YLF) with principal emissions at 1047 nm and 1053 nm, and yttrium orthovanadate (YVO_4) with principal emission at 1064 nm (see Figure 6.39). The former is especially suitable for obtaining even higher peak pulse powers than is possible with Nd:YAG. The latter is used in green laser pointers (see Figure 6.40) where a small Nd:YVO_4 crystal is pumped by 808 nm pulses from an AlGaAs laser diode. The 1064 nm radiation thus obtained is frequency-doubled by passing it through a phase-matched potassium titanyl phosphate (KTP) crystal. This process can produce several milli-watts of 532 nm green light. One can easily observe the pulsation in light from such a diode-pumped solid-state (DPSS) laser by waving the laser output at a suitable screen, when the green light appears to be broken in a train of dashes.

High power DPSS green lasers, using the same basic technology, are available with power ratings of several watts to hundreds of watts (see Figure 6.41). These lasers are used for various scientific research applications, and also for pumping other solid-state lasers.

Fig. 6.39.   Yttrium orthovanadate (YVO$_4$) crystals for Nd:YVO$_4$ lasers.

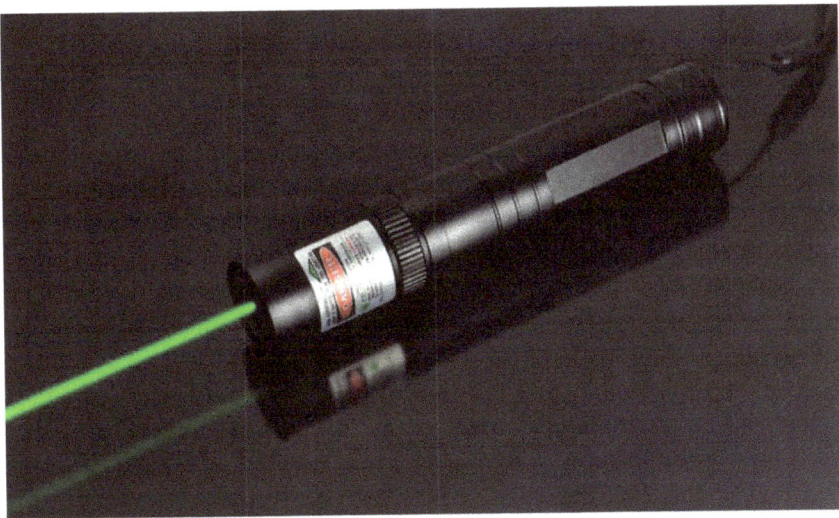

Fig. 6.40.   A diode-pumped solid-state (DPSS) green laser pointer.

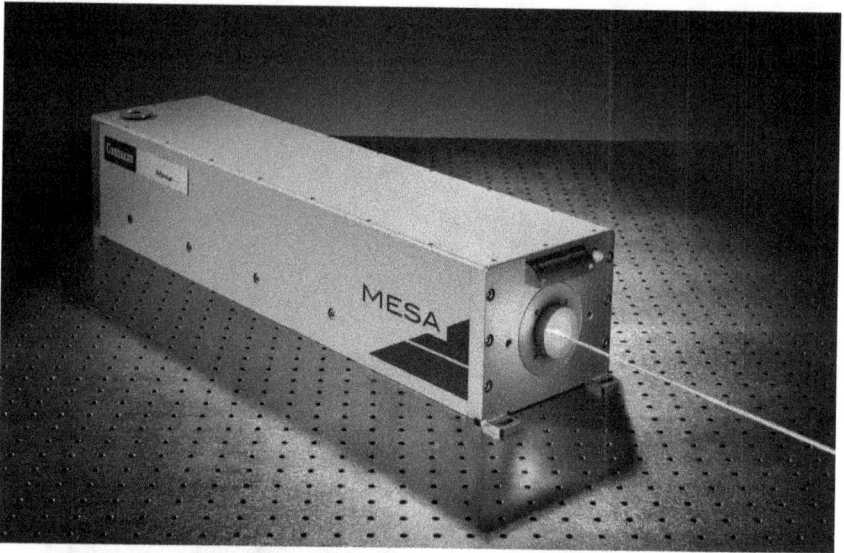

Fig. 6.41.   A commercial high-power Nd:YAG laser Mesa HP, developed by Continuum.
Courtesy: Amplitude Lasers.

Nd is also doped in silica glass (SiO_2) and phosphate glass (P_2O_5) to make Nd:Glass lasers. Glass-based lasing media, using Nd as the luminescent center, provide a much higher power capability than YAG and other crystalline materials. Glasses, being amorphous, can be uniformly doped, and to much higher concentrations than substitutional crystalline media, such as YAG. Glasses also allow laser gain media to be cast in large size and in almost any shape. These features make Nd:Glass lasers capable of achieving peak pulse powers that are much greater than can be obtained from crystalline media. The National Ignition Facility (NIF) at the Lawrence Livermore National Laboratory (LLNL) in Livermore, California, for example, uses Nd-doped phosphate-based laser amplifiers where large rectangular laser media slabs are pumped by flash lamps and used for amplifying low power laser light from the initial stages of the laser oscillator-amplifier chain. The complete system uses about 3070 42-kilogram plates of laser glass. Each glass plate has dimensions of 81 by 46 by 3.4 centimeters. Sixteen such plates are used in each of the 192 NIF beam lines (see Figure 6.42).

This Nd:Glass-based laser system produces infrared radiation at 1053 nm which is subsequently frequency-tripled (up-converted) to UV

Fig. 6.42.   Nd-doped phosphate glass laser amplifiers in the National Ignition Facility (NIF) beam lines.

radiation through the use of nonlinear optical materials. Large slabs of lab-grown potassium dihydrogen phosphate (KDP) crystals are used for this purpose (see Figure 6.43). The UV radiation is then used for creating extremely powerful bursts of X-rays for exploring laser-driven fusion energy generation.

Other than its use in pumping solid-state green lasers, Nd:YAG lasers have several other applications. Opthalmic surgery, for instance, utilizes pulsed frequency-doubled Nd:YAG lasers to perform photocoagulation in patients with diabetic retinopathy. Direct 1064 nm beam from Q-switched Nd:YAG lasers has been used for laser ablation of malignant lesions and for the removal of skin cancer growths. A number of dental procedures, such as gingivectomy, debridement, pulpotomy and frenectomy, are also frequently carried out with suitable Nd:YAG laser units. A pulsed Nd:YAG laser for dental procedures is shown in Figure 6.44. Laser radiation is delivered by fiber from where it is generated inside the main unit to the exposure head at the end of the flexible beam delivery hose. Various beam parameters, such as intensity, pulse repetition rate and pulse profile can be selected from the front touch panel user interface.

There are many industrial applications of Nd:YAG lasers, ranging from laser welding, cutting and drilling (mainly used in automotive

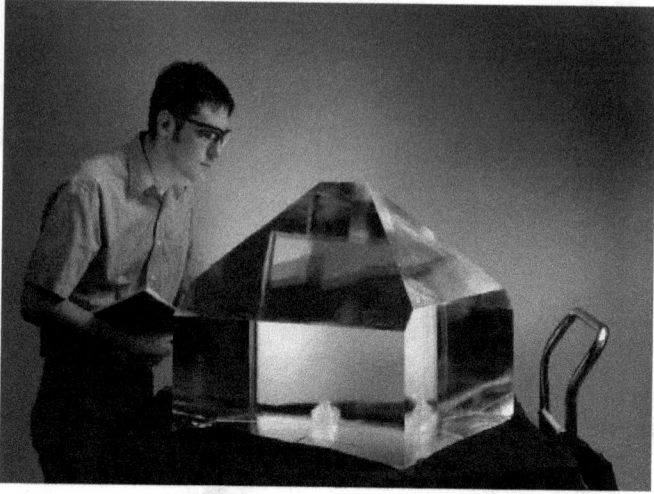

Fig. 6.43. A large lab-grown potassium dihydrogen phosphate (KDP) crystal.

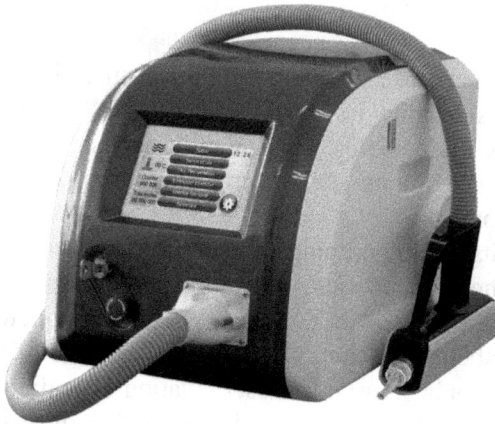

Fig. 6.44. A commercial Nd:YAG laser system for dental procedures.

manufacturing) to semiconductor processing and metal surface treatments. Large multi-kilowatt Nd:YAG lasers are specially manufactured
for such applications. Much smaller, battery-powered, portable units are
used by military forces as target designators for laser-guided ammunitions
and as laser rangefinders on guns and tanks. Relatively straightforward
construction, ability to work in both CW and pulsed modes, and an

established manufacturing base have made Nd:YAG lasers a dependable and enduring source of radiation for many applications. Despite the appearance of more modern laser systems, Nd:YAG lasers promise to remain very relevant for both research and industrial manufacturing.

6.4.6 *The dye laser*

The invention of the laser enabled studies of light–matter interaction using intense beams of laser light. With time, this gave birth to distinct sub-fields of optical sciences, such as nonlinear optics. Frequency up-conversion, through which infrared radiation is converted to visible radiation, relies on nonlinear optical processes. Solid-state green lasers, pumped by Nd:YAG lasers, as described above, make use of such frequency up-conversion in suitable crystals with pronounced optical non-linearity. Interest in exploring the interaction of strong laser beams with optically-active materials had other consequences too. In 1966, Peter P. Sorokin (see Figure 6.45) and J. R. Lankard, working at IBM Research Laboratories in York Town Heights, New York, exposed an infrared-emitting organic dye to pulses from a ruby laser and obtained strong coherent infrared pulses. They had stumbled on the first dye laser. Distinct from all the lasers that were developed before then, dye lasers were the first truly tunable lasers — capable of continuous adjustment of

Fig. 6.45.   Peter Sorokin.

output wavelength over their operating wavelength range. This makes dye lasers extremely useful wherever an intense coherent source of tunable radiation is needed. For this reason, dye lasers are often used in spectroscopy and for pumping other kinds of lasers.

All dye lasers use an organic dye as the active lasing medium. The dye may be used as a solution in a suitable solvent or it can be incorporated in a solid medium, such as a transparent polymer. The former is more traditional but the latter approach is now being widely used for making solid-state dye lasers.

At about the same time as Peter Sorokin's work in the US, Fritz Peter Schäfer (see Figure 6.46), working in the Physics Institute at the University of Marburg in Germany, also developed a very similar dye laser. Schäfer illuminated a square cuvette filled with a synthetic dye with ruby laser pulses. With low excitation, only a uniform fluorescence was seen but on increasing the incident optical power the fluorescence rapidly climbed in intensity to extremely high values. Further investigation revealed that lasing was taking place as light was reflecting back and forth between the two sides of the cuvette. This turned out to be a very

Fig. 6.46.   Fritz Peter Schäfer.

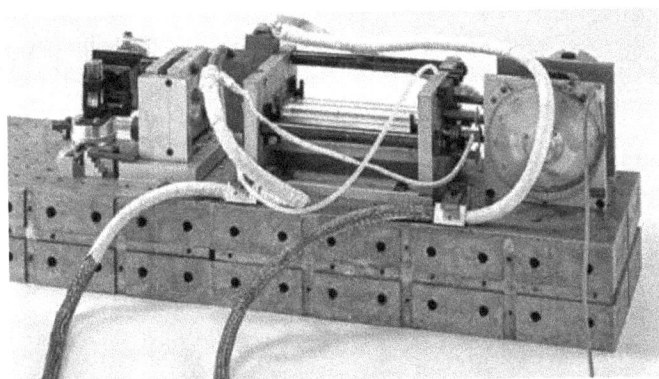

Fig. 6.47.   Peter Schäfer's original dye laser.

flexible type of laser because the lasing wavelength could be changed simply by changing the dye in solution. Schäfer's original laser appears in Figure 6.47.

Compared to atoms and simple inorganic molecules, organic dyes generally have large complicated molecules with many covalently-bonded branches. This structure endows dyes with a rich system of very closely-spaced vibrational energy levels. In solution, or other embedding medium, the energy levels get further broadened so that rich energy level continuums get formed. Electronic transitions between these wide energy bands result in spectrally broad absorption and emission bands. On the one hand, this means that dye lasers can be pumped relatively easily using broad-band light sources, such as arc lamps and flash lamps. On the other hand, the broad energy levels result in equally broad fluorescence bands and make it possible to continuously tune the lasing wavelength over wave-length intervals as wide as 100 nm, in some cases. Movable diffraction gratings are usually used for this purpose. Further wavelength selection can be affected by changing the dye itself (see Figure 6.48). This capabil-ity, at first unique to organic dye lasers, was first noticed by Peter Schäfer and made dye lasers rather special for several decades.

A range of dyes, suitable for use in dye lasers, are commercially avail-able. These differ from each other in the wavelength bands where light absorption and emission take place. This, in turn, makes certain dyes suit-able for pumping with a specific type of laser. Figure 6.49 shows a chart of laser dyes with each dye listed along with its emission range and pump source. A laser dye has to have a high damage threshold so that it is not

Fig. 6.48.   Cuvettes with different laser dyes.

adversely affected by exposure to high photon flux. Nevertheless, every dye gradually deteriorates over time due to light-induced photo-decomposition, so laser dyes are a consumable resource in dye lasers; requiring replacement after a certain amount of usage. Typical dye laser systems often use high speed dye jets to limit absorbed pump energy per unit volume of dye laser solution. Jets also simplify cooling requirements for the active medium because liquid solutions can be efficiently cooled with various flow arrangements.

Both pulsed and CW dye lasers are available. Pulsed dye lasers have been, traditionally, pumped at UV wavelengths using nitrogen or excimer lasers while CW dye lasers have employed argon ion laser pumping. Nowadays, semiconductor diode lasers are increasingly being used for pumping both pulsed and CW dye lasers.

6.4.7 *The copper vapor laser*

Lasing in neutral metal atom vapors under pulsed electric discharge was first reported by William T. Walter and his colleagues from TRG Inc. in 1966. They described a laser utilizing copper vapor as the lasing gain medium that could emit at either 510.5 nm (green) or 578.2 nm (yellow), with the former exhibiting much higher gain and efficiency. They also reported lasing at 854.2 nm and 866.2 nm in singly ionized calcium atom vapor. A year earlier, they had also reported on pulsed laser transitions in

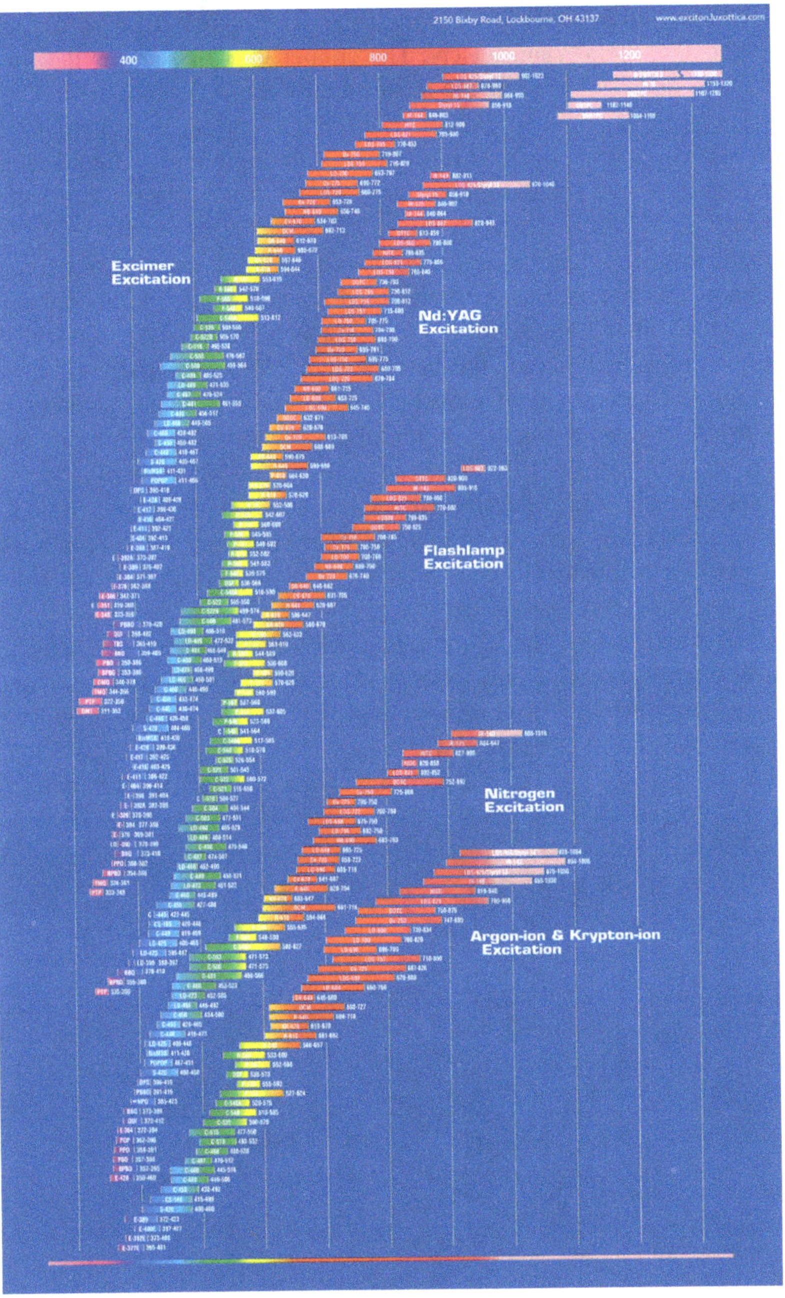

Fig. 6.49. A catalog of laser dyes.

manganese vapor. Their work started the development of a new class of lasers, called metal vapor lasers. The copper vapor laser was the first of this class, and it remains as the most important metal vapor laser.

The first copper vapor laser used direct thermal energy transfer to melt and vaporize copper. This is difficult to do, requiring high temperature sources and containment resistant to such high temperatures. Typically, 1500°C is the temperature needed to form copper vapor from metallic copper. Alumina tubes, containing a small amount of neon as a discharge gas, are used as the active medium container. Small copper pellets, placed inside the alumina tubes are vaporized when a pulsed discharge is repeatedly struck through the neon gas for around half-an-hour. This forms copper vapor at a low partial pressure of about 0.1 mbar which acts as the laser gain medium. Hydrogen is usually also added as a buffer gas to sustain electrical discharge and avoid excessive contamination from copper deposits. Flat reflective mirrors at each end of the laser tube form the laser resonator cavity (see Figure 6.50). These mirrors do not need to be wavelength selective unless oscillations at wavelengths other than the 510.5 nm high gain line are desired.

Needless to say, the requirement for high temperature heating is highly energy intensive and severely degrades the conversion efficiency of the laser device. An enormous improvement was, thus, made when copper vapor lasers operating with copper halides were demonstrated in the early 1970s. A laser operating with copper iodide was first

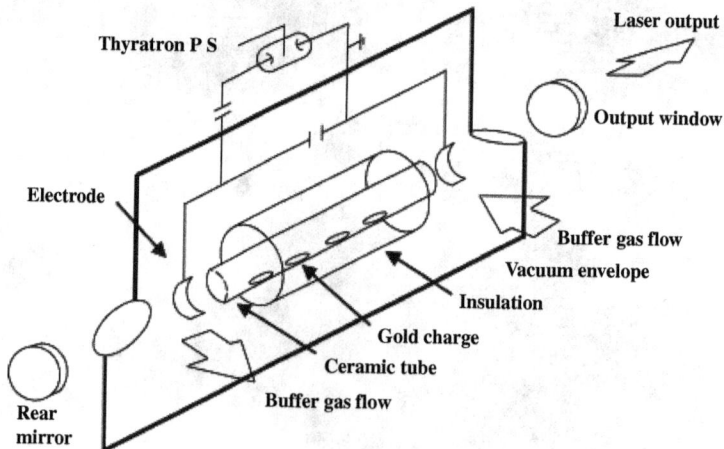

Fig. 6.50.   Schematic diagram of a copper vapor laser.

demonstrated at Westinghouse, followed by one operating with copper chloride at the Jet Propulsion Lab in Pasadena, California. In 1974, a copper bromide laser was put into operation at the Institute of Solid State Physics in Sofia, Bulgaria. All these lasers were essentially low temperature versions of the original copper vapor laser, using various copper halides as the copper source. Among them, the copper bromide laser produces the highest output power for a given electrical energy input. It operates at temperatures around 500°C which can be attained much more easily and economically when compared with close to 1500°C needed with elemental copper vapor lasers. Copper halide-based lasers attain efficiencies of 1% or more which are very high for metal vapor lasers. Other copper compounds, such as copper nitrate and copper acetylacetonate have also been used for making copper vapor lasers that can operate at still lower temperatures. Copper vapor lasers based on copper salts can use quartz tubes instead of refractory alumina tubes. They also have short warm up times of 5–10 min, compared to the more than 30 min needed by elemental copper vapor lasers. Figure 6.51 shows a copper vapor laser emitting a green beam at the Institute of Solid State Physics in Sofia, Bulgaria.

Both halide-based and elemental copper-based lasers are pulsed lasers, producing pulses that are, typically, 5–60 ns long, at repetition frequencies ranging from a kilohertz to a hundred kilohertz. Copper halide-based copper vapor lasers need two successive discharge current pulses, applied at suitable time intervals. The first pulse is a dissociation pulse, needed for producing copper and halogen atoms. The following pulse is the pumping pulse that produces population inversion. The copper vapor laser is a three-level laser, with laser transitions shown in Figure 6.52. Ground state electrons in neutral copper atoms are excited to $^2P_{3/2}$ and $^2P_{1/2}$ laser levels due to collisions with energetic electrons in the discharge pulse. Electrons in the $^2P_{3/2}$ level decay to the lower $^2D_{5/2}$ level, producing green laser light at a wavelength of 510.5 nm, while transitions from the $^2P_{1/2}$ to the $^2D_{3/2}$ level produce yellow laser light at a wavelength of 578.2 nm. The lower lasing levels are long lifetime metastable levels, and this means that the electrons in the lower lasing levels require a relatively long time to relax to the ground state. For this reason, population inversion cannot be sustained continuously and the laser action self-quenches. The laser, therefore, emits a train of very short 5–60 ns pulses with each pulse having a very high peak power, ranging from 50 to 5000 kW. As mentioned above, the lasing cycle also has a very high repeat

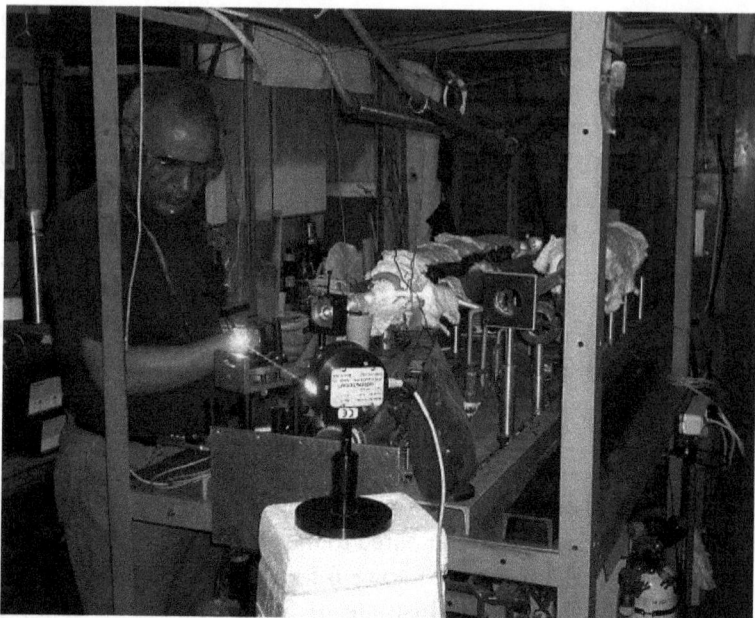

Fig. 6.51.   An operating copper vapor laser, emitting a green beam.

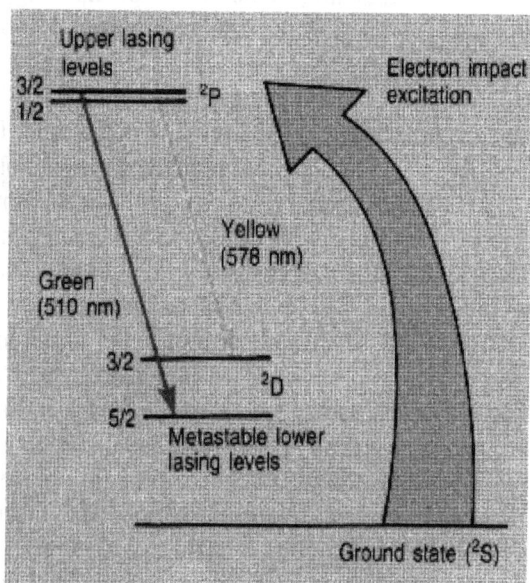

Fig. 6.52.   Energy level scheme involved in the operation of a copper vapor laser.

frequency, and pulse repetition frequencies of 2–100 kHz are common. High peak pulse powers combined with high repetition rates make copper vapor lasers one of the highest average power pulsed lasers available.

These characteristics make copper vapor lasers especially suitable for thermal materials processing applications, such as cutting, welding and drilling. The availability of high output power makes laser machining fast and economical, while the short duration of laser pulses confine heating to the intended target areas, producing high quality machined parts. Figure 6.53 shows a 20 μm hole drilled in a 160 μm diameter steel drill bit using a copper vapor laser. Oxford Lasers, based in Didcot, England, used to be a leading supplier of machining systems based on copper vapor lasers. They also produced equipment for fast imaging of high-speed phenomena, such as spray injection and ultrasonic shock wave propagation, that made use of very short visible light pulses from copper vapor lasers. In recent years, these applications have migrated to other, less expensive laser systems. The intense green light from copper vapor lasers is also used for pumping other lasers, such as the dye laser, discussed above, and the titanium sapphire laser, discussed below. Copper vapor laser-pumped dye lasers have been used in laser-based isotope separation systems. The atomic vapor laser isotope separation (AVLIS) system uses lasers emitting precise wavelengths tuned to atomic hyperfine structure transitions to separate uranium isotopes for enriching nuclear reactor fuel. Figure 6.54 shows a number of commercial copper vapor lasers operating together.

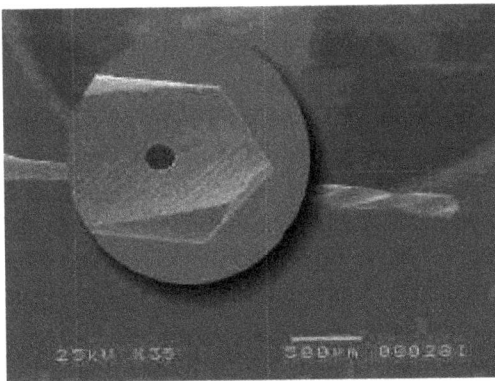

Fig. 6.53. Scanning electron micrograph of a hole drilled in a steel drill bit using a copper vapor laser.

Fig. 6.54.   Operating copper vapor lasers.

After the development of the halide variant of the copper vapor laser, there have been other advances in copper vapor laser technology. For example, it has been discovered that the addition of a small amount of hydrogen bromide (HBr) gas to the copper vapor laser, working with copper bromide, causes a very significant increase in the power output. This so-called 'kinetic enhancement', more than doubles the output power by making it energetically favorable for copper bromide to dissociate inside the laser tube and build a high concentration of neutral copper atoms. No other modification to the laser system is needed except that HBr, at a low partial pressure, is added to the neon + hydrogen buffer gas in the laser discharge tube. Kinetically enhanced copper vapor lasers show efficiencies in excess of 1.4%. In another development, porous electrodes, with reservoirs for copper and copper bromide, have been used, with advantage, in some copper vapor lasers. Porous electrodes keep the discharge tube clean, replenished with active lasing species (copper atoms), and help maintain a stable discharge.

While copper vapor lasers principally emit in the visible region, operation at ultraviolet and infrared wavelengths have also been demonstrated. Generally, the non-visible laser transitions have been obtained with the use of hollow copper cathodes. UV operation at several wavelengths in the region of 210 nm was first reported in 1976. Due to the low efficiency of hollow cathode copper vapor lasers and the availability of other more-developed laser sources in the UV region, this source has not been commercially developed.

6.4.8 *The strontium vapor laser*

Strontium vapor provides yet another lasing medium that features energy transitions in singly-charged ions. This time, in a metal vapor. Compared to copper, discussed above, strontium has a significantly lower melting point (777°C versus 1085°C). This makes strontium much easier to vaporize when compared with copper. External heaters, or ovens, placed around quartz and ceramic tubes have been used to produce strontium vapor for laser operation. For this purpose, the strontium vapor is contained in tubes that also contain a good amount of helium as a buffer gas. Strontium vapor was shown capable of lasing at infrared wavelengths several years before visible light laser action was demonstrated in this medium. J. S. Deech and J. H. Sanders from the Clarendon Laboratory of Oxford University first observed infrared laser emission at 1.03 and 1.09 μm from singly-ionized strontium ions in 1968. By 1971, lasing at twelve other infrared wavelengths was observed in neutral strontium by P. Cahuzac in France. Two years later, in 1973, Latush and Sém from Rostov-on-Don State University in Russia were the first to observe visible laser action in strontium vapor at blue wavelengths of 430.5 nm and 416.2 nm. They described their experiments in the *Soviet Journal of Quantum Electronics*. Their setup made use of a ceramic tube, 8 mm in diameter and 60 cm long. Small pieces of strontium were placed inside the tube at equally spaced intervals and strontium vapor was produced by externally heating the assembly to a temperature close to 800°C, with coils of resistive wire. Helium was used as the buffer gas, at pressures ranging from 2.5 to 35 Torr. The partial pressure of strontium, in comparison, was much lower. The gaseous medium was excited by a high current pulsed discharge. Output power was found to increase with increasing buffer gas pressure which is typical of laser excitation through collisional energy transfer. A few years later, in 1979, Latush, working with colleagues at the Colorado State University demonstrated CW operation of an infrared-emitting strontium vapor laser. This was achieved with a hollow cathode, and use of krypton as the buffer gas. Today, most strontium vapor lasers (less-commonly called helium strontium lasers) are operated as pulsed blue-emitting lasers due to their higher operating gain and efficiency at the visible blue wavelengths.

Power for operating a strontium vapor laser is delivered as short high current pulses from a high voltage capacitor bank. Either thyratrons or solid-state switches could be used for pulse generation. When a high voltage, fast rising pulse is discharged through strontium vapor, both valence electrons of neutral strontium atoms are lost, producing the doubly-charged Sr^{2+} ions. Once the discharge pulse has terminated, collisions between the Sr^{2+} ions and electrons produce the singly-charged Sr^+ ions in an excited state: $Sr^{2+} + 2e^- \rightarrow Sr^{+*} + e^-$. This ion relaxes toward the ground state in several distinct steps, with the excited electron eventually reaching the 6^2S level. Transition to the next lower lying $5^2P_{3/2}$ state is not as easy because of the large 6^2S-$5^2P_{3/2}$ energy difference. Visible laser transitions take place between these two levels.

Strontium vapor lasers have continued to develop over the years. A high-performance infrared-emitting strontium vapor laser was developed at the Tomsk State University in Russia in 2015. This laser (see Figure 6.55), emitting infrared radiation at 6.45 μm, could be used for various surgical applications.

IR-emitting strontium vapor lasers have also been used for atmospheric sounding from airborne platforms. In this role, two lasers are used for meteorological sampling of the components of the earth's atmosphere. This

Fig. 6.55.   An operating strontium vapor laser.

application benefits from the ability of strontium vapor lasers to operate at several different wavelengths. Blue-emitting strontium vapor lasers have been used for holography, lithography and spectroscopy, but those applications are now heavily dominated by helium-cadmium lasers, so strontium vapor lasers are now only used for fundamental and applied physics research, such as plasma diagnostics and investigation of photoablation etc.

6.4.9 *The titanium sapphire laser*

The titanium sapphire ($Ti:Al_2O_3$) laser is superficially similar to the ruby laser in that it makes use of a transition metal-doped sapphire crystal rod as the laser gain medium. Here, the active lasing species are triply-charged titanium ions (Ti^{3+}). A small fraction of Al^{3+} sites in the sapphire crystal are occupied by Ti^{3+} ions through isoelectronic (same ionic charge) substitution. Titanium's electronic configuration is [Ar] $3d^2$ $4s^2$. Here, [Ar] stands for a closed shell with the same electronic configuration as that of an argon atom. To form a Ti^{3+} ion, three electrons need to get lost from a Ti atom, leaving the electronic configuration as [Ar] $3d^1$. Thus, Ti^{3+} ions consist of a closed electronic shell and a lone 3d electron. d energy levels are normally five-fold degenerate but inside the sapphire crystal host the surrounding crystal field lifts the degeneracy. Each Ti^{3+} ion experiences two distinct crystal fields — a high intensity cubic symmetry field and a lower intensity trigonal symmetry field. The cubic field first splits the d levels such that these get re-distributed as three degenerate lower energy levels and two degenerate upper energy levels. These levels are referred to as the $^2T_2$ ground state and the ^2E excited state, respectively. Next, the weaker trigonal field further splits the ground state into two energy levels. The lower of these two levels is further split into two levels by the spin-orbit interaction. The two degenerate excited energy states are also split into two energy levels by a geometric symmetry breaking mechanism called the Jahn-Teller effect. The energy level scheme of titanium sapphire crystals is illustrated in Figure 6.56. Electronic energy transitions between the two upper levels and the system of three lower energy levels ($^2T_2 \rightarrow {}^2E$) take place close to 500 nm, and this is easily seen as a strong absorption band in the blue-green region. Phonon-mediated fluorescence is then observed at longer wavelengths, spanning the region from the deep red to the near infrared. With proper choice of pumping source and appropriate resonator design, Ti:sapphire crystals can be used as both oscillators and amplifiers. The wide gain region makes Ti:sapphire lasers tunable in

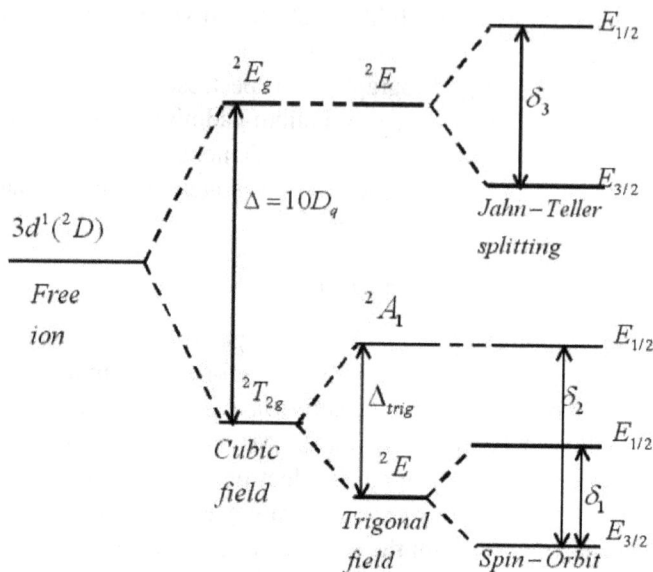

Fig. 6.56.   Titanium ion energy levels and their splitting in the sapphire host crystal.

more-or-less the same way as dye lasers. Being completely solid-state, however, Ti:sapphire lasers can be made much smaller than almost any dye laser.

Both ruby and Ti:sapphire lasers have excellent thermal characteristics which alleviate heat-related problems which often plague other solid-state lasers. But here the similarities tend to end as the Ti:sapphire laser is distinguished by its broad fluorescence band. This feature makes it possible to continuously tune its output to any desired wavelength within the 650–1180 nm range, using dispersive elements (gratings or prisms) inside the laser cavity. This laser is most efficient when producing radiation around 800 nm but is also widely used at other wavelengths within its range of tuneability. Pumping can be performed at any wavelength within the range of about 480–550 nm, which is centered at green but also extends to the bluish-green and yellowish-green regions (see Figure 6.57).

Ti:Al$_2$O$_3$ rods are made in the same way as ruby and Nd:YAG laser rods, by pulling out a crystal boule from an aluminum oxide melt doped with titanium. The crystals are a deep pink in color (see Figure 6.58) and can be cut in various orientations. Due to the popularity of this laser, many manufacturers produce Ti:sapphire laser crystals. Like rare-earth-doped

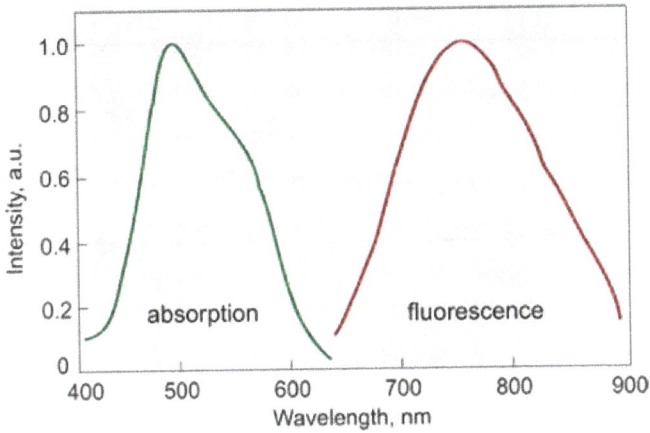

Fig. 6.57. Absorption and emission (fluorescence) spectrum of a Ti:sapphire crystal.

Fig. 6.58. Melt-grown Ti:sapphire crystals.

YAG and glass media, Ti:sapphire can also be used as either a laser to generate optical power or as an amplifier to amplify incoming radiation.

The Ti:sapphire laser was invented in 1982 by Peter F. Moulton (see Figure 6.59) at the Lincoln Laboratory of the Massachusetts Institute of Technology (MIT). He pumped his first laser using xenon flash lamps but adequate emission could only be obtained when the flash lamp pulses

Fig. 6.59.   Peter F. Moulton.

were made very narrow and intense. This greatly shortened the life of pump lamp tubes and it became clear early on that while direct flash lamp pumping works well for several lasers, it is not suitable for pumping Ti:sapphire lasers. The reason lies in the very short fluorescence lifetime of only 3.2 μs for the upper lasing level of titanium-doped sapphire crystals. This necessitates intense pumping with very short light pulses lasting around 1 to 2 μs in duration. Because flash lamps do not last long when used for generating very high brightness short duration pulses so the alternative of pumping with pulsed green-emitting dye lasers was tried next. This has the advantage that the pump light has the correct wavelength for efficient absorption in Ti:sapphire crystals whereas with xenon flash lamp light only a small portion of incident light was utilized for producing population inversion. Consequently, relatively low power pulses can pump Ti:sapphire crystals. The 503 nm output of a co-axial xenon flash-lamp-pumped Coumarin 504 dye laser was originally used for this purpose. This turned out to be a satisfactory arrangement and many characteristics of Ti:sapphire lasers were explored with such a dye laser-pumped system. Later, pulsed argon ion lasers started to be used as pump sources. However, argon ion lasers, as mentioned earlier, are bulky and have very low power efficiency. Thus, these lasers were also, in turn, replaced by frequency-doubled solid-state green lasers, once they became available in the late 1980s. Most Ti:sapphire lasers are now pumped with

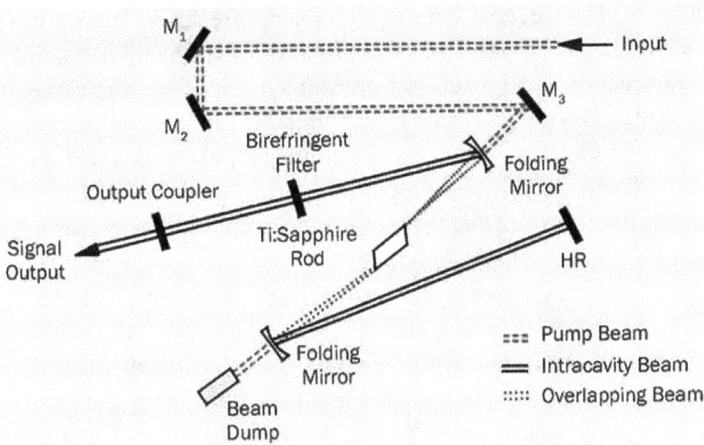

Fig. 6.60.   Schematic diagram showing the pumping and beam extraction scheme for a Ti:sapphire laser.

the 532 nm green light from diode-pumped Nd:YAG, Nd:YLF or Nd:YVO$_4$ laser. Typical, all-solid-state commercial Ti:sapphire lasers produce output in the 0.5–3 W range. As green-emitting semiconductor laser diodes become more widely available at reasonable cost, all-diode pumping of Ti:sapphire lasers will take off. Figure 6.60 shows the general arrangement used for pumping Ti:sapphire crystals. It uses a so-called folded cavity with several reflective and refractive filter elements to remove extraneous pump light from the laser cavity. The center wavelength of the high reflectance mirror HR determines the lasing wavelength.

Due to the particular energy level structure of Ti:sapphire crystals, Ti:sapphire lasers are broadly wavelength tunable. They can also operate in both CW and pulsed modes. When used in CW mode, they can be tuned to any specific wavelength, within their operating range, with very narrow line width. Use of proper wavelength-selective components and intracavity etalons can result in extremely monochromatic light with spectral width of less than 1 GHz. This makes Ti:sapphire lasers extremely attractive for spectroscopy in the red to near-IR region. Their wavelength customizability also makes them useful for pumping other existing and new solid-state lasers for which other appropriate pump sources may not be readily available.

An even more useful feature of Ti:sapphire lasers arises from their broad wavelength tuneability. The inverse relationship between wavelength bandwidth and pulse time duration means that when operating in pulsed mode, Ti:sapphire lasers can generate ultra-short pulses. Highly specialized Ti:sapphire lasers have generated pulses as short as 5 femto seconds in duration. The pulse repetition rate of such 'mode-locked' lasers depends on the length of the laser cavity. Typical pulse repetition frequencies are in the range of 70–90 MHz.

Ti:sapphire lasers are widely used in research labs — both for their highly monochromatic and tunable CW output and for obtaining ultra-short pulses of coherent light. Figure 6.61 shows one of the many commercial Ti:sapphire lasers that are available from many manufacturers. This particular unit is a CW system with very narrow tunable wavelength.

Ti:sapphire lasers are also used in ophthalmic surgery for keratectomy — clearing inflammation, scarring and other superficial, but visually significant, defects of the corneal epithelium.

Developments in Ti:sapphire laser technology have continued to this day, making it ever more useful to scientists, engineers and physicians. In recent years, multi-watt blue laser diodes emitting at 480 nm have become available and these have been used for pumping Ti:sapphire gain media. In yet another development, Ti:glass has been developed where Ti is used as a dopant in special glasses. An example is the BLG 80 glass from Schott (see Figure 6.62), which features broad emission bands, similar to

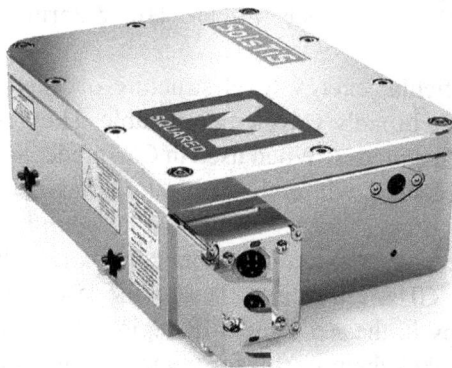

Fig. 6.61.   A commercial Ti:sapphire laser — The SolsTiS.
Courtesy: M Squared Lasers Ltd.

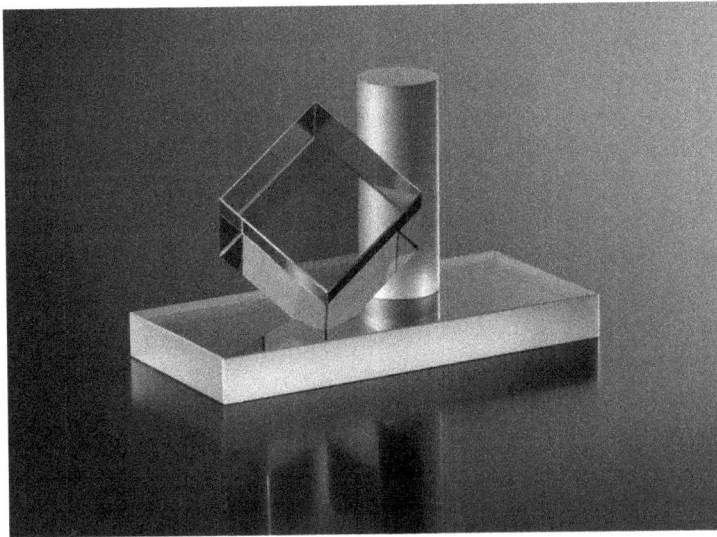

Fig. 6.62.   Titanium-doped BLG 80 glass from Schott Corporation.
Courtesy: Schott Inc.

Ti:sapphire. This material is being promoted as a cost-effective alternative to the significantly more expensive Ti:sapphire crystal.

6.4.10   *The helium–cadmium laser*

Other than the copper vapor laser, another widely used CW metal vapor laser is the helium–cadmium laser. It uses helium as a buffer gas, in the same role as it is used in the helium–neon laser. Unlike neon, cadmium is a metal so the laser tube operates at a high internal temperature to form and keep cadmium in a gaseous state. Furthermore, the active medium is not neutral cadmium atoms but singly-charged cadmium ions and, thus, the helium-cadmium laser is an ion laser. The heavy cadmium ions enable the helium-cadmium laser to operate at shorter wavelengths. There are about a dozen possible operating wavelengths, most of which are found in the UV region, but some strong emission lines are also present in the blue and green portions of the visible spectrum. Among them, the 442 nm blue line is particularly strong and is often chosen for those applications that require coherent blue radiation of high beam quality. Other possible lasing transitions are at 325 nm and 354 nm in the UV region. Because of its relatively high efficiency and compact size, this laser is often used for generating one of these wavelengths, when coherent UV radiation is needed. Its other outstanding advantages include CW operation, low noise, stable output and excellent beam quality.

Helium-cadmium lasers employ a sophisticated laser discharge tube with limited operational longevity. Typical lifetimes are in the range of 3000–5000 hours of use, after which the tube must be replaced. Having to change the very expensive discharge tube is the most conspicuous drawback of using helium-cadmium lasers. The elaborate construction of the discharge tube is dictated by the necessity to vaporize cadmium for laser operation and to keep it in its vapor phase at a relatively high pressure for lasing to take place. Furthermore, the cadmium vapor has to be kept uniformly distributed throughout the length of the tube by the application of a strong electric field. This gas phase electrophoresis process ensures that the optical field inside the laser cavity experiences a uniform cadmium vapor density — essential for achieving high gain. Nevertheless, helium-cadmium lasers show more beam noise than helium–neon lasers, because of localized fluctuations in metallic cadmium vapor concentration in the laser tube. Like the helium-neon laser, this laser also employs energy transfer from helium atoms to cadmium atoms, in vapor phase, to excite the active lasing medium. Energetic helium atoms (He*) collide with low energy neutral cadmium atoms (Cd) to generate excited singly-charged cadmium ions (Cd$^+$*) which can then de-excite for specific laser transitions: He* + Cd = He + Cd$^+$* + e$^-$. Helium atoms have a large collision cross-section which makes them particularly useful as a buffer gas for collisional excitation. Just like the helium–neon laser, the helium–cadmium laser also features an abundance of helium gas over cadmium vapor. The former is typically present at a partial pressure of 3–7 Torr whereas the latter is filled to only a few milliTorr. The laser tube, usually made of quartz, requires sophisticated construction because of the need for a cadmium metal reservoir with vaporization heater, thermionic cathode, helium pump and gas ballast, that all need to be integrated with the discharge tube. This makes the tube construction quite difficult and, consequently, there are only a few manufacturers who produce this laser commercially.

There are quite a few degradation and failure mechanisms that ultimately cause helium-cadmium laser tubes to fail. These include: gradual depletion of cadmium, which reduces its partial pressure, escape of helium buffer gas over time, non-uniform distribution of cadmium vapor along the laser tube, and the sputtering of the cathode surface.

Fig. 6.63.   Commercial helium-cadmium lasers from Kimmon Koha Corporation.
Courtesy: Kimmon Koha Co., Ltd.

Manufacturers have incorporated a number of design changes to address these issues in attempting to prolong the life of the laser tube. This is important, because just as the argon ion laser is considered an excellent source of coherent green light, in spite of its relative inefficiency, the helium-cadmium laser is highly valued for its high beam quality coherent blue and UV output.

Kimmon Koha is a leading Japanese manufacturer of helium-cadmium lasers with units available in output powers ranging from 2 to 180 mW (see Figure 6.63). Users can choose from units emitting at 325 nm, at 422 nm or even dual wavelength 325/422 nm emitters. Kimmon also offers laser maintenance and tube replacement services to the helium-cadmium laser users' community.

Helium-cadmium lasers have a number of contemporary uses. Their short wavelength, high quality, low divergence and stable output is especially suited for 3D stereolithography applications. In this 3D parts prototyping technique, a tightly-focused laser beam is used to cross-link an organic liquid into a rigid polymer which is continuously withdrawn upward as fresh solid layers are constructed underneath (see Figure 6.64). Stereolithography is often used to make complicated shapes that will be too expensive to produce by other means.

Other well-known applications of helium-cadmium laser include holography (where this laser competes with the helium-neon laser), CD and DVD mastering and the fabrication of diffraction gratings.

Fig. 6.64.   Intricately-shaped plastic parts can be produced with stereolithography.

spectroscopy and lithography are also applications that often make use of helium-cadmium lasers.

6.4.11 *The semiconductor diode laser*

Coherent light can also be produced from semiconductor light-emitting diodes. These solid-state devices are very attractive as sources of laser light due to their small size, low cost and ease of use. On the other hand, their monochromaticity, beam quality, beam divergence and coherence characteristics are quite poor when compared with typical gas lasers. In applications that are not too demanding, semiconductor diode lasers often are the preferred choice because of the benefits mentioned above. Many small, handheld and portable devices, requiring laser light, will not exist if laser diodes were not available. While semiconductor lasers are a relatively new invention, their rate of development has far out-paced that of all other types of lasers. Much of this is obviously driven by their economic potential in electronic, sensing and telecommunication applications.

Just like LEDs, semiconductor diodes that emit coherent light are also constructed from compound semiconductors. III–V semiconductors are again the most common class of semiconductors to make laser diodes from. Being *pn*-junction diodes, laser diodes have the same underlying structure as LEDs and, thus, almost all the discussion about *pn*-junction diodes presented in the previous chapter applies equally well to laser diodes. The ability to generate coherent radiation through stimulated emission, however, requires some additional structural features which are unique to laser diodes.

The basic physics of stimulated emission through population inversion remains exactly the same for laser light emanating from diodes, as that from any other laser. The only difference is that here the electron transitions take place between the conduction band and the valence band, which have been described in the previous chapter. Furthermore, the population inversion is brought about through an electric current flow, i.e. through electrical pumping — rather than optical pumping. Because, instead of sharp atomic energy levels, energy bands comprised of a continuum of very closely spaced energy levels are involved here, so light from diode lasers is necessarily not as highly monochromatic as that from other types of lasers. Creating the population inversion for enabling stimulated emission requires very high free electron population in the conduction band. Reaching the necessary number density of conduction band electrons requires injecting very large numbers of electrons through the device. This can be achieved, in practice, by establishing a very large current density (current per unit area) through the *pn*-junction. This requires a junction with very small cross-sectional area. Indeed, this was the approach that was used by the inventors of the very first diode lasers. The current density can be boosted further by making the current flow as very short duration pulses of very large current.

6.4.11.1 *History of the semiconductor diode laser*

Right after the first observations of light emission from solid-state diode junctions, several researchers began to wonder if it may be possible to make lasers from diodes that emitted light. IBM researchers, M. G. Bernard, G. Duraffourg and William P. Dumke published theoretical accounts of coherent light emission from GaAs junction diodes in the early 1960s. Several prominent physicists, including Ben Lax of the MIT, had proposed building a diode laser using silicon or germanium but it was Dumke who recognized that the indirect band gap of those semiconductors will not make it possible to make a laser out of them. He suggested using GaAs as a promising material for such a device because of its direct band gap. Due to the nascent state of the field of lasers as well as that of semiconductors, there wasn't much progress until high radiative efficiency light-emitting junctions had been demonstrated. Much of this work was being carried out in industrial research labs. Sumner Mayburg and Jacques Pankove, working separately at the GTE and RCA labs, respectively, reported in 1962 that certain semiconductor junctions, cryogenically cooled to liquid nitrogen temperature, showed very strong light emission in the infrared region. Around the same time, Robert Rediker, Ted Quist, and Robert J. Keyes of the MIT Lincoln Labs found that diffusion doping with zinc to form junctions dramatically increased the electron–hole recombination radiation from GaAs LEDs, held at liquid nitrogen temperature. They reported that their GaAs LEDs showed almost 100% radiative quantum efficiency, i.e. the conversion of electron–hole pairs to photons. With the demonstration of bright infrared-emitting GaAs LEDs, the stage was set for the next device — an infrared-emitting laser diode. That next decisive step was taken by Robert N. Hall (see Figure 6.65), together with his research team members at the General Electric R&D Labs in Schenectady, New York.

Hall was a well-recognized scientist at GE who had earlier achieved significant successes with other research projects. Knowing his experimental skills, he was teased by colleagues about not having built a semiconductor diode laser. Hall was initially skeptical of the idea of building a laser diode but after attending a conference in which Keyes described the amazingly efficient infrared LEDs fabricated at MIT, he was compelled to seriously think about the prospects for a GaAs laser diode. After conducting some literature surveys, he came to the conclusion that a diode which operated as a laser might be a real possibility, given all the materials and other resources he had available in his GE research lab. Hall asked for permission from his research manager, which was immediately granted, and with his co-workers he set upon the task of making a suitable GaAs diode structure, which will be capable of lasing.

Hall's team took several weeks to grow a proper GaAs junction diode. High doping levels (so-called degenerate doping) were used for creating large current flows during device operation. Relatively small cross-sectional area of the diode made it possible to operate with very high current densities — essential for generating a population inversion for stimulated emission. The end faces of the device were made accurately

Fig. 6.65.   Robert N. Hall.

parallel and highly polished to act as mirrors to form the laser cavity. Figure 6.66 shows the original diagram from the patent filed by Robert Hall for his device.

When this device was cooled with liquid nitrogen and driven by several microsecond long current pulses, it emitted bursts of strong coherent infrared radiation. This was the first ever demonstration of laser action through electrical current injection in a semiconductor device. Figure 6.67 shows Gunther Fenner, Robert Hall, and Jack Kingsley, from left to right, showing their experimental setup used to run and test the first infrared-emitting GaAs diode laser at General Electric. Word spread quickly and, soon after, Marshall Nathan at the IBM T. J. Watson Research Center and the MIT Lincoln Lab group also succeeded in operating their own GaAs diode lasers. Similar devices were also built at Texas Instruments and at the RCA Research Labs. Nikolay Basov in the Soviet Union also demonstrated a GaAs laser diode in early 1963. The technology was refined in later years to improve light emission and other semiconductors were also explored for making laser diodes.

A few months after the first demonstration of the infrared-emitting GaAs diode laser, Nick Holonyak Jr. (see Figure 6.68), also at GE

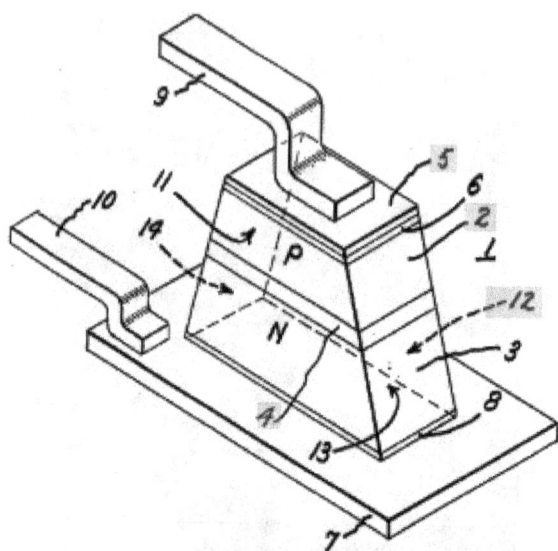

Fig. 6.66.   Schematic diagram of the first GaAs *pn*-junction laser diode.

Fig. 6.67.   Gunther Fenner, Robert Hall, and Jack Kingsley (left to right) at General Electric Research labs.

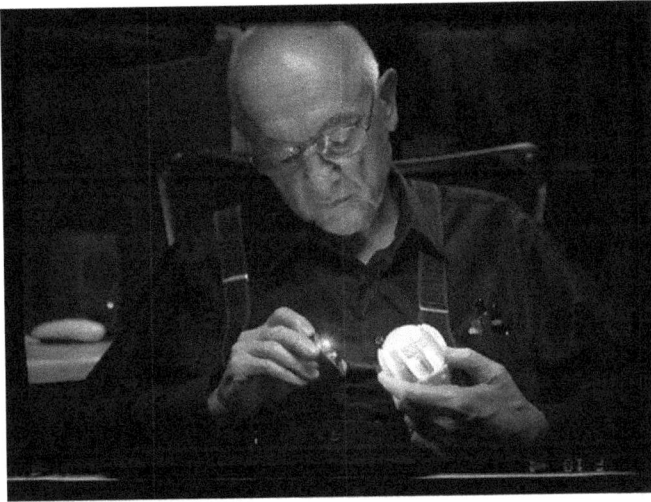

Fig. 6.68.   Nick Holonyak Jr.

Research Labs (in Syracuse, New York), added phosphorous to GaAs to make a red-emitting GaAsP diode laser. The trick was to add just enough phosphorus so that the energy band gap of the resulting ternary semiconductor became wide enough for visible light emission to take place while still remaining direct for efficient light emission. This was the first semiconductor laser that emitted visible radiation. Later, Holonyak Jr. and others experimented with other semiconductor alloys to develop laser diodes capable of emitting at several wavelengths in the red to near-infrared region. Red-emitting laser diodes are now commonly used in many everyday applications, such as supermarket barcode readers, laser-based leveling and distance measuring devices as well as CD and DVD readers.

All of the early semiconductor diode lasers were pulsed devices that required large currents and cryogenic cooling to operate. When a diode laser is operated at low current levels, it emits a mixture of spontaneous (non-coherent) and stimulated (coherent) radiation with a large spectral width. As the forward current through the diode is increased beyond a certain value called the threshold current, there is a nearly sudden transition to mostly stimulated emission with a concomitant increase in brightness and narrowing of the spectral width. The first laser diodes had quite high threshold currents running into hundreds of milli amperes. That, and the necessity to severely cool the devices for laser operation, greatly limited application possibilities. Some progress was made by improving materials and device structures but the real advance only came with the development of heterostructure *pn*-junction diodes, similar to those discussed earlier in connection with LEDs. Single heterostructures and double heterostructure quantum wells confine electrons and holes to greatly reduce the threshold current requirement of lasers diodes. Such improved devices could then operate in CW mode at room temperature. Herbert Kroemer of the University of California, Santa Barbara, and the team of Rudolf Kazarinov and Zhores Alferov of A.F. Ioffe Physico-Technical Institute in St. Petersburg, Russia, independently proposed ideas to build heterostructure semiconductor lasers. Their work later resulted in Kroemer and Alferov getting the 2000 Nobel Prize in physics. In March 1970, Alferov's group at the Ioffe Physico-Technical Institute in collaboration with Dmitri Z. Garbuzov and, independently, Mort Panish and Izuo Hayashi (see Figure 6.69) at Bell Labs made the first CW semiconductor lasers that could operate continuously at room temperature.

Semiconductor diode lasers of various designs have been developed over the past decades. Each design is particularly suited to certain applications. We now examine a few prominent types of these solid-state laser devices.

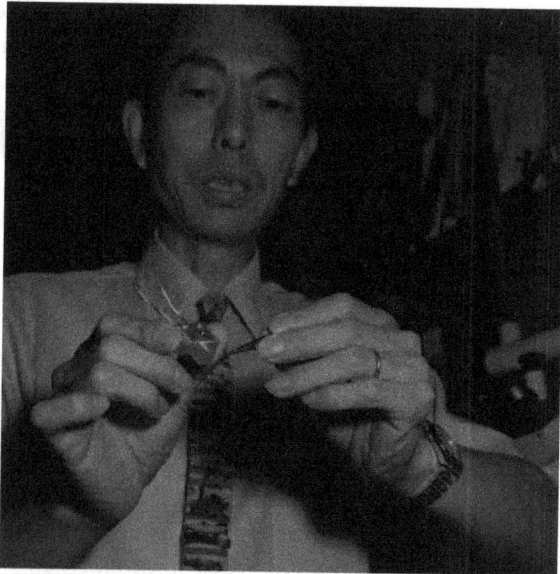

Fig. 6.69.   Izuo Hayashi.

6.4.11.2 *Fabry-Perot laser diode*

Heterostructure laser diodes have a complicated multi-layer structure with different semiconductors and doping layers. Their fabrication was, initially, enabled by liquid phase epitaxy (LPE) — a crystal growth technique developed at RCA Laboratories by Herbert Nelson. The semiconductor to be grown as an epilayer (an overlayer with the same crystal structure as that of the underlayer) is dissolved in a suitable solvent at an elevated temperature, and the growth substrate is brought in contact with the solution. To avoid residual stress in the epilayer, the thermal expansion coefficient of the substrate and the epilayer must be closely similar. Epitaxial deposition takes place at a temperature much lower than the normal melting point of the semiconductor and results in a high-quality semiconductor film forming on the substrate. Very complicated layer schemes can be deposited in this way by altering the growth solution and growth temperature in a controlled way. By sandwiching a GaAs layer between layers of aluminum gallium arsenide (AlGaAs), a heterostructure quantum well layer can be fabricated. Laser diodes using the AlGaAs/GaAs/AlGaAs structure had much lower threshold current ratings than their earlier homojunction GaAs counterparts. That enabled CW operation to be demonstrated at cryogenic temperatures, followed by CW operation at room temperature.

Figure 6.70 shows a schematic of a simple GaAs/AlGaAs laser diode. It is classified as an in-plane emission device because the light emission takes place from the plane of the device. The electrical current, in contrast, flows perpendicular to the device plane. A thin GaAs layer (usually undoped) is sandwiched between two AlGaAs cladding layers — one p-doped and the other n-doped. The AlGaAs/GaAs/AlGaAs structure forms the quantum well, with GaAs well layer and AlGaAs barrier layers. The AlGaAs layers also serve as optical confinement layers through total internal reflection, so that the radiation remains confined in the GaAs layer. Thus, these layers also act as the lower refractive index cladding layers in the device structure, where the GaAs layer is effectively fashioned into an optical waveguide by structuring the top layers in the shape of a slightly raised ridge. The region next to the long ridge may be rendered non-conductive by a technique such as proton bombardment. This makes the current flow from the top to the bottom in such a manner that the flow is confined to the region outlined by the long ridge. The radiation is generated through radiative electron-hole recombination in the

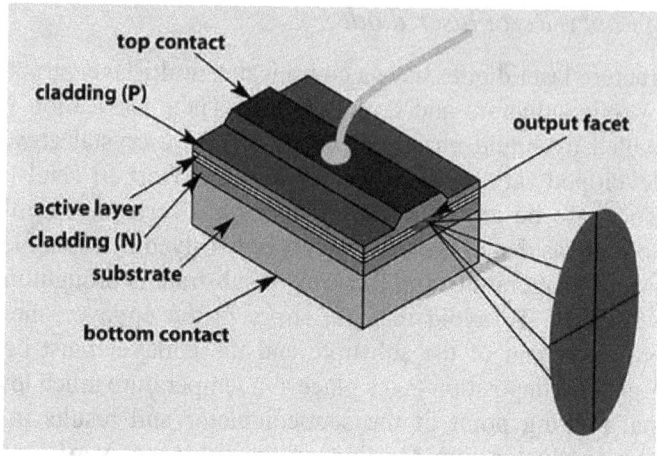

Fig. 6.70.   Schematic illustration of a typical Fabry-Perot ridge-waveguide laser diode, emitting beam with an elliptic cross-section.

stripe-like region in the GaAs layer that lies underneath the ridge. This long stripe serves as an optical waveguide confined between the cladding layers above and below. This kind of laser structure is called a Fabry-Perot laser because of its resemblance to a Fabry-Perot interferometer which is formed by an optical cavity situated between two mirrors. In this case the cavity is the stripe waveguide and the two mirrors are simply the end faces of the diode chip. Because of the entire structure being formed in an epitaxial single crystal material, cleaving the chip from the wafer automatically produces atomically flat and accurately parallel end surfaces that serve as very effective mirrors. Often these end faces are given additional dielectric coatings to improve performance. Laser action takes place as photons travel back and forth in this longitudinal cavity, gaining energy from the medium (stimulating other electron–hole pairs to recombine and emit photons). In this type of device, laser radiation can escape from one or both end faces. It is quite usual to arrange the end face reflectivity so that most of the radiation exits from a given face, but a small amount also exits from the opposite face to serve as a reference beam for real-time monitoring of device operation. In this case, a small photodiode is usually placed close to that face to measure the amount of laser emission being emitted. This signal can be used to control the laser's intensity in real time through a feedback control mechanism, in order to stabilize the intensity to a very high degree of constancy.

One somewhat undesirable feature of this kind of laser diode arises because the radiation comes out of a small rectangular aperture at the end of the long laser cavity. This makes the radiation field diffract out as an elliptical rather than a circular beam, as can be seen in Figure 6.70. The elliptical shape is quite pronounced, with the two axes of the ellipse called the 'slow' and 'fast axes'. The light needs to be shaped and collimated externally with the use of suitable cylindrical or aspheric lenses before it can be used in applications that require a round beam. This is described further in the next chapter.

Fabry-Perot semiconductor laser diodes are now widely available at many standard wavelengths, in both visible and infrared regions. Some standard peak wavelengths include 680 nm, 808 nm, 830 nm and 980 nm. The spectral width is, typically, 3 nm, so the beams are fairly monochromatic. In general, the quality of light coming out of a semiconductor diode laser is inferior to that emitted by gas lasers. This is especially true with high power laser diodes. Semiconductor lasers radiate light that has pronounced cross-sectional eccentricity, larger spectral width and lower spatial coherence than light coming from typical helium-neon or argon ion lasers. This is the reason the older gas laser technologies are still in widespread use. Diode lasers, on the other hand, excel in being much smaller in size, cheaper in price and economical to operate. Figure 6.71 shows commercially available packaged laser diodes. Note that the hermetically sealed metal cans have three pins — one for the laser diode, one for the monitor photodiode and one to serve as the common ground pin. These

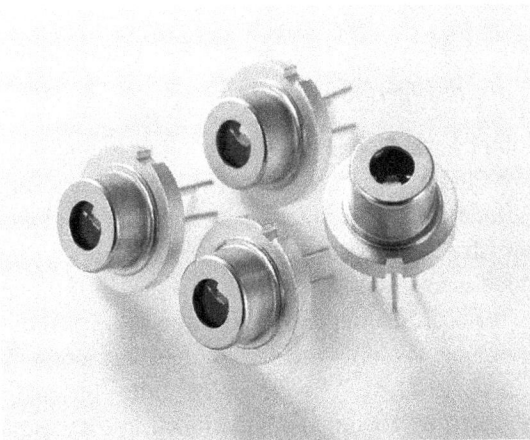

Fig. 6.71.   Commercial packaged semiconductor laser diodes.

devices can be mounted on a suitable heat sink, if needed to dissipate heat, when operated at high power levels.

While early semiconductor laser diodes (and LEDs) were made using LPE, these devices are now made using either MBE or MOCVD, in the manner described in the previous chapter. MBE provides exceptional control over doping and epitaxial layer thicknesses. It is the technique of choice for growing experimental device structures and for making devices where layer composition or thickness control is critical. MOCVD, due to its cost-effective nature, is used for the commercial manufacture of the vast majority of all modern semiconductor laser diodes. These epitaxial growth technologies have been extensively developed over the years, so that they permit the growth of high quality, low dislocation density, epitaxial, multi-layer structures which are crucial for the manufacture of these devices.

Laser diodes are often packaged with an embedded power supply and a collimating lens; forming a 'laser module' which can be readily integrated into an application. Nowadays, small, visible light emitting, low-power laser diodes are also commonly used in laser pointers, as seen in Figure 6.72. In high power applications, laser diode modules can also be cooled with either an external or an embedded Peltier cooler.

Properly collimated and circularized beams from laser diodes can be easily coupled into optical fibers. This is done with both high-power diode lasers for material processing applications and with lower power highly monochromatic lasers for optical communication applications. A fiber-coupled laser diode for use in short-haul communication equipment is seen in Figure 6.73.

6.4.11.3 *Distributed feedback laser*

The semiconductor laser diodes described so far, being of the Fabry-Perot design, have two flat parallel mirrors at each end of the laser cavity. A very different laser diode design is one where reflection originates from a longitudinal grating structure formed all along the length of the laser cavity.

Such a device is called a distributed feedback (DFB) laser. Radiation is continuously reflected from the periodic variation provided by the grating as light propagates along the length of the cavity. The grating is designed with a periodicity, Λ, which is usually one-half of the center wavelength of laser emission. The feedback is, understandably, very wavelength selective and results in considerable narrowing of the spectral

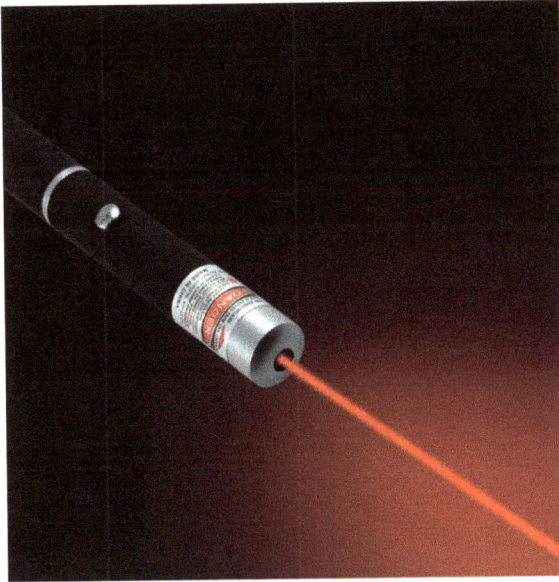

Fig. 6.72.   A laser pointer with a red-emitting semiconductor laser diode.

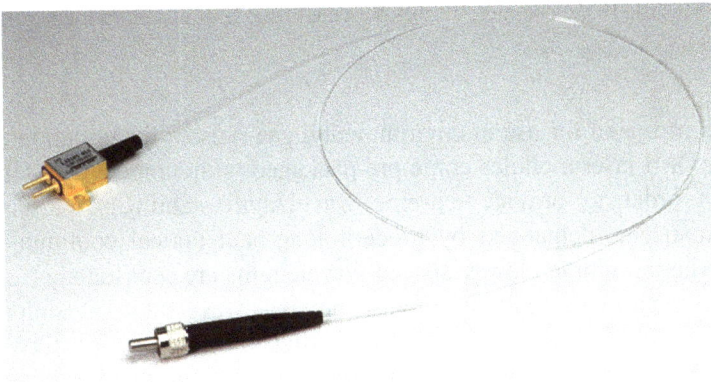

Fig. 6.73.   A fiber-coupled laser diode assembly.

width of laser emission. A schematic diagram of the DFB structure can be seen in Figure 6.74.

DFB lasers are usually designed to emit a very narrow spectral line at infrared wavelengths of importance for fiber optic communications, such as 1.55 μm and 1.3 μm. Devices emitting at even longer wavelengths have also

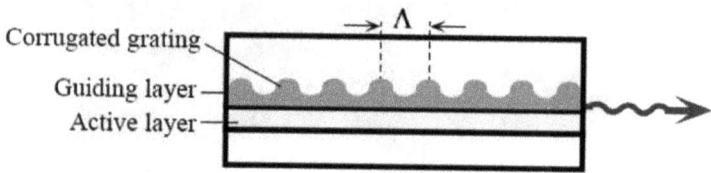

Fig. 6.74.   Schematic illustration of the structure of a distributed feedback (DFB) laser.

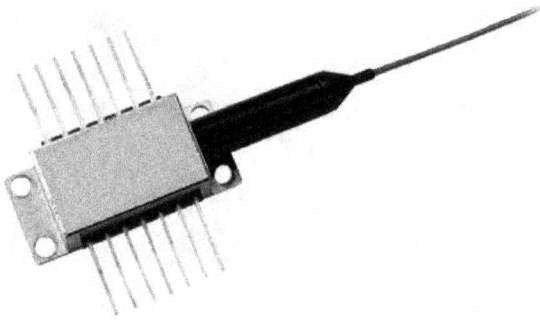

Fig. 6.75.   A 1310 nm commercial fiber pig-tailed DFB laser in a butterfly package.
Courtesy: Emcore Corporation.

been constructed for use in environmental gas sensors for pollution monitoring. DFB laser modules come pre-packaged with temperature stabilization in order to provide extreme wavelength stability. This kind of performance is demanded by modern long-haul optical communication links where multiple closely spaced wavelengths are sent together on one fiber, in the arrangements commonly known as wavelength division multiplexing (WDM) and dense wavelength division multiplexing (DWDM). Figure 6.75 shows a 1064 nm DFB laser in a so-called fiber pig-tailed butterfly package with internal thermoelectric cooler and monitor photodiode.

6.4.11.4 *Vertical cavity surface emitting laser (VCSEL)*

The laser diodes described so far all have laser cavities that lie in the plane of the semiconductor material. Unlike LEDs, that emit light perpendicular to the device plane and can be tested while still on the wafer, in-plane emission laser diodes cannot be tested for their optical emission until they have been diced and separated into individual devices. This complicates

Fig. 6.76. Vertical cavity surface emitting laser (VCSEL) chips being tested on a wafer.

their manufacture somewhat, and is one factor responsible for their relatively low fabrication yield. A very distinct type of laser diode, called a vertical cavity surface emitting laser (VCSEL) has a structure with its laser cavity oriented perpendicular to the device plane. VCSELs have a number of advantages over side-emitting lasers. One is that these devices can be tested while they are still part of the semiconductor wafer (see Figure 6.76). Another is that they emit a circular beam which does not require any corrective optics, simplifying many applications.

In a VCSEL, the semiconductor laser cavity is grown by MBE or MOCVD processes in such a way that it lies between two stacks of alternating materials that form multilayer dielectric reflector mirrors. These distributed Bragg reflectors (DBRs) sandwich the main laser cavity. The entire vertical stack, comprising of the lower DBR, the laser cavity and the upper DBR, is grown sequentially on a semiconductor wafer, from the bottom to the top. The active light-emitting layer is a quantum well or multi quantum well structure similar to that used in LEDs. The DBRs above and below the cavity, or gain region, are composed of several (typically 20–30) alternating layers of GaAs and AlGaAs. Due to the difference in their compositions, they have different refractive index and this difference causes reflections at each GaAs/AlGaAs interface. Each GaAs/AlGaAs layer pair is referred to as a period and each period has the same thickness — a quarter of the laser's peak emission wavelength. This makes the DBR highly wavelength selective, so that its reflectivity peaks at the lasing wavelength. The top stack is usually p-doped to enhance hole transport whereas the bottom stack is *n*-doped to enhance electron transport. The DBRs need to provide very high reflectivity because the laser gain region, i.e. the cavity is very short in a VCSEL device. Usually, one DBR stack has a smaller number of layers than the other and thus a lower reflectivity, in order to make the device emit from only one side (top or bottom). The GaAs–AlGaAs material system is very suitable for constructing VCSELs because the lattice parameter does not change much as the composition is changed, permitting several alternating epitaxial layers to be grown on a GaAs substrate. However, the refractive index of AlGaAs does vary quite strongly as the Al fraction is increased. This reduces the number of layers required to form an efficient Bragg mirror, compared to other candidate material systems. Furthermore, at high aluminum concentrations, an oxide can be formed from AlGaAs. The outer parts of AlGaAs layers are turned into aluminum oxide through an oxidation process which proceeds from the exposed sides. Thus, only the central portion of the structure remains electrically conducting AlGaAs; effectively forming an 'oxide aperture' which restricts current to flow through the central portion of the stack. This serves to significantly increase the current density in the device and, thus, greatly reduces the threshold current. Figure 6.77 shows the schematic structure of a VCSEL of this design. Note the mesa that isolates a single VCSEL, and the ring-shaped cathode (bottom) and anode (top) contacts.

Compared to edge-emitting semiconductor laser diodes, VCSELs offer many outstanding benefits. As mentioned before, because of their emission direction being perpendicular to the plane of the wafer, VCSELs

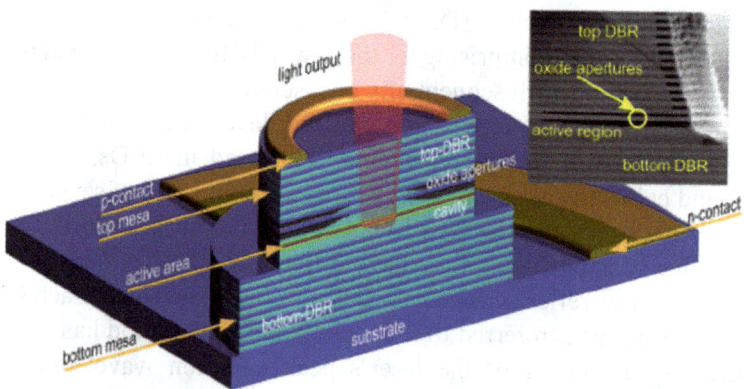

Fig. 6.77.   Schematic structure of a VCSEL.
Courtesy: SPIE.

Fig. 6.78. A packaged VCSEL array.
Courtesy: Finisar/II–VI Inc.

can be individually tested while they are still part of their substrate wafer. This greatly simplifies their testing and helps improve the ultimate device yield from a VCSEL wafer. Their vertical structure also enables two dimensional VCSEL arrays to be easily fabricated. Being circular in shape and with larger exit apertures, VCSELs produce nice low-divergence circular beams that do not require corrective optics, and can be easily coupled into optical fibers. Use of high reflectivity DBRs and current confining aperture greatly reduces the threshold current of VCSELs to values around just 1 mA. VCSELs can, thus, be used in compact, low power consumption applications. Furthermore, the low threshold current also increases their modulation bandwidth, so that these devices can be electrically modulated at very high speeds. Finally, it is possible to adjust the emission wavelength of VCSELs, to a certain extent, by changing the periodicity (thickness of a GaAs/AlGaAs pair) of the DBRs during epitaxial growth.

VCSELs are most commonly available in the near-infrared region although visible red devices can also be obtained commercially. The 980 nm is a very popular wavelength for GaAs/AlGaAs VCSELs. Blue and violet III-nitride VCSELs have also been developed, but are not widely available. As far as output optical power goes, VCSELs with power outputs from a fraction of a milliwatt to several watts have been made. Most commercial devices emit a few milliwatts per device.

VCSELs are suitable for any semiconductor laser diode application where a low threshold current and/or a circular beam is an overriding requirement. To date, VCSELs have found applications in short-haul optical communication systems, printers, hand-held and portable optical spectroscopy systems and chip-scale atomic clocks. These devices are also being used in light detection and ranging (LIDAR) systems for limited area surveillance and automobile collision avoidance systems. VCSEL arrays with hundreds or thousands of closely-spaced emitters on a single chip (see Figure 6.78) are now commercially available for 'structured-light' applications, such as in 3D object scanning and gesture sensing.

6.4.11.5 *Green laser diode*

The very first diode lasers emitted in the infrared. Further development led to red-emitting laser diodes, and then to devices capable of emitting deeper into the infrared region (longer wavelengths). Shorter wavelengths in the visible region and the UV region were not reached before developments in the 1980s and 1990s, mainly in Japan, that enabled LEDs and laser diodes to be made from GaN and related semiconductors. This has already been described in the previous chapter. Once GaN/InGaN blue-emitting LEDs were developed, laser diodes made from the same material system were not far behind. Violet-emitting GaN/InGaN laser diodes with emission centered at 405 nm are now the most widely available of all III-nitride-based laser diodes. These devices are inexpensive because they are made in enormous numbers for various applications including use in DVD players. The so-called Blu-ray players have got their name from the use of these laser diodes. A 405 nm laser diode is superior for this application compared to a red or infrared laser diode because the shorter wavelength can be focused to a much smaller diffraction-limited spot on the surface of a DVD disk. This allows much more information to be placed on a disk, and an entire high-definition movie can be easily accommodated on one standard DVD disk. Besides 405 nm, another popular GaN laser diode wavelength is 450 nm which is visibly a true-blue color. Nitride lasers have also been fabricated with operating wavelength in the UV region. These devices are expensive and not very widely available. As with LEDs, UV diode lasers make use of GaN/AlGaN quantum wells. The addition of Al to GaN causes its band gap to increase, shifting the emission wavelength to shorter values. Nitride lasers have a similar ridge-based stripe waveguide structure as described earlier for red and infrared laser diodes.

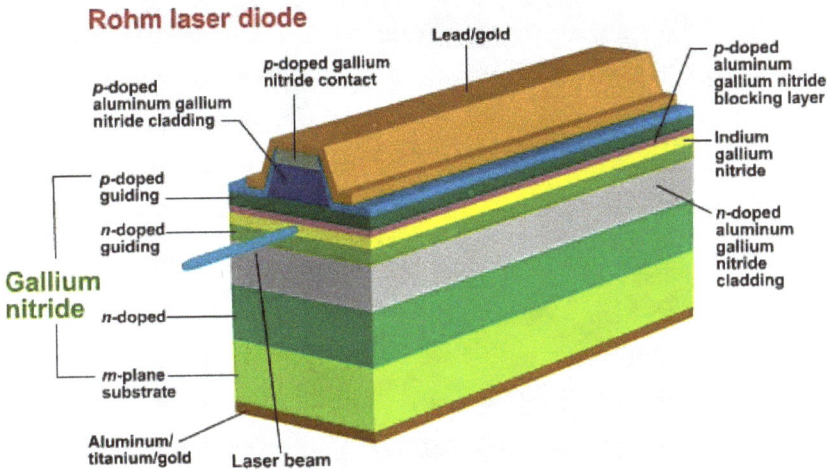

Fig. 6.79.   Schematic structure of a blue laser diode.

Courtesy: Rohm Corporation.

Figure 6.79 shows the schematic structure of a blue laser diode, developed by the Japanese ROHM Corporation

Low In content InGaN material for the active quantum wells is the easiest to grow through the MOCVD process used for commercial nitride-based laser diode production. This makes it possible to make violet and blue emitting laser diodes with high yield and, thus, low cost. AlGaN material needed for UV lasers is, comparatively, harder to grow and, thus, UV laser diodes are not widely available. However, research goes on, both to bring down the price of these devices and to push their emission wavelength deeper into the UV region. In the other direction, many years of effort directed at producing nitride-based green laser diodes culminated in the commercial availability of green-emitting InGaN laser diodes from Osram Optosemiconductors in Germany. Green laser diodes were specifically developed to serve in projection display markets that needed a replacement for DPSS green lasers which are larger and cannot be easily modulated in intensity. Osram's green laser diodes (see Figure 6.80) are now available in the wavelength range of 500–520 nm at powers of up to several watts.

The variety and uses of lasers keep increasing year-after-year. These intense sources of directional, monochromatic and coherent radiation are indispensable for many applications. From laser printers to Blu-ray

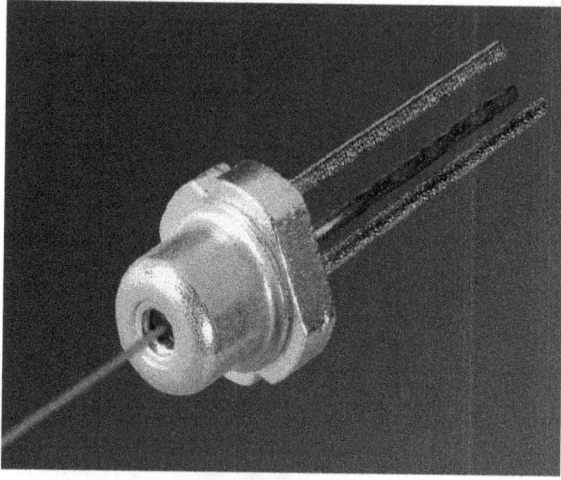

Fig. 6.80.   A packaged green laser diode from Osram Optosemiconductors.
Courtesy: Osram Optosemiconductors AG.

players and from supermarket barcode scanners to optical audio interfaces, these devices are present all around us. As the next chapter shows,
semiconductor lasers are now also being used in lighting applications —
taking on some jobs that have until now been done with LEDs, and opening up some entirely new application domains.

7

Laser-based Lighting Systems

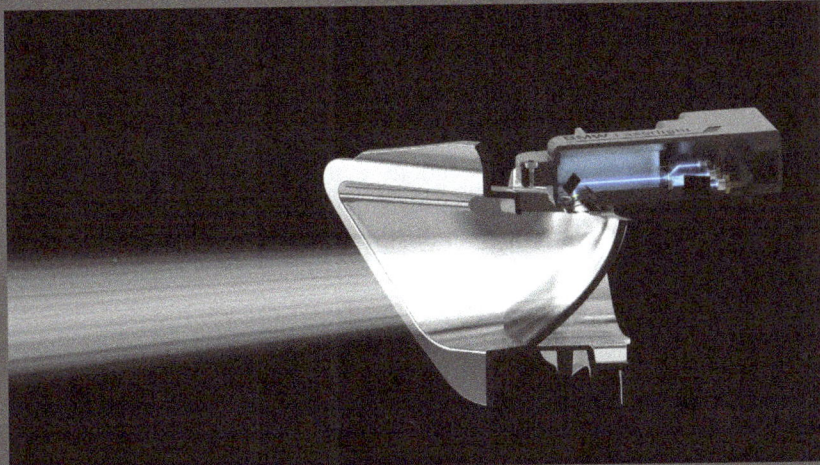

7.1 Introduction

Lasers have found their way, beyond traditional uses, into a number of illumination applications, and this trend is set to advance further as time goes by. For this reason, this chapter is devoted to the various lighting applications that are being enabled by lasers. When lasers were first developed, there was considerable excitement for many valid reasons and using them for illumination was seriously considered for some time. However, this use of lasers never then got much traction because of the peculiarities of laser emission. The light from lasers is extremely directional, highly monochromatic and shows a static scintillation pattern (speckle) when illuminating surfaces. These attributes are clearly not very desirable for most lighting applications that require a wide fan of white light. Some advances were made over the years for adapting lasers for illumination roles, especially after the invention of diode lasers, but such efforts did not result in any significant success stories. This state of affairs has now been changing, since the establishment of an entire solid-state lighting industry. While the development of solid-state lighting has been driven by innovations in LEDs, as recounted in Chapter 5, there has been increasing realization that laser diodes can also provide illumination for niche applications. This shift in thinking has come about quite naturally as LED-based lighting has matured in recent years. The focus for further innovations in solid-state lighting has, therefore, been increasingly turning toward the use of laser diodes.

Lasers did not make it on to the illumination scene initially because of their bulk, complexity and high cost. Thus, the development of laser diodes, covered in the previous chapter, was crucial for the later development of laser-powered lighting. Although even before the development of semiconductor lasers, their heftier cousins were pressed into service for applications in displays and entertainment lighting, these and many new applications have only been seriously developed during this century. Semiconductor diode lasers provide compact, low-cost light sources that can be easily integrated into various lighting systems. Both display and illumination applications, powered by laser diodes, are making rapid advances and promise to be nearly as pervasive in coming years as LED lighting is today. In fact, certain applications, such as retail lighting and movie projection, are so well matched with the capabilities of laser-based lighting that these areas will be completely taken over by sources pumped by semiconductor diode lasers.

Lasers can act as both direct light sources for displays and illumination systems, and as pump sources to generate radiation for such applications. However, display and projection (D&P) applications have mostly employed laser light directly whereas laser-based luminaires for space lighting make use of lasers to energize light-emitting phosphors. The trend, so far, has been that individual red, green and blue semiconductor lasers are being used for high color-purity image D&P applications. Lower cost displays and projectors have been developed that use only a single blue laser diode and obtain other colors through the use of luminescent materials. For illumination applications, on the other hand, blue- or violet-emitting semiconductor lasers are used to pump a broad-spectrum phosphor to generate white light.

While laser-based lighting, projection and display systems have seemingly evolved from their LED-based counterparts, laser diodes are by no means a drop-in replacement for LEDs. Light coming out of a laser diode has characteristics that are much different from that of LED light. Laser light has very high power per unit area, is highly directional, very narrow in spectral width, and has high coherence. All of these make developing a laser-based lighting system a very different engineering challenge. High brilliance results in high thermal loads as well as possible phosphor saturation. Beam directionality may require proper optical design if wide output beams are needed. Coherence can leave visual artefacts that may require elimination or suppression. All of these quirks are being addressed to come up with practical light sources to serve a variety of markets.

7.2 Laser-based Space Illumination Systems

LEDs serve very well for many lighting applications. Their small size, low cost, high efficiency (at low power levels) and long life are all very desirable attributes for building lighting systems. Laser diodes, on the other hand, are significantly more expensive and their cost escalates rapidly with increase in output power. They require more carefully-designed drive circuits and can fail catastrophically if proper care is not taken during their operation. For these reasons, laser diodes are not the first natural alternative that comes to mind when one considers replacing traditional LEDs with a superior alternative. Their cost and availability, in particular, have kept them from widespread use in solid-state lighting applications.

Along with their pertinent disadvantages, however, they also offer some outstanding benefits. Due to these features, using laser diodes instead of LEDs for certain kinds of luminaires makes a great deal of sense.

7.2.1 *Laser diodes versus light-emitting diodes for solid-state luminaires*

To begin with, we have the absence of 'droop' in semiconductor laser diodes. Droop, as discussed in Chapter 5, refers to a reduction in operating efficiency of LEDs when driven with increasingly higher drive currents. Its causes have been debated and investigated for several years now but it appears to be a fundamental feature of LED devices. This then means that very high brightness single chip LED sources will not be available any time soon, if ever. Multi-chip single encapsulation chip-on-board (COB) devices on ceramic substrates have been developed (see Figure 7.1) to somewhat offset this shortcoming but even these have size and brightness limitations. Such composite LED devices can go up to several tens of watts. Still, it is not possible to find single LED packages with electrical power ratings of 50 W and above. Laser diodes offer an attractive alternative as these very high brightness sources are free from droop; with their efficiency only suffering slightly at high forward currents (within their maximum current rating and with adequate cooling). The droop effect severely limits the maximum attainable wall-plug efficiency of GaN LEDs so that both the maximum attainable brightness and the efficiency are capped by the diminution of internal quantum efficiency at high drive

Fig. 7.1.   A multi-chip chip-on-board (COB) LED on a ceramic substrate.

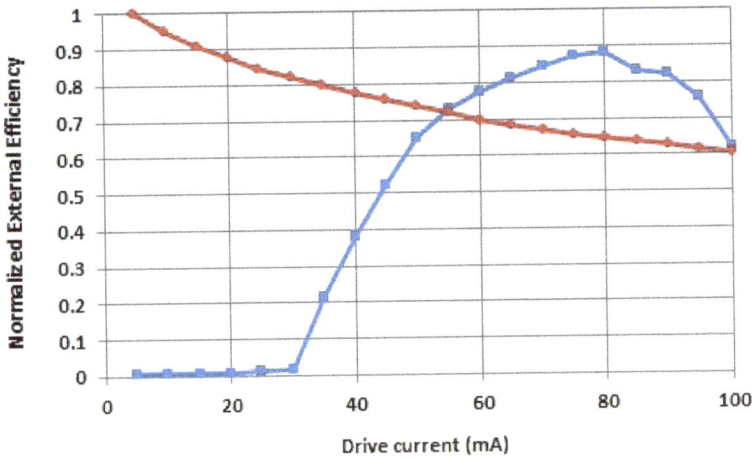

Fig. 7.2.   External conversion efficiency variation of an LED (red) compared with a laser diode (blue), as a function of drive current.

currents. This is clearly seen in a comparison of the external efficiency of a typical pumped-phosphor LED and a laser diode, as a function of drive current (see Figure 7.2). The external efficiency of an LED, operated at a constant temperature, falls steadily with increasing drive current whereas that of a comparable laser diode, operating beyond its threshold current of 30 mA, rises quickly to a large value and then drops slowly until the maximum allowable current rating is reached. The decrease in laser diode efficiency seen in this figure, after reaching a peak value at 80 mA, is due to heating effects, not droop. With better thermal management this decrease can be minimized and, thus, efficiency maintained at even larger drive currents.

The quality of light emitted by LEDs shows considerable device-to-device variation, which is controlled by 'binning' LEDs at the point of manufacture, i.e. by grouping them in different intensity and peak wavelength classes. Laser diodes are, comparatively speaking, much more uniform in their output radiation characteristics. This greatly simplifies the manufacture of laser-powered luminaires.

With phosphor-converted LEDs there is also the shift in color coordinates to contend with, as an LED warms up and its color point changes. While LEDs rarely suffer from catastrophic instantaneous failure, they do lose their brightness over time due to deleterious effects going on in their phosphor coatings. These effects are only accentuated at high temperatures so that devices with poor thermal management show a more pronounced reduction in emission intensity. The emission wavelength from laser diodes shifts little with changes in temperature. Furthermore, phosphors used with laser diode-pumping are used in a remote phosphor arrangement where the phosphor is not in contact with the diode. This results in no or very little heating of the phosphor from contact with a warm device, avoiding both in-use chromaticity shifts and long-term phosphor degradation.

Another benefit offered by laser diodes is the possibility of obtaining very broad spectral output when used with appropriate phosphor mixtures. Blue LED-pumped phosphor-converted sources are always deficient in the deep blue and violet portions of the visible spectrum. Not so with light sources pumped by 404 nm laser diodes — the cheapest and most widely available diodes because of their use in Blu-ray players. As space illumination lamps do not need directional emission so the laser light can be broadened by a concave lens before striking a large phosphor-coated plate or a phosphor-in-glass (PiG) disk. This arrangement removes thermal loading, phosphor saturation and speckle that can affect other laser-pumped systems. Luminaires, where these laser diodes pump a mixture of red-, green- and blue-emitting phosphors can emit a wavelength-rich bright white light that is eminently suitable for applications like retail illumination that require wavelength-rich light. LED lamps can neither provide similar brightness levels nor the spectral richness that is so needed for accurate color rendering.

7.2.2 *Laser diodes for solid-state lighting*

Laser diodes, at a number of different emission wavelengths, can be used for pumping down-conversion phosphors. Traditional, blue LED down-conversion phosphors (designed for excitation in the 445–470 nm region) can be pumped by blue-emitting laser diodes having a peak wavelength around 450 nm. These, and near-UV excitable phosphors, can often also be pumped with violet-emitting LEDs. These latter devices, emitting at 405 nm wavelength (see Figure 7.3), are often well-suited to this task because they are the most mature and cheapest of all 'blue' laser diodes.

A further advantage of these laser diodes is that pumping at this wavelength also provides the violet light component in white light which is so

Fig. 7.3.   405 nm violet laser diodes in metal can packages.

missing in light from white LEDs. Sub-400 nm laser diodes (such as devices emitting at 390 nm) can also be used for exciting near-UV phosphors. In this case, all the three primary color components have to be generated from individual phosphor components in a trichromatic phosphor blend. This allows precise spectral shaping for obtaining high color rendering index (CRI) light sources. However, sub-400 nm laser diodes are considerably more expensive. The 405 nm violet laser diode pumping approach is very attractive because of the wide availability of low-cost devices and the availability of phosphors that can be excited in the 400–420 nm wavelength region. As seen in Figure 7.4, both LEDs and laser diodes are available with peak wavelengths around 405 nm. The differences between these two pump sources are in the different spectral full width at half maximum (FWHM), angular beam profile, wavefront structure and coherence. Light from a laser diode is significantly narrower in wavelength spread (~2–3 nm) compared to LED light (~12–30 nm). It is also highly coherent which can lead to interesting effects in certain situations. A number of companies, such as Sony, Nichia and Osram produce laser diodes suitable for solid-state phosphor pumping applications. Most of them come packaged in TO-style metal cans for good heat dissipation. Such through-hole leaded packages can be mounted directly on a printed

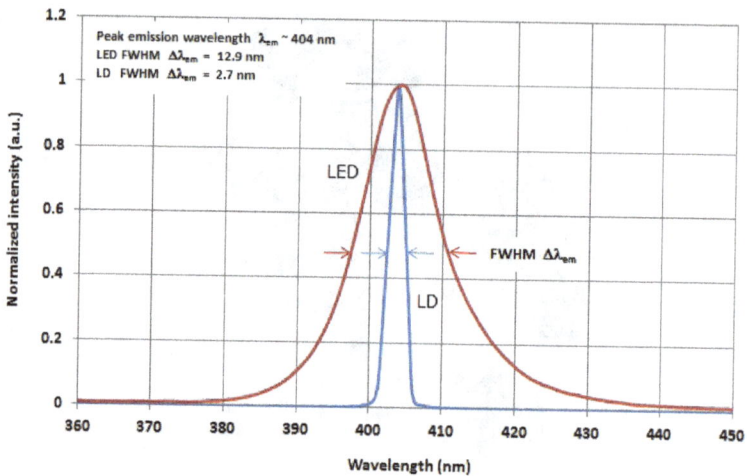

Fig. 7.4.   Spectra of LED (red) and laser diode (blue) with the same peak emission wavelength of 404 nm.

circuit board (PCB). Suitable heat sinking must be provided to remove the heat generated during laser diode operation. The metal can packages have a transparent glass window at the top through which the laser light escapes. Other than that, no beam shaping optics are present. Optical elements needed to appropriately shape the beam for efficient phosphor pumping must be provided external to the laser diode package. Devices emitting in the UV region may have a fused silica window instead of an ordinary float glass window. Usually, a photodiode is packaged with the laser diode chip inside the same package. This photodiode is used to monitor the emission intensity of the laser diode so that closed-loop feedback control can be used to keep the emission intensity stable against changes in emission intensity due to temperature change etc. Thus, most laser diode packages have three leads that come out of the metal can. Two of them connect to the photodiode and laser diode anodes whereas the third one is a common cathode lead. Some laser diodes also come mounted on a Peltier cooler but this form of device is, generally, not used for phosphor pumping.

A number of parameters are important for selecting laser diodes suitable for exciting phosphors. These include the emission wavelength, threshold current, operating current, operating voltage and power conversion efficiency. All of these are carefully measured by laser diode manufacturers and listed on datasheets. Violet light-emitting laser diodes, as mentioned above, are the most common for pumping phosphors because of their wide availability and low price. These devices usually emit at 405 nm (±1%) with a narrow spectral width of around 2 nm FWHM. Their beam quality is often not very good, but this is not important for phosphor pumping applications. As long as their operating current is within the range bounded by the threshold current and the maximum allowable current, these devices perform very well.

Unlike LEDs, laser diodes are more susceptible to gradual degradation through mechanisms related to dislocation multiplication in the active region. Threading dislocations that, inevitably, exist in laser diodes, tend to multiply at high current densities and elevated operating temperatures. These dislocations are non-radiative recombination sites, i.e. electron-hole pair recombination at their locations does not result in the emission of photons. Instead, the energy from carrier recombination appears as heat. Linear threading dislocations appear as, so-called, dark line defects (DLDs). With prolonged device operation, especially, at high current levels, DLDs tend to proliferate and the laser diodes can gradually lose their brightness. This is a serious degradation mechanism for many types of semiconductor lasers which can place an upper limit on the longevity of laser diode-pumped phosphor-converted white light sources. Short wavelength visible light laser diodes have been much improved over the past several years, but more work needs to be done to further enhance their long-term reliability. Growing laser diode device layers on native (free standing) GaN wafers, which are now commercially available, can go a long way toward reducing the density of threading dislocations and, thus, of DLDs.

Light coming out of semiconductor laser diodes is almost always divergent and with non-Gaussian intensity profile. The reason for this is the shape of the resonant cavity in typical side-emitting laser diodes. The long narrow cavity, with rectangular cross-section, causes the light output to diffract asymmetrically. The wavefronts diverge more along the rather small thickness of the multi-quantum well (MQW) structure and less along the wider lateral rib that defines the width of the lasing cavity. Light exiting such a rectangular aperture assumes a divergent fan shape with elliptical cross-section (see Figure 7.5). This is far from the structure of the nice round TEM_{00} beams that come out of most gas lasers. The two elliptical axes of a laser diode light beam are called 'slow' and 'fast'. Rounding and collimating such a laser beam can be very challenging.

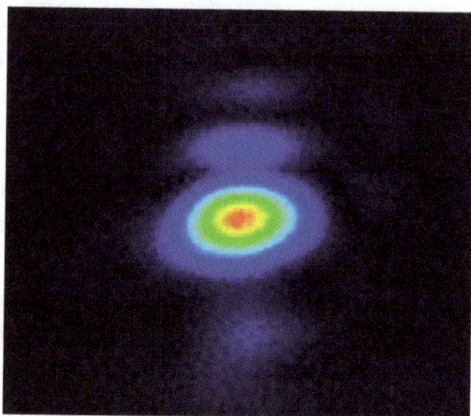

Fig. 7.5.   Far-field emission profile of light from a diode laser without any beam shaping optics.

Nevertheless, effective techniques have been developed to shape this type of laser light beam. The first task is to circularize the cross-section of light from side-emitting laser diodes. A common way to achieve this is to use cylindrical lenses. These lenses have cylindrical shapes at right angle to the optical axis and tend to shape light fields such that the light is spread out into a line at right angle to the cylinder's axis (see Figure 7.6). Two separate plano-convex cylindrical lenses with different focal lengths can be used to circularize the output of a laser diode by shaping light along both the slow and the fast axis. Cemented and mounted cylindrical doublets are available for this purpose. Once the beam has attained a circular cross-section, any curvature of the wavefront can be removed through the use of a collimating lens assembly. An appropriate plano-convex or plano-concave lens can achieve this. A well-collimated laser light beam has maximally parallel rays that imply plane parallel wavefronts. Such a light beam is in a sense 'well-behaved' and can be further shaped to suit a given application.

7.2.3 *Drive electronics for laser diode-based illumination systems*

Diodes that emit light — both LEDs and laser diodes — are usually operated in a constant current mode such that they are driven by a source that appears as a current source with high internal (Norton) resistance. An

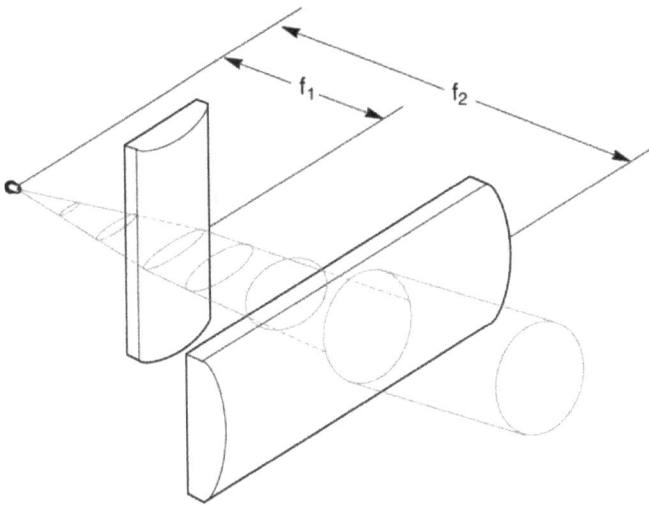

Fig. 7.6.   Circularizing an elliptical diode laser beam using two cylindrical lenses.

ordinary voltage source with an appropriate zener diode can make a tolerable current source but current-mode transistor-controlled power supplies are much preferable. Such supplies are now also available as integrated circuits that require only a few external components to make excellent LED and laser diode drivers.

Pulse width modulation (PWM) is the most widely used scheme for controlling the intensity of light emission from all kinds of LEDs and laser diodes. By varying the duty cycle of a pulse waveform, used to drive a laser diode, between 0% and 100%, it is possible to control the light output from the diode. Typical PWM frequencies are around a few kilo Hertz and these waveforms can be generated by a variety of devices such as microcontrollers, complex programmable logic devices (CPLDs), field programmable gate arrays (FPGAs) and special LED-driving integrated circuits. Normally, the PWM output is filtered with a simple one-pole RC filter before it is used to drive LEDs, either directly or with a power transistor. A better alternative has now become possible with the availability of LTC 2645 from Linear Technology Inc. This IC is a PWM-to-DC level converting chip which Linear Technology calls a PWM to V_{OUT} DAC. This device takes in, up to, four channels of PWM waveforms and produces corresponding DC voltage levels. Unlike an RC filter, this is accomplished with no latency and very high accuracy. Linear single channel PWM circuits can control the brightness of phosphor-converted laser diode-pumped white light sources whereas multiple channel PWM circuits can individually control the red, green and blue channels of a color-mixing type white light source, pumped by three (or more) laser diodes. The latter technique allows the spectral profile of the resulting mixed light to be tuned as desired. The combination of multiple laser diodes (either emitting band edge light or down-converted light) with individual PWM controls enable the generation of full-spectrum white light with desired chromaticity point and correlated color temperature (CCT).

Several semiconductor companies also make special laser diode drivers that can be used to drive laser diodes in both continuous wave (CW) and pulsed modes. iC-Haus in Germany, for example, makes a range of CW and pulsed laser diode drivers that are suitable for operating blue and violet GaN laser diodes for illumination applications. Typical driver ICs provide spike-free switching of laser diode current, appearing as voltage-controlled current sinks to connected laser diodes.

In addition to the laser diode driver, a complete laser diode-based illumination system also needs a power supply to convert mains power to low voltage DC power that can be used by the laser diode driver. This is accomplished by a switch-mode power converter. These systems use step-down converter ICs with a few external components and can convert any voltage in the 80–300 V AC range to a low DC voltage. Such 'universal' power supplies are small, light and fairly inexpensive. While it is easy to develop and implement the electronics needed for laser diode-based white light systems, their long-term reliability is usually inferior to that of other components in the system. Thus, a light source can fail because an electronic component, instead of a light path component, has degraded, gradually or catastrophically. This kind of reliability issue is also seen in LED-based white luminaires.

7.2.4 *Approaches to laser-based white light generation*

There are two principal ways in which laser diode-based white lighting systems can be implemented. These are very similar to approaches used with LEDs to build sources of illumination-quality white light. The conceptually simpler, but operationally more complicated, strategy is to simply combine light from separate red-, green- and blue-emitting lasers to generate light that appears white to our eyes. This color mixing approach has been used with color LEDs as well. This route, whether implemented with laser diodes or with LEDs, offers the great benefit of the ability to tune the color or shade of the resulting light by altering the relative proportions of red, green and blue lights. Thus, this is the preferred approach where partial or full color control is needed over illumination. The downside of this approach is the need for properly mixing the different lights to generate fully-mixed output light. This is quite hard to do with high efficiency, i.e. without losing a big fraction of the light in the mixing process. The simplest light mixing elements consist of translucent polymer or glass plates that randomize the direction of light rays that travel through them, and in the process mix different light components. However, such devices, necessarily, also absorb a good fraction of light propagating through them and, thus, are only used in applications where low efficiencies can be tolerated. Other, more efficient mixing devices consist of mirror baffles and multifaceted mixing chambers that take more volume but lose less of the lights that are being mixed. It has been shown that red, yellow, green and blue light from four different lasers can be mixed together using chromatic thin-film beam combiners. After further mixing in a volumetric spatial diffuser, the resulting light appears white in color. Very significantly, for illumination applications, it has been found that, in spite of the resulting light's spectrum being composed of four narrow spectral lines, it is able to render the color of everyday objects quite faithfully. This is an important observation for laser-based illumination systems, especially those that employ only lasers (no phosphor-based down-conversion) or make use of narrowband phosphors. It turns out that due to the nature of the spectral responsivity function of the human eye, even white-appearing light with 'spiky' spectrum is able to render actual physical colors quite effectively. Of course, use of broadband phosphors leads to spectrum-filling white light with much better color rendering characteristics, but 'spiky' white light also performs reasonably well; somewhat contrary to expectations. True RGB laser diode-based white light sources have now become available with the development of green-emitting laser diodes from Osram and other companies. Typical PL 520 type green laser diodes from Osram, for example, seen in Figure 7.7, emit 30 or 50 mW of output optical power in the 510–530 nm range, and can be combined with red and blue laser diodes emitting about 100 mW of optical power to make compact white light sources.

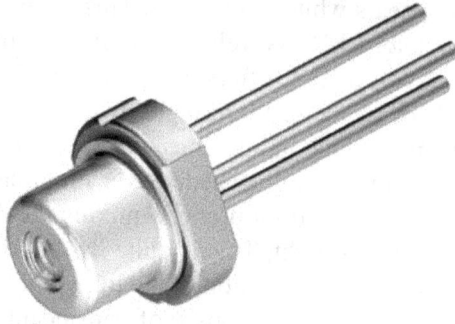

Fig. 7.7.   Osram PL 520 green laser diode.
Courtesy: Osram Optosemiconductors AG.

The other scheme for generating white light from laser diodes is more recent but is now the dominant approach. This is through the down-conversion of laser radiation, used as a pump source, to one or more longer wavelengths through the use of single or multi-component phosphors. This is very similar to the approach used with phosphor-converted white LEDs. The only difference here is that a laser diode instead of an LED is used as the pump source. Just like the case with LEDs, laser diode-pumped phosphor down-conversion is a much simpler scheme for generating white light. In contrast to LED-based systems, this approach is exclusively based on remote phosphor arrangements so that the phosphor is not in contact with the laser diode chip or package. This is discussed in greater detail in the following sub-section.

There are also other strategies for generating white light from monochromatic laser diode light. Because of system cost and complexity, these approaches are much less prevalent and are only met with in research environments. One of these approaches is through spectral stretching of optical energy through the use of special photonic crystal optical fibers. A number of nonlinear effects can convert short intense pulses of monochromatic laser light into broad spectrum light. This, so-called, supercontinuum effect was first observed in 1970 and is now a valuable method for generating intense broadband white light. Systems available from NKT Photonics in Denmark, for example, provide continuum light in the 400–2400 nm region (see Figure 7.8).

Another supercontinuum source, providing coverage in the 1.3–4.5 μm wavelength region is the SC4500 from Thorlabs (see Figure 7.9).

Fig. 7.8. Supercontinuum laser from NKT Photonics, Denmark.

Courtesy: NKT Photonics, Denmark.

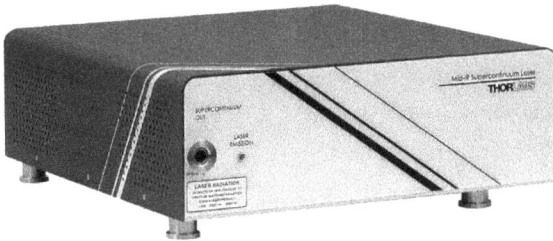

Fig. 7.9. Supercontinuum laser from Thorlabs Inc.

Courtesy: Thorlabs Inc.

The other approach makes use of plasma generation by focusing high power laser diode light to a tight spot (~100 μm diameter) inside a rarefied gas (usually, Xe) atmosphere. This produces a small bright broadband source of light that can be coupled to both free space and to optical fibers. Figure 7.10 shows a commercial laser plasma-based white light source from Energetiq Inc. (now a part of Hamamatsu Corporation).

7.2.5 *Phosphors for laser diode-based illumination systems*

Since the advent of LED-based solid-state lighting, around two decades ago, inorganic phosphors of various compositions, pumped by blue LEDs, has remained the primary scheme for generating quasi-white light. Almost every commercial white LED on the market today employs a GaN/InGaN blue LED that pumps a coating of one or more phosphors, placed either

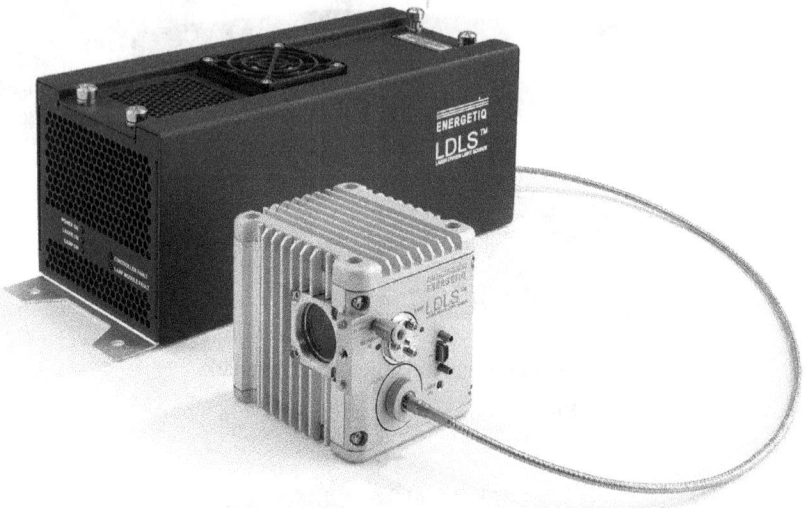

Fig. 7.10.   Laser-driven broadband light source from Hamamatsu Corporation.
Courtesy: Hamamatsu Corporation.

directly on top of the LED (proximity pumping) or some distance away from it (remote pumping). Most LED phosphors can be used for laser diode-based lighting as well, as long as they have high absorption at the laser diode emission wavelength. Another important consideration for suitable laser diode-pumped phosphors is high saturation intensity so that phosphor particles do not stop down-converting photons, as the incident optical flux increases. This is an especially important concern because phosphor pumping by laser diodes results in much higher aerial power densities than are achieved with even very high brightness blue LEDs.

Laser diode-pumped white light sources can be produced with either a single broadband yellow-emitting phosphor or a mixture of red, green and blue phosphors — a so-called trichromatic phosphor. This latter approach is preferred for its better spectrum-filling light quality. A trichromatic white-emitting phosphor capable of pumping with laser diodes emitting at 405 nm can, for example, be made by mixing three rare-earth-doped phosphors: (1) Potassium europium tungstate phosphor ($KEu(WO_4)_2$, emission color: red, λ_{max} = 616 nm; (2) Europium and manganese-doped barium magnesium aluminate phosphor ($BaMg_2Al_{16}O_{27}$:Eu,Mn), emission color: green, λ_{max} = 515 nm; and (3) Europium-doped strontium magnesium silicate phosphor (($Sr,Mg)_2SiO_4$:Eu) — emission color: blue, λ_{max} = 460 nm. By mixing appropriate amounts of these three primary color phosphors, it is possible to produce balanced white light with chromaticity point near the center of the chromaticity diagram. In order to obtain a good white point, phosphors have to be mixed in inverse proportions to the amount of radiometric energy they contribute to the white light's spectrum. Thus, a significantly larger amount of red phosphor, of the above-mentioned type, is needed in the phosphor mix due to its relatively low conversion efficiency. Even for narrowband phosphors it is possible to produce very good white light by proper mixing of individual phosphors. The chromaticity points of the above-mentioned phosphors and of the white light from their mixture, on excitation with a 405 nm diode laser, appear in Figure 7.11. The RGB phosphors generate their characteristic

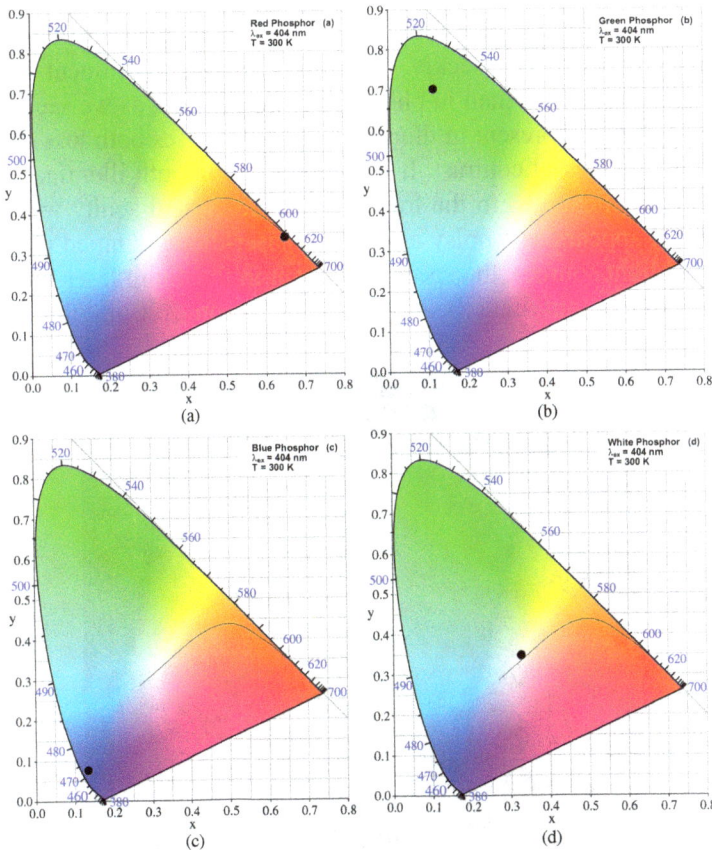

Fig. 7.11. Chromaticity coordinates of emission from (a) red, (b) green, (c) blue, and (d) mixed RGB phosphors, pumped by 405 nm laser radiation.

colors from respective atomic transitions in Eu^{3+} (red phosphor), Eu^{2+} (blue phosphor) and Mn^{2+} (green phosphor) ions. In the case of the green phosphor, the Eu^{2+} ions act as sensitizers — transferring energy gained from the absorbed laser pump photons efficiently to luminescent Mn^{2+} ions. The Mn^{2+} ions are themselves hard to pump directly because their electronic transitions are forbidden on both spin and parity grounds. The presence of Eu^{2+} ions in the same host (phosphor crystal) allows first their pumping, followed by efficient radiation-less transfer of energy to Mn^{2+} ions, causing them to fluoresce.

7.2.6 *Speckle in laser diode-based illumination systems*

Laser light, reflecting off from plane surfaces, displays a static scintillation pattern, called speckle (see Figure 7.12). Laser-based illumination systems can also show speckle. This is much more prominent in direct laser-based luminaires than in phosphor-pumped ones, as we see below.

Speckle, when present in illuminating light, causes both loss of visual definition and loss of contrast. It is troublesome in both illumination and projection applications. In the former it reduces visual acuity whereas in the latter it appears as a noisy artefact. In laser-diode pumped phosphor-converted light sources, the speckle inherent in the pump laser light reappears in the output light beam. Partly, this comes from the residual

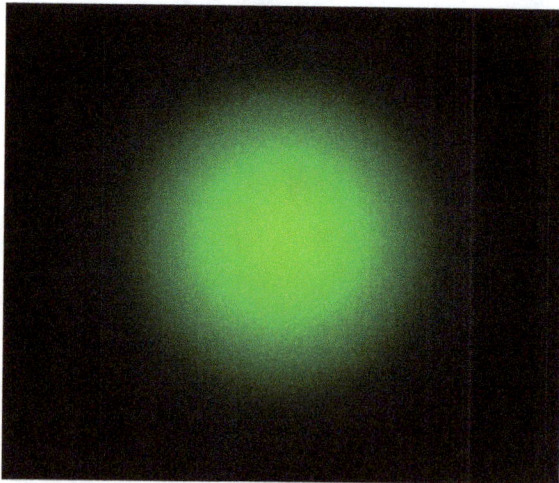

Fig. 7.12.   Laser speckle seen in green laser light reflecting from a plane white surface.

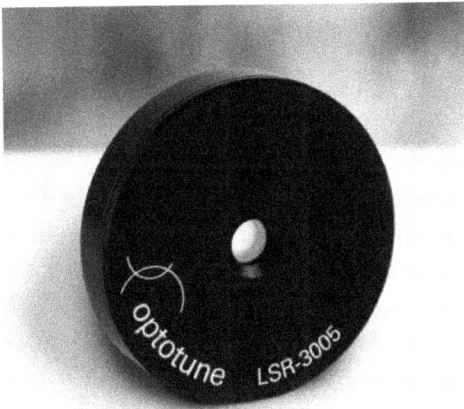

Fig. 7.13.   Optotune laser speckle reducer.
Courtesy: Optotune AG, Switzerland.

(un-converted) pump light and the rest from stimulated emission from wavelength up-conversion from intense pumping. It has been shown that luminaires that make use of powdered phosphors exhibit very little speckle. This is because of multiple light scattering events within the phosphor volume which destroy spatial optical coherence.

Where speckle can be a problem, it can be eliminated or reduced with various devices that serve to effectively reduce the coherence of laser radiation. One such device is based on an electrically-driven electro-active polymer membrane (see Figure 7.13). This type of laser speckle reducer is a dynamic light diffuser which reduces optical coherence, without any change in peak wavelength and spectral width of the light passing through it. The light does become un-collimated after passing through this device, but it can be collected and refocused with a lens system. Figure 7.14 shows the effect of the laser speckle reducer; with the upper panel showing the distribution of local intensity pattern with the speckle reducer switched off and the lower panel showing the distribution with the speckle reducer switched on. The reduction in speckle is easily seen. Reduction of speckle causes the far field pattern of laser light to become more uniform which avoids saturation of phosphor particles. Both saturated and under-illuminated phosphor particles produce reduced down-converted light which adversely affects system performance.

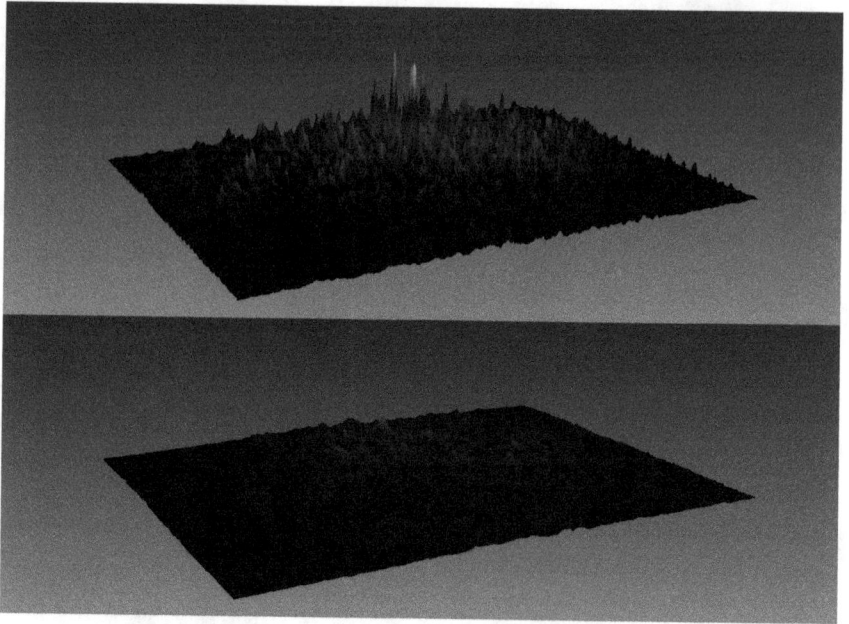

Fig. 7.14. Spatial intensity distribution of laser light passing through an Optotune speckle reducer with the device switched off (upper panel), and with the device switched on (lower panel).

7.2.7 *Luminaire design for laser diode-based illumination systems*

Depending upon application requirements, various types of laser diode-pumped white light sources can be designed. An illustrative example of a domestic-type lamp is described here but most recent lamps have been designed for such high intensity applications as automobile headlights which are described in the next section. This is because laser diode-pumped luminaires have an immediate advantage over LED-pumped luminaires in that they are unaffected by LED droop and, thus, can easily achieve higher brightness levels at respectable conversion efficiencies. The laser-driven white light source described here was designed to have approximately the same size and form factor as a traditional MR-40 style tungsten halogen lamp. The device comprised a 405 nm violet laser diode module with a Sony laser diode, attached to an interior-aluminized plastic reflector shell. A phosphor-coated borosilicate glass disc was attached to

Fig. 7.15. Prototype 3D-printed body of a laser diode-pumped white lamp with phosphor-coated glass plate seen at the front.

the wide end of the reflector. Figure 7.15 shows the lamp body from the front. This component had a uniform wall thickness of 2 mm, and it was made through 3D printing on a Stratys 60 printer using a hard polymer. The optimized profile of the reflector was first created as a 2D sketch. This profile was subsequently revolved using SolidEdge software to produce a suitable 3D model. The .stl file thus created was directly used by the printer to build the reflector.

Geometric ray tracing simulations of the luminaire assembly were carried out, using Lambda Research Corporation's TracePro software. For this purpose, a geometric model was set up. It consisted of a parabolic-figured reflector with aluminized inner surface with 95% reflectance (ALCOA BriteCoat 85 Specular), a phosphor plate modeled by a diffuse scattering coating on its reflector-facing surface, a double concave glass lens with 12 mm focal length, and a cylindrical base/adapter. Laser light was simulated by a circular 1 mm diameter grid source with a Gaussian intensity distribution. Typically, 271 rays per wave were simulated and the results are shown here in Figure 7.16.

For simulation visualization, the external surface of the phosphor plate was made opaque in order to clearly see the behavior of light inside the laser-reflector-phosphor plate assembly. The expansion of the bundle of rays (red) comprising the incident light from the laser diode module is clearly seen on the left. The middle image shows the expanded light cone striking the inside (phosphor-coated) surface of the phosphor plate. Some light is backscattered from the plate (blue) and travels back to be reflected by the specular inner surface of the reflector. This light

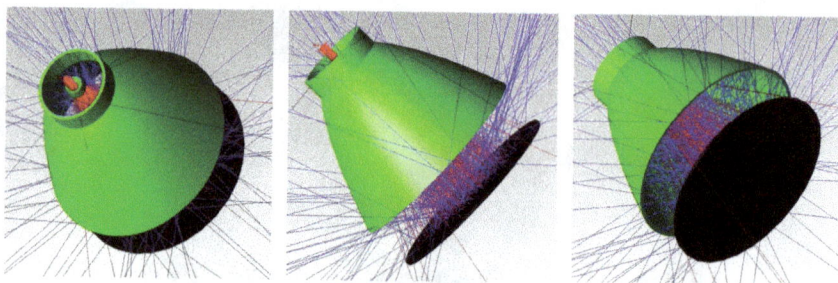

Fig. 7.16.   Geometric ray tracing simulation of a laser diode-pumped white lamp.

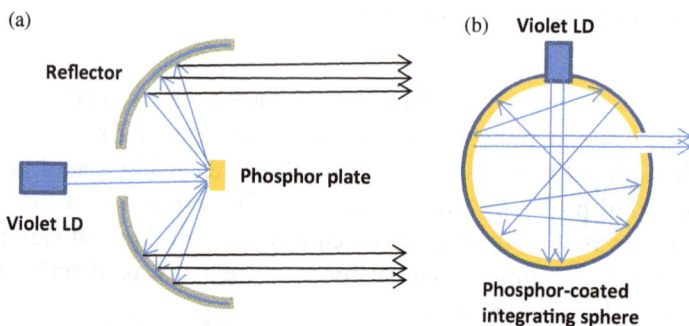

Fig. 7.17.   (a) A unidirectional reflector-based laser luminaire design. (b) An integrating sphere-based laser luminaire design.

is then re-incident on the phosphor plate. Note that only very few rays come out of the back end of the reflector. These rays are wasted as they are absorbed by the laser diode module housing. Rays seen escaping the gap between the reflector and the phosphor plate are a result of deliberately leaving a gap between the two, during system simulation, in order to see inside the assembly. In actual practice, the phosphor plate is flush with the end of the reflector and no light escapes there. The actual behavior of the luminaire was found to be very close to the simulation results depicted here.

Another possible arrangement for making laser diode-pumped white light source is shown in Figure 7.17(a). In this case, light coming out of a hole in the center of a concave mirror strikes a much smaller opaque phosphor plate and the down-converted light is completely directed outwards by the reflector. Although not shown, a beam expansion lens is needed here too, in order to avoid saturating the phosphor coating.

Figure 7.17(b) shows yet another arrangement where down-conversion takes place in a special enclosed spherical phosphor chamber that serves as an integrating sphere. Here, as shown, the pump light and the down-converted light have a 90° angular separation but other separations are also possible due to the symmetry of the integrating sphere.

In extremely high-power luminaires, such as powerful headlights and searchlights, multiple laser diodes need to be used to pump phosphors effectively. In such cases, simple mirrors can be used to combine light from multiple diodes. Alternatively, polarization beam combiners can also be used which combine orthogonally-polarized laser beams into a single higher power beam. A possible concern in this and similar high intensity lamps is related to damage to the phosphor due to the use of high incident optical flux. This problem can be mitigated by scanning the incident beam on the phosphor plate with a small electrically-driven scanning mirror that produces a raster pattern on the phosphor plate. An even more practical arrangement is to use a spinning phosphor-coated disk, as is described in the next section.

Phosphor plates directly bonded to heat sinks are also available for making high power laser diode-based luminaires (see Figure 7.18). In this case, heat generated by multi-watt laser beams impinging on the phosphor is dissipated into the metal heat sink through a low thermal resistance path. The finned heat sink can be cooled through either natural or forced convection.

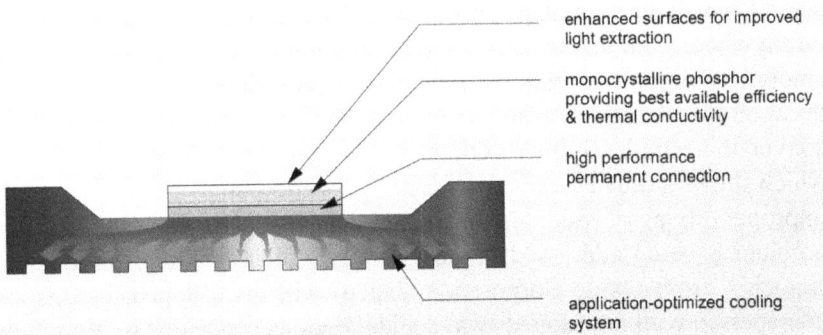

enhanced surfaces for improved light extraction

monocrystalline phosphor providing best available efficiency & thermal conductivity

high performance permanent connection

application-optimized cooling system

Fig. 7.18.   Monocrystalline phosphor plate bonded to a metal heat sink, intended for use in reflective laser diode-pumped white lighting systems.

7.3 Low-divergence Laser-based Illumination Systems

A number of lighting applications require fairly directional light beams. Examples are provided by automotive headlamps, search lights, spot lamps, medical endoscopy, rear projection displays and cine projection. These kinds of applications can be very well performed by laser-based lighting systems where narrow and parallel laser beams can be usefully utilized. Laser light is characterized by low intrinsic divergence and if the low angular spread can be maintained as the light passes through a color-converting medium then highly directional light of a color different from that of the pump laser can be obtained. Ordinary powder phosphors, as used in LEDs, are often not suitable for this purpose because multiple light scatterings between phosphor particles scramble the original direction of light beam propagation, making the color-converted light exit as a wide cone. For this reason, very directional phosphor-converted light sources employ single crystal phosphors.

The very high intensity of typical laser beams poses immediate challenges when trying to pump wavelength-conversion materials. Like almost any other material, phosphors can get damaged by very high photon flux. High optical damage threshold phosphors are, thus, attractive for laser pumping. Fortunately, most phosphor host crystals are materials like silicates, nitrides, oxides and tungstates that have respectable damage thresholds. YAG:Ce, grown from a cerium-doped YAG melt, for instance, is available as single crystals for such applications (see Figure 7.19). Typically, a crystal measuring 5 mm × 5 mm is used, where most of the crystal face is lit by a pump laser beam. The thickness of the crystal can be anywhere from less than 1 mm to several millimeters, depending on the amount of color conversion needed. However, due to the short transit distance for laser light inside a single crystal, the amount of wavelength conversion remains limited. Often, this can result in a converted beam which shows a blue center surrounded by a yellow ring. This is caused by improper mixing of residual pump beam with phosphor fluorescence. This problem is being addressed in various ways, such as composite crystals (Epoch™ from Oxide Corporation, Japan) where Ce-doped regions are interspersed with non-doped 'waveguide' regions consisting of aluminum oxide (sapphire) which serve to effectively mix pump and converted light (see Figure 7.19). Sapphire has high thermal conductivity which serves to effectively dissipate heat generated by Stoke's shift and non-radiative transitions in the Ce-doped regions.

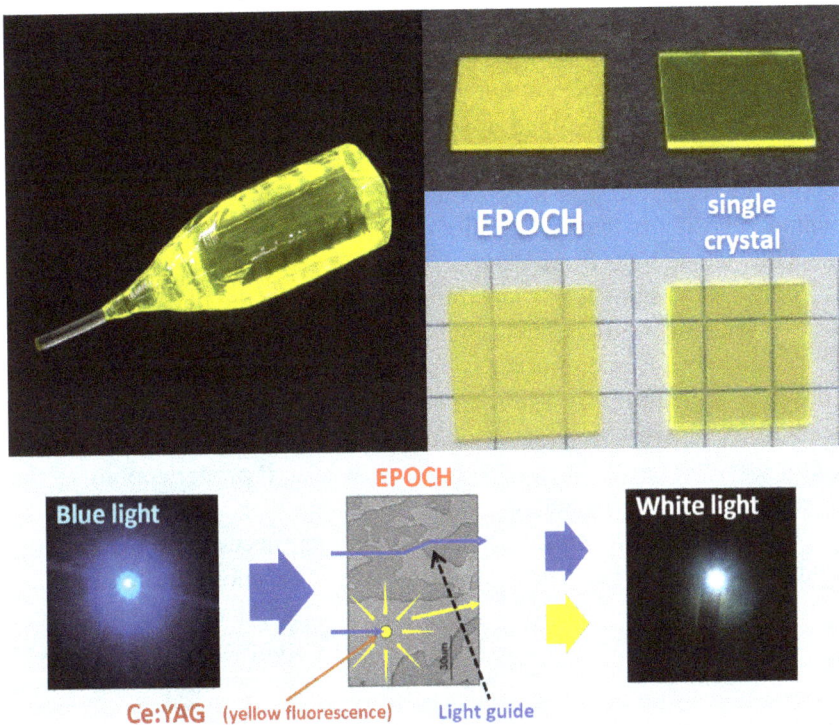

Fig. 7.19.   (Left) Melt-grown YAG:Ce boule. (Right) Cut and polished plates of EPOCH and single crystal YAG:Ce material. (Bottom) Scheme for converting blue laser diode light into white light using EPOCH wavelength-conversion material.

Courtesy: Oxide Corporation.

Other suitable laser phosphor materials which have also received attention due to their high tolerance to thermal loads include Ce-doped aluminum nitride (AlN:Ce) and alumina-YAG:Ce ceramics.

A more serious issue arises from the sheer number of photons per unit time present in laser emission. A given volume of phosphor can absorb a lot of photons but plenty still may be left unabsorbed due to the depletion of electrons in the ground state. As it takes some time for absorbed photons to result in the emission of a longer wavelength photon so the phosphor can easily get saturated and stop absorbing any more photons. This saturation effect is obviously a more serious problem with long emission time phosphors, because their luminescent centers remain in their excited state for longer times, and, thus, short luminescence decay times are desirable for laser-pumped phosphors. A comparison between cerium and europium ions — two of the most commonly used luminescent centers in commercial phosphors — shows that at room temperature the former has an excited state lifetime a hundred times shorter (~24 ns) than that of the latter (~480 ns), when doped in strontium thiogallate crystals. This comparative difference remains when other host crystals are used. For this reason, cerium (Ce^{2+}) is a better luminescent ion for LED- and laser-pumped phosphors than is europium (Eu^{3+}).

The intense light from lasers also causes phosphors to heat up. If not cooled quickly enough, this heat buildup can cause the phosphor's wavelength conversion efficiency to plummet (thermal quenching). It can also result in thermal deterioration of phosphors over longer time periods. Phosphor degradation is also seen in LEDs but where it takes months to years for the decrease in LED brightness to become noticeable, with lasers useful lifetimes could be as short as a few weeks, if proper cooling of the phosphor is not implemented.

When very low beam divergence is not necessary, a reflective arrangement where laser pump light reflects off a phosphor coated plate is the configuration of choice. Both optical saturation and thermal loading can be largely eliminated in this case by using a spinning phosphor-coated wheel, like the one shown in Figure 7.20. When a laser spot illuminates a

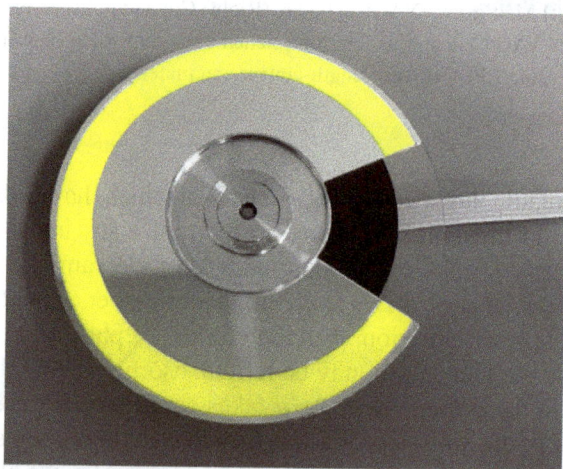

Fig. 7.20.   Phosphor disk that can be spun to distribute laser pump flux around the disk's periphery.

Courtesy: A-Tech System Company Ltd.

phosphor-covered area near the periphery of the wheel then thermal loading does not remain confined to one localized spot. As the wheel spins, heating gets distributed all around it, greatly easing the thermal burden on the phosphor. This also helps with phosphor saturation because fresh phosphor keeps coming under the pump beam as the wheel rotates.

High intensity beams of colored light are just what are needed for D&P applications. Here, however, instead of wide wavelength coverage, very narrow spectral widths are desirable. Semiconductor diode lasers are ideally suited for this. Red laser diodes have been available for decades, blue ones have also been available for many years, the remaining color — green — took a long time coming but is now finally commercially available. Thus, all three laser diode sources can now be used in RGB D&P systems. Projection equipment manufacturers have been increasingly replacing bulky, short-lived and power-hungry xenon lamps with RGB laser units. Projection light source modules built with laser diodes last for more than 30,000 hours before requiring replacement, which is vastly longer than the few hundred hours lifetime of xenon projector lamps. These sources produce bright pictures on large screens with nearly full color saturation, thanks to the brilliance and monochromaticity of laser light. 3D cine projection especially benefits from this because the glasses required for viewing 3D images absorb a significant amount of light, so the screen image needs to be brighter than that needed for 2D projection. Going even further, because of the ability of semiconductor laser diodes to provide extreme narrowband emission at precise wavelengths, some 3D projection systems have been implemented using six instead of three primary colors. These, so-called 6P systems, make use of two distinct sets of RGB laser units. One set emits light for the right eye image whereas the other set emits light, at slightly different wavelengths, for the left eye image. Color-filtering glasses separate out the two sets of images so that each eye only sees the image that was intended for it. Furthermore, the pictures from RGB laser systems are free from motion streaking and frame breakup because of the ability of semiconductor lasers to be directly modulated in intensity. Figure 7.21, on the left, shows a pure RGB laser-based 45,000 lumen D4K 40-RGB projector from Christie Digital Systems. RGB lasers are also being used for rear-projection 'video walls' manufactured by Barco Inc. where they offer significantly increased screen brightness over LED-powered systems.

RGB laser projectors are still expensive but a lower cost laser projection option is provided by phosphor-converted laser sources, similar to those described above, where blue laser diode light illuminates a spinning

Fig. 7.21.   (Left) RGB laser-based 45,000 lumen D4K 40-RGB projector from Christie Digital Systems. (Right) Barco DP4K-36BLP 35,000 lumens laser phosphor projector.
Courtesy: Christie Digital Systems and Barco Inc.

broadband phosphor wheel. Dichroic filters are used to separate green and red components from the resulting light which is then used for image projection. Some systems, such as BoldColor™ from Christie Digital Systems, even employ a hybrid arrangement where a red LED or laser diode is added to make up for the deficiency of red component in YAG:Ce-converted light. Due to their lower cost, laser-pumped phosphor-converted sources have now become the illumination of choice in many home and office projectors, as well as professional cinema projection systems. Figure 7.21, on the right, shows a Barco DP4K-36BLP 35,000 lumens laser phosphor projector. All professional laser projectors can display high contrast, high frame rate images on large screens, with minimal maintenance requirements.

Many laser-based illumination and projection systems now make use of a phosphor wheel to make good use of the high photon flux present in laser diode beams. This is seen here in Figure 7.22 which shows the arrangement used in some NEC laser projectors. A high-power blue diode laser illuminates a spot on a spinning phosphor wheel. Unconverted blue laser light simply passes through the wheel. Wavelength up-converted radiation with green and yellow components comes off the front surface of the wheel. The various color components are mixed with red light from either an LED or a red laser diode at a color selection filter wheel, and then used for image projection.

As far as wavelength-converted projection light sources are concerned, even better performance can be obtained in the future from using a blue laser diode whose light is converted to other required primary

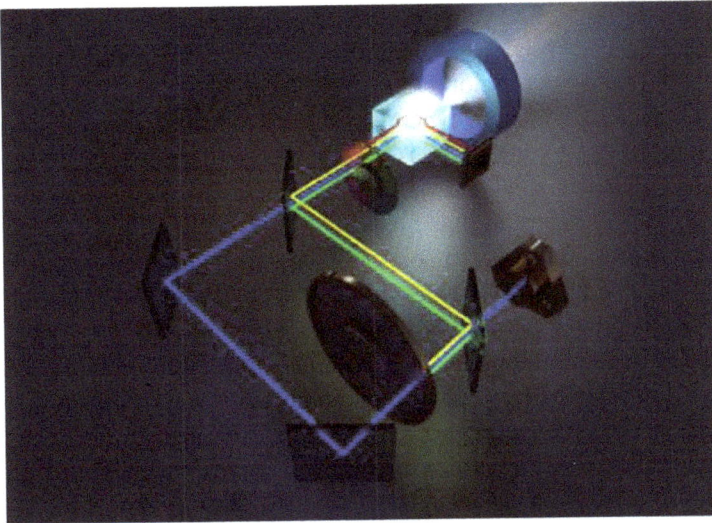

Fig. 7.22.   The light engine used in some NEC projectors — showing the use of a spinning phosphor-coated wheel for producing yellow and green light from wavelength up-conversion of blue laser diode light.

Courtesy: NEC Corporation, Japan.

colors through separate wavelength conversion materials. This approach will still not give the same high performance as a trio of band gap RGB lasers, but it promises to be significantly cheaper, and is, thus, an area of active research. Blue diode laser is the obvious candidate for this, because blue light can be wavelength up-converted to both green and red emissions. This, however, requires luminescent materials capable of wavelength conversion to very narrow spectral width emissions. Most conventional phosphors produce wide emission spectra which are much more suited for space illumination needs than for display applications. Whereas typical semiconductor-based color LEDs emit light with spectral widths of around 20 nm (measured as FWHM), phosphor-converted color LEDs pumped by blue LEDs can give broadband emissions measuring 60 nm or more. For D&P applications, though, the narrower the emission the better the display performance. Narrow emission phosphors are needed and are being developed for this purpose. One example is $K_2SiF_6:Mn^{4+}$ TriGain™ red-emitting (631 nm) phosphor from General Electric (GE). This is currently being used in TVs and portable display devices, but may also finds its way into laser color conversion systems.

Quantum dots offer yet another route to quasi-monochromatic emissions that are essential for wide color gamut D&P applications. Now widely used in 'quantum dot' televisions, these luminescent materials upconvert wavelengths to narrow emission lines that are typically 5 to 10 nm wide. Indium phosphide (InP) is commonly used in consumer quantum dot liquid crystal displays. This material is far less toxic than the more traditional II–VI quantum dots made with cadmium sulfide (CdS) or zinc selenide (ZnSe), though with somewhat lower photon conversion yields. At least one company (TCL) has also tried using Perovskite quantum dots which produce even narrower emissions than most non-Perovskite quantum dots. These materials have, so far, only been used for small screen displays. For laser pumped sources, it is far harder to use quantum dots as luminescent color conversion materials. This is because of the lower damage threshold of most quantum dots, as well as their higher susceptibility to degradation from atmospheric oxygen and water vapor. Well-encapsulated quantum dots need to be developed. An interesting possibility is the use of opal crystals for containing quantum dots. Artificial opal structures made from sub-micron silica spheres (see Figure 7.23) can provide a suitable host for quantum dot particles.

The structure of the synthetic opal as well as the size of the constituent spheres can be controlled to increase the effectiveness of the laser

Fig. 7.23. Scanning electron micrograph of a sample of synthetic opal, showing the regular arrangement of spherical silica spheres. The spaces between the spheres can accommodate luminescent species, such as quantum dots.

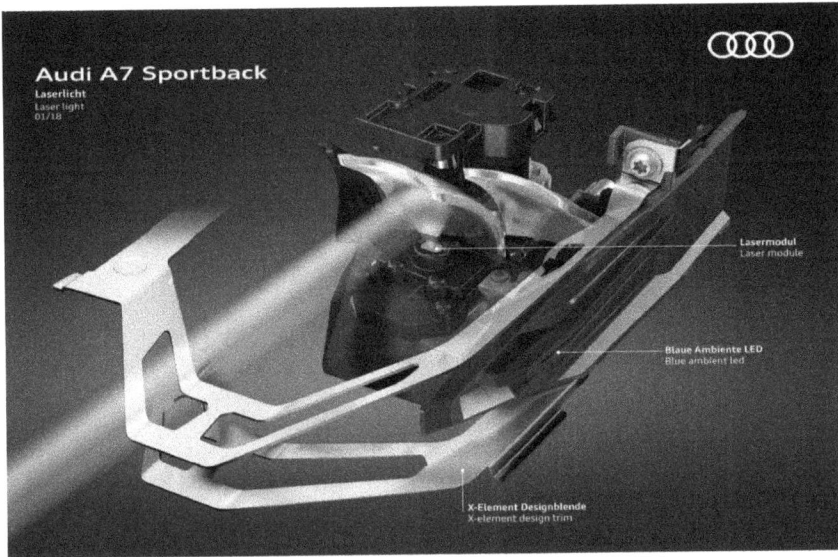

Fig. 7.24.   Audi 'Laserlicht' headlamp assembly.
Courtesy: Audi AG.

radiation field pumping resident quantum dots. This kind of control is very attractive and is exercised by engineering the stop band and pass band of the opal, considered as a 3D photonic crystal. Opal crystals for this application can be grown in both bulk and thick film forms using controlled evaporation techniques. Once grown, quantum dots could be infiltrated into the interstices of an opal by capillary suction from a quantum dot suspension in a suitable liquid.

From a commercial point-of-view, laser diode-pumped phosphor-converted light sources have been gaining popularity in niche applications other than just D&P industries. Automotive headlights have been slowly making a transition to laser-powered luminaires. An example is provided by the 'Laserlicht' headlamp assembly from Audi shown in Figure 7.24. This is reminiscent of the emergence of LED-based lighting in automobiles, several years ago. With the passage of time, more vehicles will come out with laser-based headlight systems.

Solid-state lighting powered by semiconductor diode lasers has a bright future. At the present time, this technology is at the leading edge of artificial lighting developments. There is much room to grow and the coming years, in all likelihood, will see it making its presence felt in a big way.

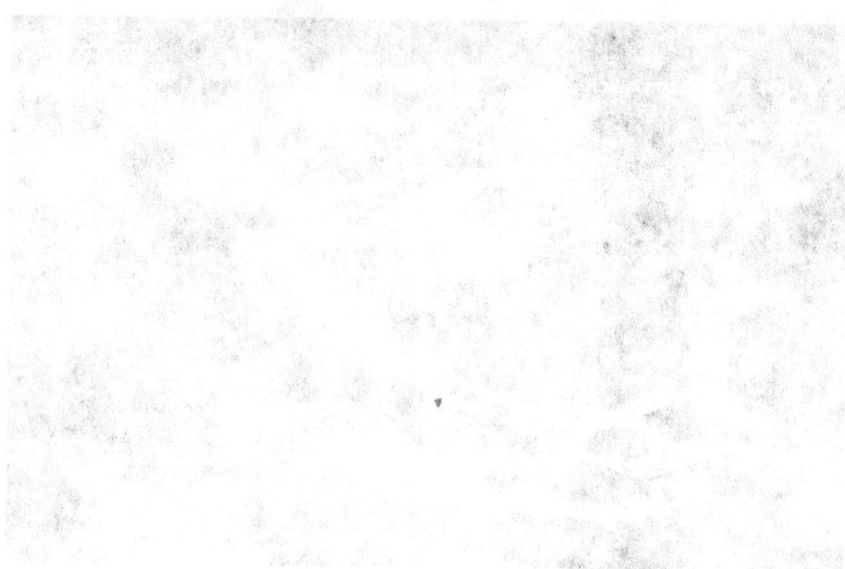

8 Other Light Sources

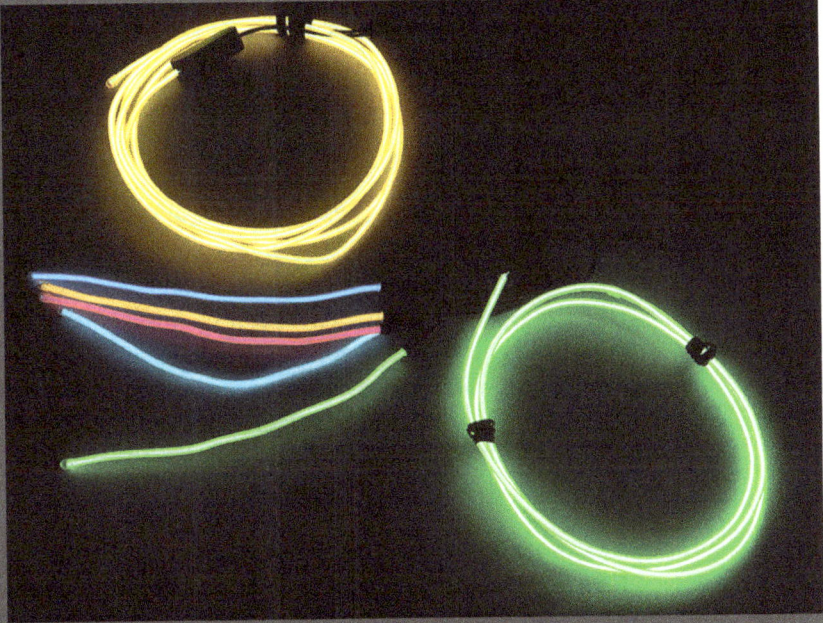

8.1 Introduction

In this chapter, we take a look at some light-emitting materials and devices that have not been covered elsewhere in this book. The four main technologies covered in this chapter are electroluminescent devices, cathodoluminescent devices, chemiluminescence and radioluminescent sources. In the overall scheme of light sources, these are topics more of scientific interest than commercial importance (except for electroluminescent and cathodoluminescent devices which have been used in commercial devices for a long time).

8.2 Electroluminescent Devices

The term 'electroluminescence' can, in principle, stand for any process where light is produced through electrical means. Practically, it is almost exclusively used in situations where solid-state devices are involved. Here too, the term, electroluminescence, can be broadly applied to light emission from all types of light-generating devices, including LEDs. In this chapter, we will use this term in a narrower sense where light is generated through electron impact-based mechanisms in solid materials, often in the form of a thin film. In such cases, *pn*-junctions are not present and, thus, LEDs are excluded from consideration. Use of the term, electroluminescence, has become quite common in this sense and here we will also use it as such.

8.2.1 *Historical development of electroluminescent devices*

The term 'electroluminescence' was first used by the French scientist and inventor, Georges Destriau, in the year 1936, when he described the emission of light from zinc sulfide (ZnS) powder, doped with traces of copper ions, under AC electrical excitation. Doped ZnS particles were suspended in oil for electrical insulation and a high voltage was applied by placing it between the plates of a parallel plate capacitor. A steady blueish glow was observed, coming from ZnS particles. For some time, this phenomenon was known as the Destriau effect. This remained a scientific observation until 1951 when the first commercial electroluminescent lamps, employing doped ZnS powder, were developed by Keith Butler at Sylvania Corporation. A few years later, in 1954, the first electroluminescent display with a two-dimensional array of pixels was also developed at

Sylvania by Sandford Peek. In later years, during the period from 1956 to 1958, electroluminescent light source technology was further improved by Elmer Fridrich of the General Electric Corporation in Cleveland, Ohio. Fridrich also made very important contributions to several other lighting technologies and was central to the development of the tungsten halogen lamp, as has been described in Chapter 2. Still later, around the late 1960s, Aron Vecht, at the University of Greenwich in London, UK, developed DC electroluminescent devices for watches and displays. By that time, electroluminescent devices were being made from thin film phosphors rather than powdered phosphor material, as was the case during the 1950s and 60s. Thin film devices were significantly superior to the earlier powdered phosphor devices in terms of their brightness and longevity. The next significant development took place in 1974 when Tuomo Suntola (see Figure 8.1), in Finland, developed atomic layer epitaxy (now called atomic layer deposition — ALD), for depositing extremely uniform thin films on large substrates. This was a game changer because it enabled economical manufacture of uniform thickness, pin hole-free, thin films for large, high reliability electroluminescent display panels. At the same time as this development, Toshio Inoguchi, working at Sharp Corporation in Osaka, Japan, developed the first bright and long-life monochrome thin

Fig. 8.1. Tuomo Suntola — Finnish physicist and inventor who among his other contributions, developed atomic layer deposition (ALD) technology.

film electroluminescent (TFEL) displays. This was a major development which raised the profile of Sharp Corporation as a leading producer of display equipment. In later years, TFEL display technology was further developed by both academic and industrial researchers to create high-resolution displays for use in applications ranging from ATM machines to airport information boards. Much of the development during the 1980s was spearheaded by Planar Systems, based in Hillsboro and Beaverton, Oregon, where researchers including Christopher King, Jim Hurd, John Laney and Eric Dickey did extensive work to improve large format TFEL displays. Still later, during the 1990s, Xingwei Wu, of iFire Technology in Oakville, Ontario, Canada, developed a thick film display and named it thick dielectric electroluminescent (TDEL) display. Capable of much higher brightness than TFEL devices, these displays used a 'color by blue' approach to achieve good RGB light generation. TDEL was the first full color-capable electroluminescent display technology.

8.2.2 *Operation of electroluminescent devices*

The physical mechanism of light generation in *pn*-junction-based devices has been examined in Chapter 5. It involves carrier (electrons and holes) injection through *n*- and *p*-type semiconductor regions into a charge-depleted junction region where electrons and holes recombine, radiatively, to release energy as photons. This process is quite efficient and could be made even more efficient with the proper choice of materials and structural features, such as the use of quantum wells. Almost all efficient semiconductor light emitters are based on *pn*-junction devices. While junction-based electron–hole recombination is the main mechanism utilized in modern solid-state light emitters, a different mechanism, based on kinetic impact excitation is used in impact excitation (also called impact ionization) electroluminescent devices. Most impact excitation-based devices are relatively low luminosity light emitters that are used in displays, illuminated signs, backlights and night lights. This reflects the fact that impact excitation is significantly less efficient in generating light than is electron-hole pair recombination. The latter is, therefore, preferred over the former, wherever an appropriate device structure could be realized. Carrier impact-based devices which, in accepted parlance, we will refer to as electroluminescent devices have the advantage of architectural simplicity. These devices are further differentiated from LEDs due to the presence of high voltages to accelerate electrons in the luminescent material.

Light emission

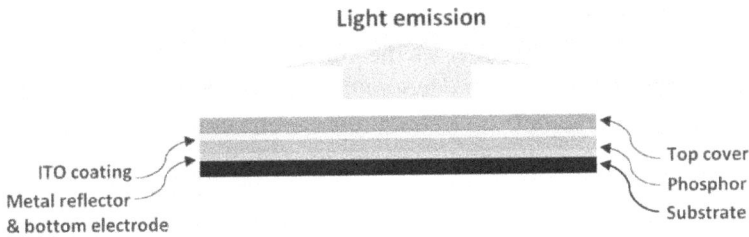

Fig. 8.2. Simple schematic structure of a DC electroluminescent device. AC devices have dielectric layers on each side of the phosphor layer to block the direct flow of electric current.

Application of high voltages, anywhere in the range of approximately 80–300 V, produces high electric fields of the order of 10^6 V/cm in the luminescent material. Among other attributes, the luminescent material needs to have sufficiently high dielectric breakdown strength to survive electric fields of this magnitude.

Fabricating a good *pn*-junction involves a significant engineering challenge. Where, for one reason or another, *pn*-junctions could not be constructed, it may be possible to achieve electroluminescence with the use of a capacitor-like arrangement. Here, an active light-emitting material is sandwiched between two electrodes (see Figure 8.2). One of these electrodes is, generally, made reflecting while the other is transparent. The transparent electrode is most frequently made of a thin indium tin oxide (ITO) or fluorine-doped tin oxide (FTO) film, but other transparent conductive materials could be used as well. A high voltage is applied between the device electrodes. This voltage can be DC, in certain cases, but is much more frequently an AC voltage. AC electroluminescent devices have a dielectric coating on both sides of the phosphor layer to block the direct flow of electric current. The amount of light emission usually increases with increase in the peak-to-peak voltage, until certain values are reached. At very high voltages, the device can suffer catastrophic dielectric breakdown which constrains the maximum usable voltage. The amount of light emission also increases with rise in frequency of the drive waveform, but with continued increase in frequency the light output tends to saturate.

The operating principle behind electroluminescent devices is quite straightforward. The electric field between the parallel electrode plates of an electroluminescent device causes naturally present free electrons in the luminescent material to move from the (instantaneously) negative electrode to the positive electrode. As the electrons move under the influence of the electric field they gain energy and get accelerated to high velocity. The ultimate energy gained is limited by the mean distance traveled ballistically before a collision with a material atom, impurity or defect takes place. It is during the inter-collision transit time that electrons gain energy from the electric field. The kinetic energy thus picked up is dissipated when a collision takes place. These collisions transfer energy from high-energy non-thermal electrons to the atom's electrons which become excited to non-equilibrium configurations. It is the decay of, thus, excited electrons to their ground state energy level that results in light emission. In other cases, an electron can be released completely free of the atom, causing the atom to become ionized. This impact ionization process creates a free electron and a positively charged ion, bearing an electron vacancy, i.e. a hole. In this way, impacts from field-accelerated electrons can generate electron-hole pairs. Their recombination can then produce photons that correspond in energy to the band gap of the luminescent material.

Many of the same phosphors, as used with LEDs, can be used in electroluminescent devices and one in particular has been extensively used for a long time. This is manganese-doped zinc sulfide (ZnS:Mn). It emits an orange-yellow light with fairly good efficiency. Another similar electroluminescent phosphor is copper-doped zinc sulfide (ZnS:Cu). The physics of impact excitation is such that several potentially luminescent materials that will be impossible or hard to turn into *pn*-junction-based devices can be utilized in electroluminescent devices. For example, boron nitride (BN) is a promising material for making deep ultraviolet (DUV) emitters, but it is very challenging to grow device-quality BN films and even harder to make BN *pn*-junctions. Impact ionization-based luminescent devices may provide an alternate way to make DUV emitters from this material. Due to their simpler capacitor-like structure, complicated doping strategies are not required when making electroluminescent devices. The principal downside of almost all electroluminescent sources remains their poor efficiency and, consequently, low output powers.

8.2.3 *Examples of electroluminescent devices*

Simple electroluminescent devices have been used as small flat light sources in night lights (see Figure 8.3 (left)). 'Panelescent' night lamps,

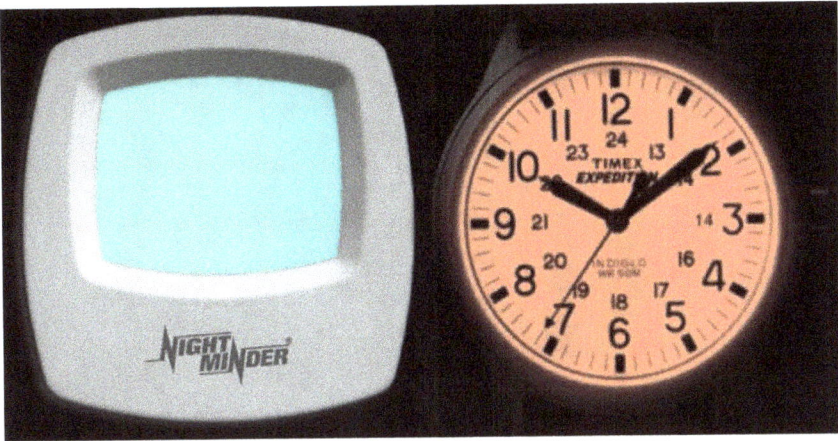

Fig. 8.3.   (Left) Sylvania 'Panelescent' night light, (Right) Timex 'Indiglo' wrist watch.

manufactured by Sylvania have been the longest available devices of this kind. Consuming only 20 mW during operation from the mains AC line, this greenish-blue night light has long been a favorite in many Western countries. Various types of non-self-emissive displays have also used flat electroluminescent light-emitting panels as backlights. The blue light in 'Indiglo' Timex watches also utilizes an electroluminescent backlight where the front face of the watch can be illuminated from the back at the push of a button (see Figure 8.3 (right)).

The high voltage needed for the electroluminescent backlight is generated by a switch-mode embedded power supply. One can hear the feeble 'hum' generated by high-frequency current flowing through the inductor used in the high voltage generator circuit each time the watch dial illumination button is pressed.

Electroluminescent backlights are also used with some LCD displays to make them readable in low light conditions. In addition to use as passive backlight sources, self-emissive displays based on electroluminescent technologies have also been developed and commercialized. These TFEL devices feature pixels that are formed as individual electroluminescent capacitors (see Figure 8.4). Initially, TFEL displays were only monochrome, with usually an orange-on-black color scheme, but three-color RGB displays have also been built. Full color displays make use of broadband white-emitting phosphors whose light is separated into red, green and blue colors using filters on each sub-color pixel. TFEL displays are used in demanding applications where their outstanding advantages can

Fig. 8.4. Electroluminescent information display at Helsinki international airport.

be fully utilized. These advantages include fully solid-state architecture, vibration immunity, insensitivity to temperature variations, wide viewing angle, high contrast even in bright ambient light conditions, fast video refresh rate, and long lifetime. Due to such desirable characteristics, flat panel TFEL displays have found use in industrial applications, military display systems, aerospace vehicles and other hazardous environments. Flexible TFEL displays have also been demonstrated. Such devices have been built with organic luminescent and charge conducting materials coated on flexible substrates.

Electroluminescent lamps for illumination purposes have also been built but did not find wide acceptance because of their relatively low brightness. With typical brightness in the range of 2–10 lumens/W, electroluminescent devices, despite their many advantages, such as low power consumption, soft glow, mechanical robustness and long life, have only been used in applications where their relatively low light output is not an overwhelming disadvantage. Displays remain the primary application for electroluminescence technology with new applications targeting hobby, illuminated signage and illuminated garment markets. Flexible,

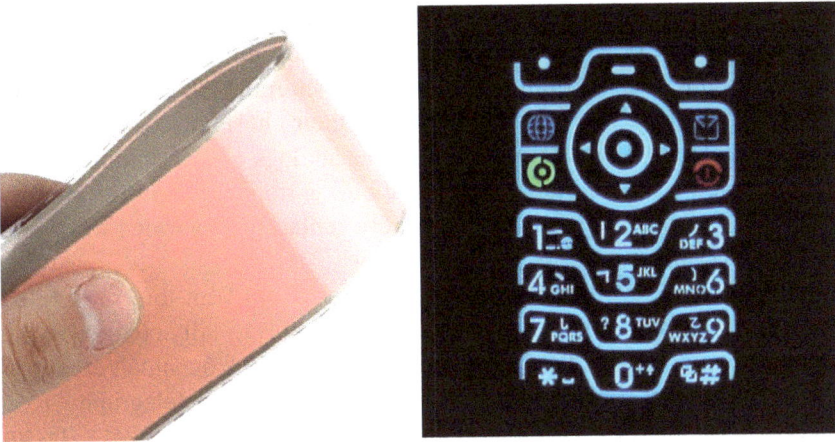

Fig. 8.5.   (Left) Flexible electroluminescent strip, (Right) mobile phone keyboard backlit with electroluminescent backlight.

inexpensive, thin-film display panels that can be bent, cut into desired shapes and used as keypad backlights are now available from a number of manufacturers (see Figure 8.5).

With the development of new organic and inorganic light-emitting materials, the prospects of electroluminescent technology being used with novel phosphor materials are also looking promising. It is certain that with the passage of time electroluminescent light sources will get brighter and also become available at wavelengths in the infrared and UV regions.

8.3 Cathodoluminescent Sources

The electrons in atoms of luminescent materials can not only be excited by collisions with electrically injected electrons, but also through collisions with electrons that bombard the material's surface. This is achieved by producing a free space electron beam in a vacuum. Systems that rely on this mode of material excitation often emit electromagnetic radiation as the excited atoms decay back to their ground states. The radiation emitted due to this process is called cathodoluminescence. This term implies luminescence due to cathode rays which is what electron beams were called during the initial years of their discovery in evacuated discharge tubes.

The most familiar example of cathodoluminescence is provided by cathode ray tube (CRT) based TV displays and oscilloscope screens. They produce light when a beam of accelerated electrons, emanating from an electron gun at the rear of the tube, strikes their phosphor-coated screens. In a sense, these devices are the free space analogs of electroluminescent devices. Being vacuum tubes, CRTs are bulky, fragile and power-hungry. For these reasons, they have been largely replaced by various flat panel display technologies.

CRTs were developed during the Second World War, to serve as displays for radar systems. However, their invention is usually credited to the German engineer Karl Ferdinand Braun (see Figure 8.6), much earlier in the 19th century. Braun was an early pioneer of radiotelegraphy who received a Nobel Prize with Guglielmo Marconi. In a traditional CRT, an 'electron gun' at the narrow base of the funnel-shaped glass tube emits a narrow beam of electrons. These are generated by heating a tungsten filament, by passing a current through it. The filament is coated with a material with a low work function, such as thorium oxide. In some cases, the heated filament heats up a metal cap over it which emits electrons. Once emitted, the electrons are confined to a beam and then accelerated through a set of cylindrical apertures (anodes) set at increasingly higher voltages. The highest accelerating voltage could be in the range of 10–12 kV. Either

Fig. 8.6. Karl Ferdinand Braun — German physicist and engineer who developed the first cathode ray tube displays.

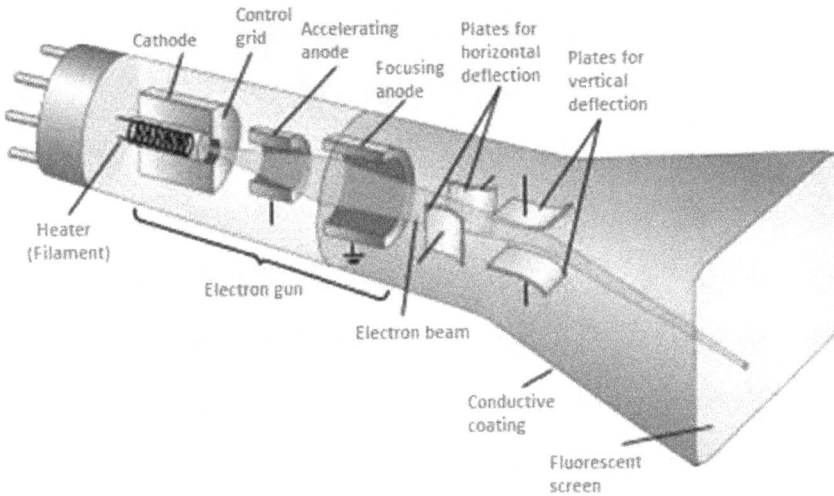

Fig. 8.7. Schematic illustration of a cathode ray tube display, as used in oscilloscopes. Various components inside the tube generate, accelerate, focus and deflect the electron beam which causes the tube's screen to glow where it impacts the phosphor coating.

electrostatic capacitor-like or magnetic coil-based electron-optic lenses are used to focus and then deflect the electron beam. The inside surface of the broad end of the tube is coated with a 'TV phosphor'. This is somewhat similar in composition to the phosphors used in fluorescent lights. When the high-velocity electrons strike the phosphor surface their kinetic energy causes the phosphor to get excited locally and emit light. A conductive coating on the inside surface of the tube is used to make sure that the tube does not accumulate any charge as a result of electron bombardment (see Figure 8.7).

For monochrome TVs, white light-emitting phosphors, similar to that used in fluorescent lights, were used whereas for oscilloscope and radar displays a higher efficiency green phosphor was used. TV phosphors were chosen to have short luminance decay times so that their light emission was synchronous to electron beam impact. This made sure that no visual artefacts, such as motion streaks, were visible during fast-changing scenes. On the other hand, slow, long persistence phosphors were often used for oscilloscopes and radar displays. Later, when color CRT displays were developed, three-color phosphor screens were used where the RGB phosphors were printed in either a dotted or a striped pattern on the inside

Fig. 8.8.   A television CRT, seen from the side. The electron gun is visible at the left and magnetic deflection coils in the middle.

surface of the tubes. CRT-based TVs (see Figure 8.8) held sway until the 1990s when flat panel displays began to edge them out.

The use of cathodoluminescence is not limited to uses in CRTs. The detectors inside a scanning electron microscope (SEM) work by converting secondary electrons, given off by a sample being imaged, into flashes of light in a suitable scintillator. The scintillator emits cathodoluminescence when secondary electrons strike it. Cathodoluminescence is also used for studying materials ranging from geological samples and minerals to semiconductors. This is done by analyzing the intensity and spectral make-up of light that is emitted when a beam of accelerated electrons hits a specimen. Special equipment can be installed inside SEMs to collect the cathodoluminescence emitted when an electron beam hits a sample under observation. Prominent in this setup is a large parabolic or ellipsoidal mirror for collecting the cathodoluminescence and focusing it on a sensitive light detector. Many materials luminesce when impacted by accelerated electrons, and, thus, this technique is very valuable for studying their composition and structure.

Electrons traveling through free space can excite phosphors much more strongly than electrons moving through a solid material under the influence of an electric field. In the latter case, the short mean free path, caused by frequent collisions inside the material, limits the maximum energy that can be attained by electrons, and, thus, the maximum energy that these electrons can transfer to the atoms of the luminescent material. This means that cathodoluminescent displays can be significantly brighter than electroluminescent displays. Although, electroluminescent displays have been greatly improved in recent decades so that their brightness levels are comparable to that of CRT displays, cathodoluminescent sources are capable of being much brighter. This potential has not gone unnoticed. Researchers in Russia, at the Moscow Institute of Physics and Technology (MIPT) and the Lebedev Physical Institute, have developed illumination-quality cathodoluminescent lamps. These can be retrofitted in existing lamp installations with screw-in bases (see Figure 8.9).

Capable of delivering up to 250 lumens while consuming 5.5 W of electrical power, these lamps can operate continuously at high temperatures and in hostile environments. The electron gun, located at the base of the lamp unit, features a carbon fiber-based field-emission electron source (see Figure 8.10). The electron source is made of a cathode modulator (1), a carbon fiber field-emission tip (2) and a beam concentrator (3). The emitted electrons (4) travel forward to hit the phosphor layer (5) which is backed by an evaporated aluminum reflector (6). Electrons can drain out of the system through the charge collector anode connection (7). The potential difference between the carbon fiber field emitter and the aluminum charge collector is 10 kV. Relying on field-emission rather than thermionic emission increases the energy efficiency of the lamp and also

Fig. 8.9.   An experimental cathodoluminescent lamp.

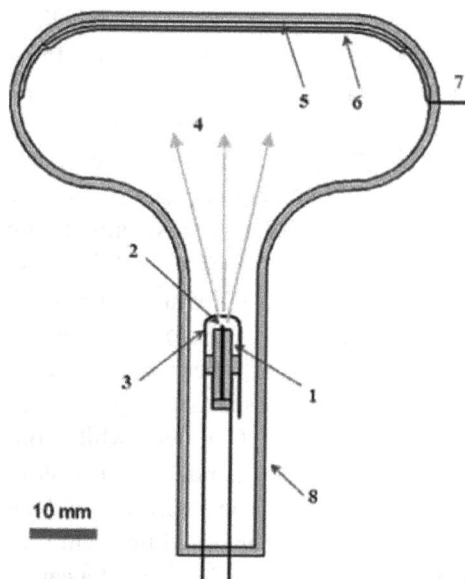

Fig. 8.10.   Schematic diagram showing different parts of the cathodoluminescent lamp.

enables it to turn on instantly when power is applied. The lamp envelope, made of glass (8), is evacuated to around 10^{-6} torr pressure, in order to provide unimpeded travel for electrons.

While cathodoluminescent lamps are unlikely to become common-place because of their cost, bulk and fragility, they do illustrate the potential of free space electron impact-based devices for generating substantial amounts of light.

8.4 Chemiluminescence

Many chemical reactions result in energy output, in addition to reaction products. Reactions where the energy appears as heat are called exo-thermic reactions. In a very small minority of cases, the energy released as a result of a chemical reaction, serves to excite one of the reaction products which then releases this electronic excitation as visible radia-tion. Light produced in this manner is called chemiluminescence (see Figure 8.11). This is completely different from combustion reactions where hot flames are produced. Chemiluminescence reactions generate no or very little heat, and, thus, the light is considered 'cold'. Here, the

Fig. 8.11. Chemiluminescence from a chemical reaction taking place inside a glass test tube.

reaction product (or one of the products if there is more than one product) gets created in an electronically or vibrationally excited state, which then quickly decays to the ground state by emitting visible or infrared photons. If the emission is in the visible region, then the reaction stands out due to the luminosity that it generates. Because light is produced as a direct result of a chemical reaction taking place so chemiluminescence is almost the opposite of a photochemical reaction where external light is provided to drive a chemical reaction. A chemiluminescence reaction can be regarded as a chemical reaction where a 'fuel' is oxidized to create an excited state molecule that, by itself or through energy transfer to another molecule, emits light.

Perhaps the most well-known example of chemiluminescence is the reaction between luminol and hydrogen peroxide. Also known as 3-Aminophthalic hydrazide (see the chemical structure in Figure 8.12(a)), luminol ($C_8H_7N_3O_2$) is a yellowish crystalline solid which is insoluble in water, but is soluble in most polar organic solvents. It shows bright blue chemiluminescence when mixed with appropriate oxidizing agents, in the presence of a catalyst, such as copper or iron.

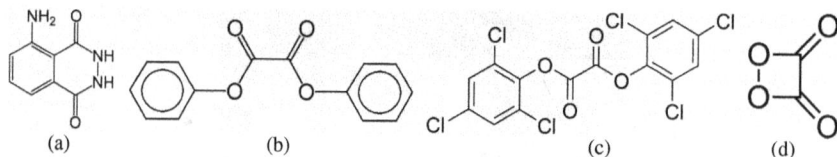

Fig. 8.12. Chemical structures of luminol (extreme left), diphenyl oxalate (DPO) (second from left), bis(2,4,6-trichlorophenyl)oxalate (TCPO) (third from left) and 1,2-dioxetanedione (C_2O_4) (extreme right).

The reaction product is 3-aminopthalate which is formed in an excited state but decays rapidly to the ground state with the emission of blue light. In actual practice, not all product molecules in any chemiluminescence reaction are produced in excited state, capable of emitting light. The fraction of 'active' light-generating molecules produced, is the quantum efficiency of chemiluminescence. In most cases it is quite poor and hardly exceeds about 2%.

Chemiluminescence has a number of applications. The most visible is likely to be its use in toys and entertainment. Glow sticks, also called light sticks, are a popular kids' toy item. These sticks contain two different chemicals which when mixed together, give off light for an extended period of time. The light is sufficiently bright at first to be seen even in brightly lit rooms, but gradually fades away over a period of hours. Glow sticks are also used as a utilitarian light source in situations where electrically-powered light sources are either not available or are not desirable. Various designs for purely chemical light sources were developed during the decades of the 1960s and 1970s, with a number of patents issued to quite a few inventors. The most practical among these were designs based on the mixing of two chemicals contained in containers, such that one was nested inside the other. This has become the basis of inexpensive glow sticks which are now widely used for recreational purposes (see Figure 8.13). These devices contain an inner thin-walled glass ampoule containing a suitable oxidizing agent (usually hydrogen peroxide). The long tubular outer plastic container is filled with a chemiluminescent precursor, such as diphenyl oxalate (DPO) (see Figure 8.12(b)) or *bis*(2,4,6-trichlorophenyl)oxalate (TCPO) (see Figure 8.12(c)), together with a fluorescent dye, and sodium salicylate as a catalyst. On bending the tube, the inner glass ampoule breaks, causing the chemicals to mix together and immediately generate a chemiluminescent glow.

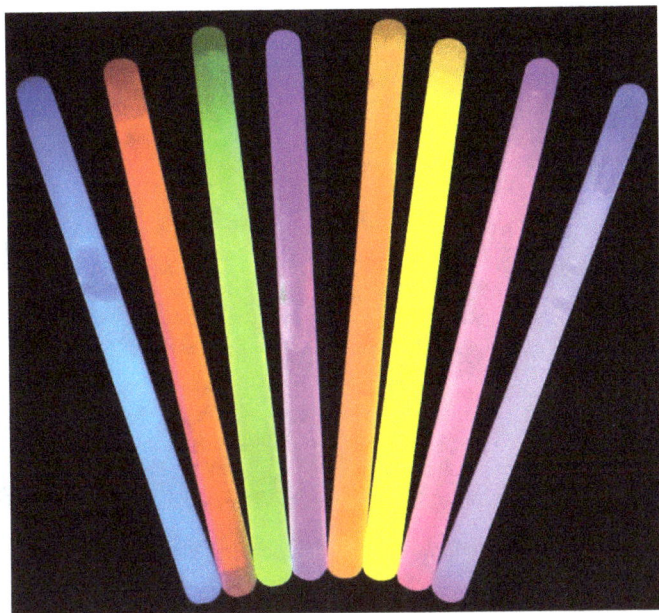

Fig. 8.13.   Toy glow sticks in various colors.

The chemiluminescent reaction takes place with the decomposition of the oxalate ester into phenol (C_6H_5OH) and 1,2-dioxetanedione (C_2O_4) (see Figure 8.12(d)) which is an unstable intermediate product. The dioxetanedione — more commonly known as peroxyacid ester — is an extremely unstable and energetic compound which can be viewed as an exotic oxide of carbon. Structurally, it resembles two carbon dioxide molecules stuck together. It rapidly transfers energy to the fluorescent dye molecules (usually an anthracene derivative) and itself decomposes into carbon dioxide. The excited dye molecules then glow with a color that is determined by their chemical composition. By selecting the right dye and the color of the outer plastic container, a range of glow stick colors can be obtained. Proper choice of chemicals and their concentrations also affects the emission profiles of glow sticks, such as a relatively dim glow over a long period of time or a bright glow for a shorter time period.

Chemiluminescence has a number of applications in biochemistry and life sciences. A particularly important application is for the forensic detection of blood. This uses the same luminol-based chemiluminescence

Fig. 8.14.   Chemiluminescent (bioluminescent) glow from a firefly.

reaction as was mentioned earlier. Blood detection through this method relies on using the iron present in blood's hemoglobin to catalyze the oxidation of luminol with hydrogen peroxide. There are many other applications of chemiluminescence in analytical chemistry and biochemistry for the detection of a large number of proteins, antibodies and other biologically-active compounds. Because even a few photons can be detected with sensitive optical detectors so such methods are generally very sensitive and can detect even trace amounts of compounds which will be impossible or very difficult to do with other methods.

Biological organisms often exhibit chemiluminescence which is then usually called bioluminescence. The most well-known example is the glow given off by fireflies (see Figure 8.14). Fireflies, which, despite their name, are actually beetles, have evolved specialized bioluminescent organs called 'lanterns' on the rear underside of their bodies from where their bioluminescence emanates. Other species of light-emitting beetles are known where the lanterns are located in the front part of their body, behind their heads. In addition to beetles, a variety of bacteria, fungi and marine animals also show bioluminescence. Rotting meat, for example, has been known from ancient times to glow with a very weak green light.

This glow is now known to be caused by concentrations of bioluminescent bacteria that cover decomposing flesh.

In most cases, the bioluminescent glow is green or blue in color with very few known cases of red light emission. There are a variety of reasons for light emissions given off by living organisms. These include luring of prey, various kinds of warning signals, attracting mate, camouflage, and illuminating the environment. Most bioluminescence is a result of a chemical reaction between the enzyme luciferase and the luminescent pigment luciferin. Luciferins, which are small organic molecules, can undergo an enzyme-catalyzed oxidation reaction with the formation of excited state intermediates that emit light upon decaying to the ground state. The molecular intermediates are produced in electronically excited states as luciferin breaks down into smaller molecules. In fireflies, for example, the oxidation of luciferin with both adenosine triphosphate (ATP) and molecular oxygen is responsible for light emission. As another example, certain fungi species, found on rotting wood, generate green bioluminescence as luciferase catalyzes luciferin secretions (see Figure 8.15). Luciferin is not the only bioluminescent material. Another protein, aequorin, found in certain jellyfish, produces blue light in the presence of calcium.

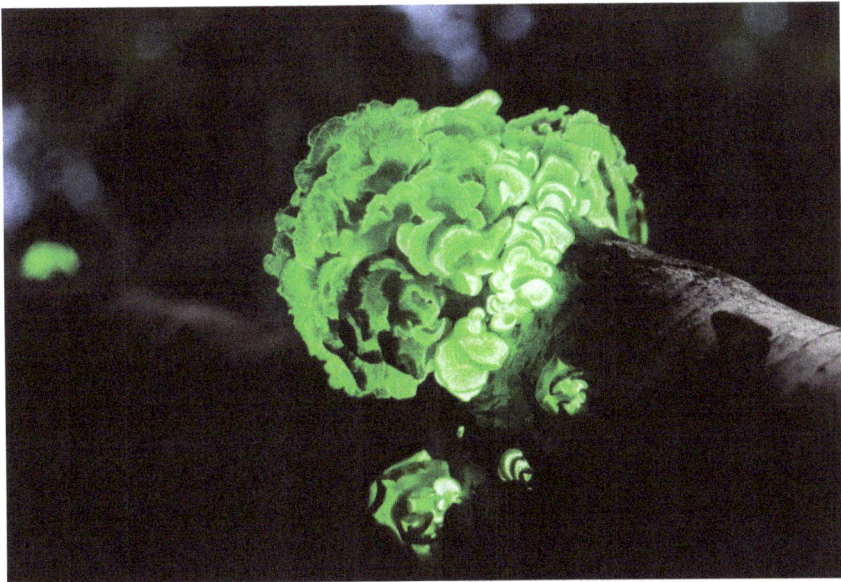

Fig. 8.15. Bioluminescent bacteria on rotting wood.

Glowing jelly fish are well-known to inhabit the deeper parts of the world's oceans. Their glow, in their otherwise completely dark environment, is thought to attract prey and deter predators.

The artificial generation of light through chemical means is not just a preserve of the entertainment and biomedical industries. Certain types of lasers derive energy for achieving population inversion from purely chemical means. The most well-known of these is the chemical oxygen iodine laser (COIL). This laser emits CW radiation in the near-infrared at 1.315 μm. Its very scalable design makes it possible to produce anywhere from several watts to several megawatts of output power. This feature has made it of particular interest to military establishments where laser-based directed energy weapons have long been viewed with interest.

The COIL system derives energy from a chemical mixture that produces singlet oxygen — a reactive form of molecular oxygen in a spin-flipped excited state. Where ordinary oxygen ($X^3\Sigma$, in spectroscopic notation) has a pair of electrons with parallel spin orientation on each oxygen atom, singlet oxygen atoms ($a^1\Delta$, in spectroscopic notation) have these electrons oriented in anti-parallel fashion. This unusual electron spin configuration raises the overall energy level of singlet oxygen molecules. Because relaxing to the ordinary oxygen's parallel electron configuration would require spin flips — forbidden under quantum mechanical transition rules — so the singlet state is long lived, with a half-life of around 45 min. The key to the working of a COIL system is the fact that iodine atoms in diatomic molecular iodine (I_2) have a set of energy levels that are almost perfectly aligned with the energy levels of singlet oxygen molecules. Thus, singlet oxygen molecules can efficiently transfer their energy to iodine molecules and relax to the ground state. Stimulated emission then takes place in iodine atoms with the emission of photons at 1.315 μm (see Figure 8.16).

Chemically-driven generation of active singlet oxygen makes COIL a purely chemical laser. In its most common manifestation, singlet oxygen for a COIL system is produced by pumping an aqueous mixture of hydrogen peroxide and potassium hydroxide with gaseous chlorine into a reaction chamber. The reactants react together to form potassium chloride, water and molecular oxygen in its excited triplet state:

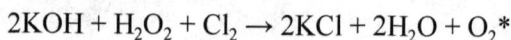

$$2KOH + H_2O_2 + Cl_2 \rightarrow 2KCl + 2H_2O + O_2*$$

Here, O_2* denotes a singlet oxygen molecule.

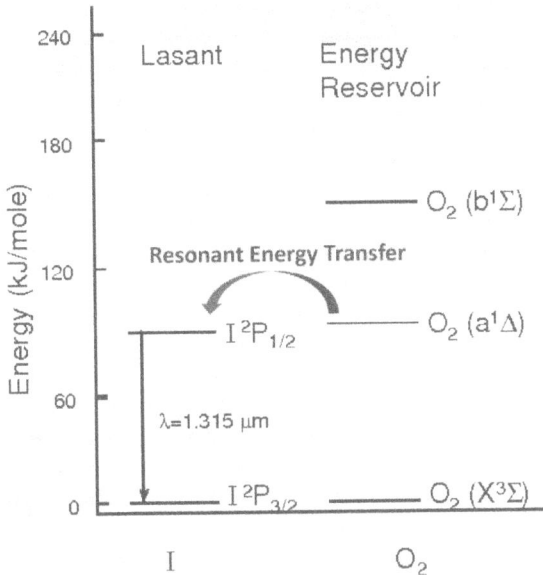

Fig. 8.16. Energy level schemes involved in the operation of COIL lasers.

The excited singlet oxygen molecules are mixed with iodine in a resonant laser cavity where energy transfer to iodine atoms, stimulated emission and laser oscillations take place.

The entire COIL system is quite bulky with all the chemical storage containers, tubing and pipework, as well as gas handling systems and various interaction chambers. On the other hand, the laser is capable of delivering very high CW powers in a concentrated invisible laser beam that can be used for both defensive and offensive military operations. Since the mid-1970s, COIL has been under development by the US Air Force, with a 20-kW laser tested in 1998. It forms a component of the United States' military airborne laser and advanced tactical laser programs. On February 11, 2010, an airborne COIL system (see Figure 8.17) was successfully deployed to shoot down a missile off the coast of central California.

The extensive hardware requirements for chemical-fed oxygen iodine lasers have limited their further development for operational laser weapons, especially those carried aloft by aircrafts. Such systems appear more suitable for deployment on warships where their weight and bulk can be more easily accommodated. In the quest for fieldable airborne laser weapons, an alternative form of a hybrid chemical-electric oxygen

Fig. 8.17.   United States air force aircraft-mounted COIL system with omnidirectional targeting turret.

iodine laser called an electri-COIL or e-COIL has been investigated. In this system (see Figure 8.18), triplet oxygen is produced through an RF electric discharge and is then fed to a laser cavity where iodine vapor is injected for lasing to take place. This system has, so far, not demonstrated comparable performance to the purely chemical version of oxygen iodine lasers. Work continues on its further development. More recently, even newer designs of defense system using COIL and similar lasers have been studied to realize systems for boost-phase interception of ballistic missiles.

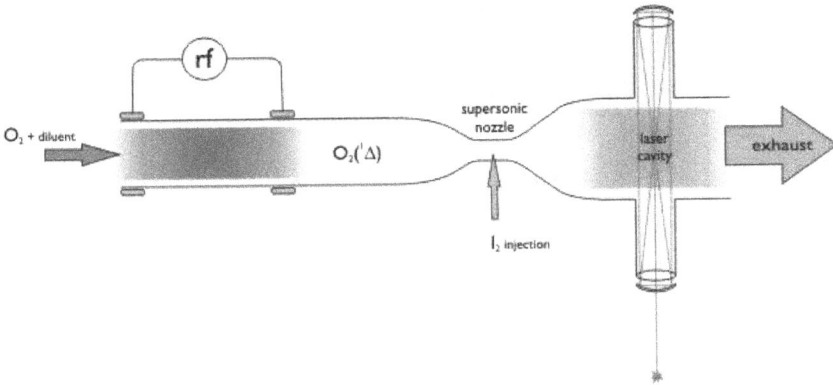

Fig. 8.18.   Schematic diagram of an electrically-operated COIL system.

8.5 Radioluminescence

Radioluminescence refers to visible light emission that is created by energy input from a radioactive source. When nuclei of radioactive isotopes decay, they either eject a helium nucleus (an alpha particle) or one of the neutrons transforms into a proton, accompanied by the formation of an electron (a beta particle) and an electronic neutrino that are ejected from the parent nucleus. Some nuclei decay through alpha particle emission whereas others decay through beta emission. Often one radioisotope decays into a daughter isotope which is also radioactive. In some cases, a radioisotope decays through a succession of radioisotope 'daughter products' until a stable isotope is reached. In such cases of radioactive isotope chains, both alpha and beta particles are produced simultaneously, as a given sample of the material will contain various amounts of different daughter nuclei, some of which may be alpha emitters while others may be beta emitters. The ejected alpha and beta particles carry sufficient kinetic energy to cause luminescence in suitable materials. Any such luminescence is called radioluminescence and can be used in practical applications, as long as the radioactivity itself is deemed safe. Generally, this requires the ejected particles to have low penetrating power so that they are unable to enter living tissue.

The emission of light from nuclear reactor-generated neutrons traversing moderator water, outside the reactor's core, has been described in Chapter 1. The bluish glow of Cerenkov radiation thus generated is a

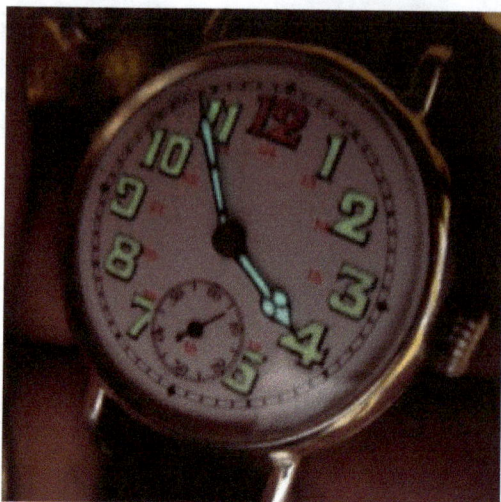

Fig. 8.19. Dial of an old wristwatch with numerals and hands painted using a radium salt-containing luminescent paint.

result of neutrons traveling at speeds in excess of the speed of light in water. While Cerenkov radiation can be considered as a kind of radioluminescence, the term is, generally, applied where radioactive radiation from a radioactive element or compound is used in some way to generate artificial light for some specific application.

One example of radioluminescence is provided by the 'radium' dials on some older wristwatches (see Figure 8.19). Radium salts give off a visible glow in the surrounding air as fast-moving alpha particles ionize ambient air and cause light emission. In order to make the numerals and hands of a watch glow in the dark, it was once a common practice to use a mixture of a suitable phosphorescent material with a radium salt as a luminescent paint. Radium chloride ($RaCl_2$) was in particularly widespread use. The energetic particles (alpha particles, beta particles and gamma rays) given off during the radioactive decay of radium atoms (^{226}Ra), and its daughter products, energized the phosphorescent material and caused the paint to emit a fairly bright green glow in the dark.

Beginning around 1910, radium-based luminescent paints were also used on the dials of various instruments to make them readable under low light conditions. By the 1920s their use had become widespread and one company was making a fortune by selling 'Undark' — a glow-in-the dark paint incorporating radium compounds — and touting its product widely in advertisements (see Figure 8.20).

Fig. 8.20. A newspaper advertisement, circa 1921, for 'Undark' — a radium salt-containing luminous paint.

The adverse health effects from highly radioactive radium compounds were discovered later. The dangers were recognized, most notably, from severe illnesses, and even deaths, caused in the so-called 'Radium Girls' who were employed to paint watch and instrument dials with radium paint. Their plight led to a complete ban on the use of radium-based luminous paints for any application, and by 1970 radium's use had completely ceased. Radium is particularly dangerous because its main isotope, ^{226}Ra, has a half-life of 1600 years, which is short enough to cause radium to be highly radioactive but long enough for the radioactivity to remain strong for hundreds of years. Furthermore, radium has a long radioactive decay chain, which results in simultaneous production of alpha and beta particles, as well as gamma radiation. It is now recognized that the presence of any radium compound not only poses dangers from that element's radioactivity but also because its decay product is radon — an even more radioactive gas, which can easily diffuse through enclosed spaces. Due to these reasons, any use of radium, except for strictly medical applications, is now strictly forbidden. Until some time since the worldwide ban on the use of radium came into effect, promethium — another radioactive element — continued to be used in place of radium. A completely synthetic element, its ^{147}Pm isotope was used for years as a replacement for ^{226}Ra for making self-luminous paints. Compared to radium, promethium is much safer, but with more stringent radiation safety rules gradually gaining ground, its use in commercial products was also banned.

Where radium and promethium are now completely off-limits as sources of energy for radioluminescent devices, the heaviest isotope of hydrogen — tritium (^{3}H) — offers a near-perfect alternative. Of the three isotopes of hydrogen: ordinary hydrogen (^{1}H), deuterium (^{2}H) and tritium, this is the only radioactive isotope — and the one which is not found naturally on earth. Tritium can be made by irradiating lithium in nuclear reactors. It has a half-life of about 12.3 years which makes it highly radioactive. Where it differs from radium is in that it decays through the emission of a very low energy (5.7 keV) electron (beta particle) and the decay product is the stable ^{3}He isotope. The energy of a tritium-emitted electron is even lower than that of electrons used in ordinary CRT displays. These low energy electrons can barely travel through 6 mm of air and are unable to penetrate even the outermost layer of human skin. Furthermore, the ^{3}He nucleus is created in its ground state so there is no gamma ray emission to accompany the ejection of a beta particle. These two characteristics of tritium's decay make its radioactivity very safe from a biohazard

Fig. 8.21.   Tritium-filled key chains.

point-of-view, making it suitable for a range of applications. One of these is powering up tritium-based radioluminescent devices, such as key chains (see Figure 8.21), firearm sights, self-illuminated signage and even watch dials.

Tritium's use in self-luminous devices was patented by Edward Shapiro in 1953. A company was started to develop and market gaseous tritium light source (GTLS) devices. Nowadays, many companies produce tritium-based light-emitting devices for a range of applications. Because the beta decay of tritium is the source that powers these devices so they are also known as 'betalights'.

Compared to radium-based luminescent devices, tritium-powered devices are overwhelmingly safe. Their outer plastic enclosures provide more than enough protection from the low energy beta emission of tritium that their use is completely safe. While tritium does not glow visibly by itself, all tritium luminescent devices contain phosphor coatings that emit light by the impact of electrons coming from the decay of tritium.

By using different phosphors, several different colors can be produced. The use of a beta particle emitter, like tritium, is much better than using a source, such as radium, which also emits alpha particles. This is because alpha particles are helium nuclei (^4He) which, due to their substantial mass, can cause lattice damage in phosphors, shortening the phosphors' useful lifetime.

Although tritium-based radioluminescent devices are very safe and do not require any electrical power to operate, a potential hazard comes from the accidental escape of tritium. While tritium is not dangerous externally, there can be potential radiation hazards if it is ingested or inhaled. However, this is considered an acceptable risk because each device contains only a minute amount of tritium, and care is taken in making the enclosure mechanically robust. Borosilicate glass is generally used for containing tritium because it is hard, impervious to the gas, and is resistant to breakage.

The one significant downside of any tritium-powered luminescent device comes from the relatively short half-life (~12.3 years) of tritium. In just over 12 years, tritium's radioactivity gets half as strong as before.

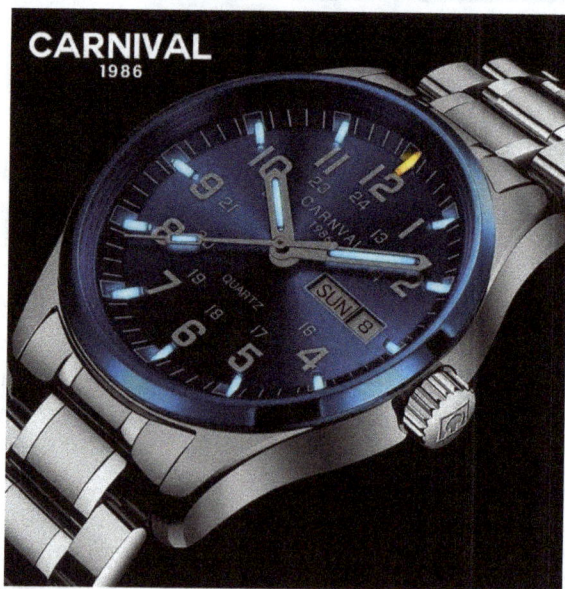

Fig. 8.22.   A wristwatch with night-time luminescence provided by phosphor-coated tritium-filled tubes.

This means that radioluminescent devices using tritium as the source of phosphor excitation get visibly dimmer over the course of a few years. There is no way around this shortcoming, other than replacing the entire device because refilling with tritium is not a practical proposition. Due to this limitation, tritium luminescent devices have found limited applications, with uses in decorative items being the most widespread. In applications that do require lifetimes of 20 years or more, a partial solution is to use a higher than needed tritium fill level. This makes the device start out in a brighter than needed state, giving it a longer useful life. Tiny, super-fill tritium tubes have even been used to provide illumination for the hands and numerals on wrist watches (see Figure 8.22), providing a much safer alternative to radium-powered luminescence of yesteryears.

At present, around 400 grams of tritium, sourced from a number of nuclear reactor facilities around the world, is used up annually to make tritium-based radioluminescent devices. By providing a vastly safer alternative to radium-based devices, tritium, thus, serves a small but important market niche.

Index

www.ingramcontent.com/pod-product-compliance
Lightning Source LLC
Chambersburg PA
CBHW070741220326
41598CB00026B/3718